Workshop Processes, Practices and Materials

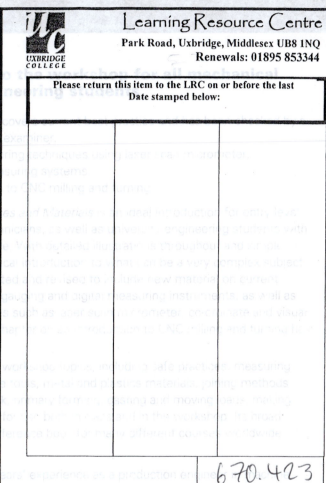
An essential guide to the workshop for all mechanical and production engineering students

▶ Health and safety chapter covers current health and safety legislation by a certified health and safety examiner.
▶ Addition of modern measuring techniques using laser, solid micrometer, co-ordinate and visual measuring systems
▶ Addition of an introduction to CNC milling and turning

Workshop Processes, Practices and Materials is an ideal introduction for entry-level engineers and workshop technicians, as well as novice, for engineering students with little or no practical experience. With extensive illustrations throughout and in full, clear language, this is a practical introduction to what can be a very complex subject. It has been significantly updated and revised to include new material on current health and safety legislation, gauging and digital measuring instruments, as well as modern measuring techniques such as laser, solid micrometer, co-ordinate and visual measuring systems. A new chapter on an introduction to CNC milling and turning has also been added.

This book covers all standard workshop topics, including safe practices, measuring equipment, hand and machine tools, metals and plastics materials, joining methods (including welding), presswork, primary forming, casting and moving loads, making it an indispensable handbook for use both in college and in the workshop. Its broad coverage makes it a useful reference book for many different courses worldwide.

Bruce J. Black has over 40 years' experience as a production engineer in the machine tool and aircraft industries, as well as lecturing in production engineering, culminating as workshop director (wood, metal and plastics) at the then Gwent College of Higher Education in South Wales, UK. Now retired, he works as a freelance technical author.

Workshop Processes, Practices and Materials

Fifth edition

Bruce J. Black

Routledge
Taylor & Francis Group

LONDON AND NEW YORK

Fifth edition published 2015
by Routledge
2 Park Square, Milton Park, Abingdon, Oxon OX14 4RN

and by Routledge
711 Third Avenue, New York, NY 10017

Routledge is an imprint of the Taylor & Francis Group, an informa business

First edition published by Newnes 1979
Fourth edition published by Routledge 2010

Trademark notice: Product or corporate names may be trademarks or registered trademarks, and are used only for identification and explanation without intent to infringe.

British Library Cataloguing-in-Publication Data
A catalogue record for this book is available from the British Library

Library of Congress Cataloging in Publication Data
Black, Bruce J.
Workshop processes, practices & materials / Bruce Black. – 5th edition.
pages cm
Includes index.
ISBN 978-1-138-78472-7 (pbk. : alk. paper) – ISBN 978-1-315-76822-9 (ebook) 1. Machine-shop practice.
I. Title. II. Title: Workshop processes, practices, and materials.
TJ1160.B52 2015
670.42'3–dc23
2014034244

ISBN: 978-1-138-78472-7 (pbk)
ISBN: 978-1-315-76822-9 (ebk)

Typeset in Univers LT by
Servis Filmsetting Ltd, Stockport, Cheshire

Printed and bound by CPI Group (UK) Ltd, Croydon, CR0 4YY

Dedication

To my wife Gillian, children Susan and Andrew, and grandchildren, Alexander and Thomas Hattam, Darcey, Sophie and Bailey Black

Contents

Contents

Preface to the first edition

I have written this book to cover the objectives of the Technician Education Council standard units Workshop processes and materials I (U76/055, which is based on and intended to replace U75/001), for students of mechanical/production engineering, industrial measurement and control, and polymer technology, and Materials and workshop processes I (U75/002), for students of electronic, telecommunication, and electrical engineering. These two units contain a great deal of common material, and by covering them both I hope that the book will be useful to a larger number of students and teaching staff.

From my own experience I have found the content of these units too great to be covered in the design length of 60 hours while still leaving time for adequate practical involvement by the student, which can best be carried out under the guidance of the lecturer. In writing this book, my aim has been to set out in detail the theoretical aspects of each topic, with appropriate line illustrations, in the hope that, by using the book as a course text and for assignments, more time can be spent by the student in practical work where machines and equipment can be demonstrated, handled, and used. Questions at the end of each chapter are directly related to the chapter content as a means of reinforcing the learning process.

An extensive coverage of health and safety has been included, as I feel very strongly that anyone involved in an industrial environment should develop a responsible awareness of the hazards which could affect not only his own health and safety but also that of his fellow workers.

Although specifically written for the TEC standard units, the content is also suitable for the syllabus requirements of the GCE ordinary-level examinations in Engineering workshop theory and practice (AEB) and Engineering workshop practice (WJEC), as well as a considerable amount of first-year work in the higher national diploma in mechanical engineering.

Finally, I would like to thank my wife for her patience and understanding throughout the period of writing the book, my colleagues for their assistance, and Mrs Brigid Williams for her speedy and skilful typing of the manuscript.

Preface to the second edition

Preparing the second edition has enabled me to update a number of areas and to increase the scope of the book by including additional material. It has also afforded the opportunity of resetting to current popular book size and format.

In this second edition I have increased the content to cover a wider range of topics in order to make the book even more comprehensive by providing additional chapters on processes to include sand casting, rolling, extrusion, drawing, forging, presswork, investment casting, shell moulding and die casting.

I have updated the Safe Practices chapter to include current Health and Safety Regulations and the chapter on Measuring Equipment to include electronic instruments. A section on bonded abrasive grinding wheels has been added to the chapter on Surface Grinding and moulding processes has been included in the chapter on plastics.

Preface to the third edition

The third edition has enabled me to increase the scope of the book by including additional material to cover much of the units in the Performing Engineering Operations syllabus.

Where required, chapters have been updated in line with current developments, e.g. lost foam casting and metal injection moulding.

New chapters have been added to cover Standards, measurement and gauging as well as Moving loads and Drawing specifications and data.

I have also taken the opportunity to include review questions at the end of each chapter.

Preface to the fourth edition

In this fourth edition I have included, by request, additional material including gas welding and the dividing head. I have updated to the current health and safety legislation in Chapter 1 and in relation to abrasive wheels, power presses and manual handling and pictures of machine guards have been added. The section on adhesives has been enhanced and the section on protective coatings has been enlarged to include plasma electrolytic oxidation, electrocoating, powder coating and coil coating. Chapter 14 has additions to include the high-performance polymers such as polyimides and PEEK as well as the recycling of plastic materials.

Preface to the fifth edition

This fifth edition has enabled me to further increase the scope of the book. I have updated to current health and safety legislation. Chapter 5 has been extended by adding the use of length bars and angle gauge blocks as well as updating some of the text. Some alterations have been made to Chapter 6 and I have included the latest digital instruments used in measurement and added a section on modern measuring techniques using the laser scan micrometer, co-ordinate and visual measuring machines. Cutting tool materials has been re-written in line with current practice and the use of high-pressure coolant has been included. A description of hand tapping on the lathe and knurling has been included by request. In the materials section I have been requested to include a description of non-destructive testing. Chapter 16 has been enhanced by the addition of the vacuum casting technique. Finally I have added a completely new chapter (Chapter 12) as an introduction to the computer numerical control of machine tools.

Acknowledgements

The author and publishers would like to thank the following organisations for their kind permission to reproduce photographs or illustrations.

JSP Ltd. (Fig. 1.7); Crown Copyright (Fig. 1.8); Chubb Fire Ltd. (Fig. 1.18); Desoutter Brothers Ltd. (Fig. 2.22); Neill Tools Ltd. (Figs. 3.10, 3.11, 3.14, 6.18, 6.23, 10.7, 10.8); Mitutoyo Ltd. (Figs. 3.15, 5.33, 6.6, 6.9, 6.10, 6.13–17, 6.22, 6.26–28, 6.31–34, 6.36, 6.37, 6.39–41, 6.44, 6.45, 6.47); A.J. Morgan & Son (Lye) Ltd. (Figs. 4.2, 4.8); Walton and Radcliffe (Sales) Ltd. (Fig 4.3); Q-Max (Electronics) (Fig. 4.4); TI Coventry Gauge Ltd. (Figs. 5.2, 5.5, 5.7, 5.16, 5.19, 5.21, 5.22); L.S. Starrett & Co./Webber Gage Division (Figs. 5.11, 5.12); Verdict Gauge Sales (Fig. 5.30); Rubert & Co. Ltd. (Fig. 5.32); Thomas Mercer Ltd. (Figs. 6.38, 6.42); Faro Technologies UK Ltd. (Fig. 6.46); Draper Tools Ltd. (Figs. 7.17, 7.19); Sandvik Coromant UK (Figs. 7.22, 7.23); W.J. Meddings (Sales) Ltd. (Fig. 8.1); Procter Machine Guards (www.machinesafety.co.uk) (Figs. 8.3, 9.8, 9.9, 11.6); Clarkson-Osborne Tools Ltd (8.18, 11.11); TS Harrison (Fig. 9.1); Pratt Burnerd International Ltd. (Figs. 9.10, 9.12–15); Elliot Machine Tools Ltd. (Figs. 10.1, 11.1, 11.2); Jones & Shipman Hardinge Ltd (Figs 10.5, 10.6); Saint-Gobain Abrasives (Fig. 10.10); Engineering Solutions Ltd. (Figs. 15.11, 15.12); Hinchley Engineering Co. Ltd. (Fig. 15.13); Sweeny & Blockside (Power Pressing) (Fig. 17.2); Verson International Ltd. (Fig. 17.3); P.J. Hare Ltd. (Fig. 17.4); Lloyd Colley Ltd. (Fig. 17.13); P.I. Castings (Altrincham) (Figs 18.1–18.7); Dennis Castings (Fig. 18.10); Ajax Machine Tools International Ltd. (Figs. 12.3, 12.11); Lloyds British Testing Ltd. for information on lifting equipment; Fig. 15.10 is based on an ICI technical service note by permission of Imperial Chemical Industries; Figs. 5.6, 5.7, 5.10 and 15.15 were photographed by John Kelly; Figs. 1.8 and 20.20 are reproduced by kind permission of HMSO. Figures 11.15 and 16.14 were drawn by Graeme S. Black.

The following are reproduced from or based on British Standards by kind permission of the British Standards Institution from whom copies of the complete standards may be obtained.

Table 1.2 and Figs. 1.12–1.16 (BS ISO 3864); Tables 5.4 and 5.5 (BS 1134–2010); Table 5.3 (BS EN ISO 286–1:2010).

CHAPTER 1

Safe practices

Almost everyone working in a factory has at some stage in his or her career suffered an injury requiring some kind of treatment or first aid. It may have been a cut finger or something more serious. The cause may have been carelessness by the victim or a colleague, defective safety equipment, not using the safety equipment provided or inadequate protective clothing. Whatever the explanation given for the accident, the true cause was most likely a failure to think ahead. You must learn to work safely. Your workplace will have its own safety rules so obey them at all times. Ask if you don't understand any instruction and do report anything which seems dangerous, damaged or faulty.

1.1 Health and Safety at Work Act 1974 (HSWA) (as amended)

This Act of Parliament came into force in April 1975 and covers all people at work except domestic servants in a private household. It is aimed at people and their activities, rather than at factories and the processes carried out within them.

The purpose of the Act is to provide a legal framework to encourage high standards of health and safety at work.

Its aims are:

▶ to secure the health, safety and welfare of people at work;
▶ to protect other people against risks to health or safety arising from the activity of people at work;
▶ to control the keeping and use of dangerous substances and prevent people from unlawfully having or using them;
▶ to control the emission into the atmosphere of noxious or offensive substances from premises.

What the law requires is what good management and common sense would lead employers to do anyway, i.e. to look at what the risks are and take sensible measures to tackle them.

1.2 Health and safety organisation (Fig. 1.1)

The HSWA established two bodies, the Health and Safety Commission and the Health and Safety Executive (HSE). These were merged in 2008 to form a single body, the HSE. The HSE is a statutory body established by the HSWA (as amended) which consists of a chairperson and between 7 and 12 executive directors.

The HSE Board is responsible to appropriate ministers for the administration of the 1974 Act.

Workshop Processes, Practices and Materials, Fifth Edition. 9781138784727.
© 2015 Bruce J. Black. Published by Taylor & Francis. All rights reserved.

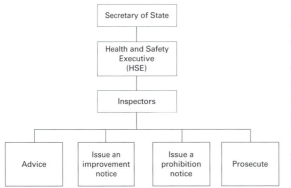

Figure 1.1 Health and safety organisation

The HSE's mission is: 'the prevention of death, injury and ill health to those at work and those affected by work activities'.

HSE's aims are to protect the health, safety and welfare of people at work, and to safeguard others, mainly members of the public, who may be exposed to risks from the way work is carried out.

HSE's statutory functions include:

▶ proposing new and updated laws and standards;
▶ conducting research;
▶ providing information and advice;
▶ making adequate arrangements for the enforcement of health and safety laws.

In recent years much of Britain's health and safety law has originated in Europe. Proposals from the European Commission may be agreed by member states who are then responsible for making them part of their domestic law. Where HSE consider action is necessary to supplement existing arrangements, their three main options are to provide:

1. guidance;
2. approved codes of practice (ACOPs);
3. regulations.

Guidance – can be specific to health and safety problems of an industry or of a particular process used in a number of industries.

The main purposes of guidance are to:

▶ help people understand what the law says;
▶ help people comply with the law;
▶ give technical advice.

Following guidance is not compulsory, but if followed, will normally be enough to comply with the law.

Approved codes of practice – offer practical examples of good practice. They give advice on how to comply with the law by, for example, providing a guide to what is 'reasonably practicable'.

Approved codes of practice have a special legal status. If employers are prosecuted for a breach of health and safety law, and it is proved that they have not followed the relevant provisions of the approved code of practice, a court can find them at fault unless they can show that they have complied with the law in some other way.

Regulations – are law, approved by Parliament. These are usually made under HSWA following proposals from HSE. This applies to regulations based on EC directives as well as 'home-grown' ones.

Health and safety law is enforced by inspectors from HSE.

Local authorities also enforce health and safety law in the workplace allocated to them – including offices, shops, retail and wholesale distribution centres, leisure, hotel and catering premises.

Inspectors may visit a workplace without notice at any reasonable time. They may want to investigate an accident or complaint or examine the safety, health and welfare aspects of the business. They can talk to employees and safety representatives and take photographs and samples.

Inspectors may take enforcement action in several ways to deal with a breach in the health and safety laws.

In most cases these are:

▶ *Informal*: where the breach of the law is relatively minor – give advice both face to face and in writing.
▶ *Improvement notice*: where the breach of the law is more serious – requires remedial action within a time period.
▶ *Prohibition notice*: where the activity involves risk of serious personal injury – prohibit the activity immediately until remedial action is taken.

▶ *Prosecution*: where individuals or corporate bodies fail to comply with the regulations – can lead to a substantial fine or imprisonment or both.

The Health and Safety (Fees) Regulations 2012 puts a duty on the HSE to recover costs from companies who are found to be in breach of health and safety law, referred to as Fee for Intervention (FFI) cost recovery scheme. The fee is based on the amount of time an inspector has had to spend in identifying the breach, helping to put it right, investigating and taking enforcement action. Companies who comply with the law will not pay a fee for any work that the HSE does with them.

1.3 Employer's responsibilities (Fig. 1.2)

Employers have a general duty under the HSWA 'to ensure, so far as is reasonably practicable, the health, safety and welfare at work of their employees'. The principle of 'so far as is reasonably practicable' applies to all the following areas and means that an employer does not have to take measures to avoid or reduce the risk if they are technically impossible or if the time, trouble or cost of the measures would be grossly disproportionate to the risk. The HSWA specifies five areas which in particular are covered by the employer's general duty.

1. Provide and maintain machinery, equipment and other plant, and systems of work that are safe and without risk to health. ('Systems of work' means the way in which the work is organised and includes layout of the workplace, the order in which jobs are carried out or special precautions to be taken before carrying out certain hazardous tasks.)
2. Ensure ways in which particular articles and substances (e.g. machinery and chemicals) are used, handled, stored and transported are safe and without risk to health.
3. Provide information, instruction, training and supervision necessary to ensure health and safety at work. *Information* means the background knowledge needed to put the instruction and training into context. *Instruction* is when someone shows others how to do something by practical demonstration. *Training* means having employees practise a task to improve their performance. *Supervision* is needed to oversee and guide in all matters related to the task.
4. Ensure any place under their control and where their employees work is kept in a safe condition and does not pose a risk to health. This includes ways into and out of the workplace.
5. Ensure the health and safety of their employees' working environment (e.g. heating, lighting, ventilation, etc.). They must also provide adequate arrangements for the welfare at work of their employees (the term 'welfare at work' covers facilities such as seating, washing, toilets, etc.).

1.4 Safety policy

The HSWA requires every employer employing five or more people to prepare a written statement of their safety policy. The written policy statement must set out the employer's aims and objectives for improving health and safety at work.

The purpose of a safety policy is to ensure that employers think carefully about hazards at the workplace and about what should be done to reduce those hazards to make the workplace safe and healthy for their employees.

Another purpose is to make employees aware of what policies and arrangements are being made for their safety. For this reason you must be given a copy which you must read, understand and follow.

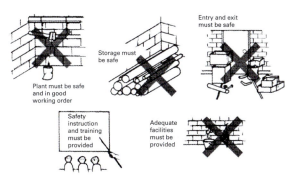

Figure 1.2 Duties of employers

1

The written policy statement needs to be reviewed and revised jointly by employer and employees' representatives as appropriate working conditions change or new hazards arise.

1.5 Safety Representatives and Safety Committees Regulations 1977 (as amended)

1.5.1 Safety representatives

The Regulations came into force on 1 October 1978 and provide recognised trade unions with the right to appoint safety representatives to represent the employees in consultation with their employers about health and safety matters of the organisation.

Employers have a duty to consult on the following matters:

▶ the introduction of any measure at the workplace which may substantially affect the health and safety of employees;
▶ arrangements for getting competent people to help them comply with health and safety laws;
▶ the information they must give their employees on the risks and dangers arising from their work, measures to reduce or eliminate these risks and what the employees should do if exposed to a risk;
▶ the planning and organisation of health and safety training;
▶ the health and safety consequences of introducing new technology.

An employer must give safety representatives the necessary time off, with pay, to carry out their functions and receive appropriate training.

The functions of an appointed safety representative include:

▶ investigating potential hazards and dangerous occurrences in the workplace;
▶ investigating complaints relating to an employee's health, safety or welfare at work;
▶ making representations to the employer on matters affecting the health, safety or welfare of employees at the workplace;
▶ carrying out inspections of the workplace where there has been a change in conditions of work, or there has been a notifiable accident

or dangerous occurrence in a workplace or a notifiable disease has been contracted there;
▶ representing the employees he or she was appointed to represent in consultation with inspectors or any enforcing authority;
▶ attending meetings of safety committees.

1.6 Health and Safety (Consultation with Employees) Regulations 1996 (as amended)

The law is different if there are employees within the organisation who are not represented under the Safety Representatives and Safety Committees Regulations 1977.

For example:

▶ The employer does not recognise trade unions.
▶ There are employees who do not belong to a trade union and recognised trade unions have not agreed to represent them.

Where employees are not represented under the Safety Representatives and Safety Committee Regulations 1977, the Health and Safety (Consultation with Employees) Regulations 1996 will apply.

If the employer decides to consult the employees through an elected representative, the employees have to elect one or more within their group to represent them. These elected health and safety representatives are known as 'representatives of employee safety' in the Regulations. The range of functions of union-appointed representatives and elected representatives within each Regulation is similar and it is good practice for employers to give equivalent functions, where they agree to them.

1.6.1 Safety committees

If two or more union-appointed health and safety representatives request, in writing, that the employer set up a health and safety committee, then it must be done within 3 months of the request.

Although there is no such requirement if the employer consults health and safety representatives elected by the workforce, it is good practice to do so.

The main objective of such a committee is to promote co-operation between employers and employees in setting up, developing and carrying out measures to ensure the health and safety at work of the employees.

The committee can consider items such as:

► statistics on accident records, ill health and sickness absence;
► accident investigation and subsequent action;
► inspection of the workplace by enforcing authorities, management or employee health and safety representatives;
► risk assessments;
► health and safety training;
► emergency procedures;
► changes in the workplace affecting the health, safety and welfare of employees.

If the health and safety committee is discussing accidents, the aim is to stop them happening again, not to give blame. Committees should:

► look at the facts in an impartial way;
► consider what precautions might be taken;
► recommend appropriate action;
► monitor progress of any action taken.

Consulting employees about health and safety can result in:

► a healthier and safer workplace – employees can help identify hazards, assess risks and develop ways to control or remove risks;
► better decisions about health and safety – based on the input and experience of a range of people including employees who have extensive knowledge of their own job;
► a stronger commitment to implementing decisions or actions – since employees have been actively involved in reaching these decisions;
► greater co-operation and trust – employers and employees who talk and listen to each other gain a better understanding of each other's views;
► joint problem solving.

These can then result in real benefits for the business, including:

► increased productivity;
► improvements in overall efficiency and quality;
► higher levels of workforce motivation.

1.7 Employees' responsibilities (Fig. 1.3)

Under the HSWA it is the duty of every employee while at work:

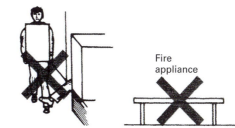

Every employee must take reasonable care of the safety of himself and others

Fire appliance

Never interfere with anything provided in the interests of safety

Figure 1.3 Duties of employees

► To take reasonable care for their own health and safety and that of others who may be affected by what they do or don't do.

This duty implies not only avoiding silly or reckless behaviour but also understanding hazards and complying with safety rules and procedures. This means that you correctly use all work items provided by your employer in accordance with the training and instruction you received to enable you to use them safely.

► To co-operate with their employer on health and safety.

This duty means that you should inform, without delay, of any work situation which might be dangerous and notify any shortcomings in health and safety arrangements so that remedial action may be taken.

The HSWA also imposes a duty on all people, both people at work and members of the public, including children, to not intentionally interfere with or misuse anything that has been provided in the interests of health, safety and welfare.

The type of things covered include fire escapes and fire extinguishers, perimeter fencing, warning notices, protective clothing, guards on machinery and special containers for dangerous substances.

You can see that it is essential for you to adopt a positive attitude and approach to health and safety in order to avoid, prevent and reduce risks

1

at work. Your training is an important way of achieving this and contributes not only to your own but also to the whole organisation's health and safety culture.

1.8 New regulations for health and safety at work

Six new sets of health and safety at work regulations came into force on 1 January 1993. The new regulations implement European Community (EC) directives on health and safety at work in the move towards a single European Union. At the same time they are part of a continuing modernisation of existing UK law.

Most of the duties in the new regulations are not completely new but clarify and make more explicit what is in current health and safety law. A lot of out-of-date law will be repealed by the new regulations, e.g. many parts of the Factories Act 1961. Some of these have been updated since 1993.

The six regulations are:

▶ Management of Health and Safety at Work Regulations 1999;
▶ Provision and Use of Work Equipment Regulations 1998;
▶ Workplace (Health, Safety and Welfare) Regulations 1992;
▶ Personal Protective Equipment at Work Regulations 1992;
▶ Health and Safety (Display Screen Equipment) Regulations 1992;
▶ Manual Handling Operations Regulations 1992.

1.9 Management of Health and Safety at Work (Amendment) Regulations 2006

These Regulations set out broad general duties which operate with the more specific ones in other health and safety regulations. They are aimed mainly at improving health and safety management. Under these Regulations employers are required to assess the risks posed to workers and any others who may be affected by the work or business.

They focus on risk assessments and how to use them affectively to identify potential hazards and

risks, preventive measures that can be applied and the management and surveillance of health and safety procedures that should be followed in the event of serious or imminent danger.

The Regulations require employers to:

▶ assess the risk to health and safety of employees and anyone else who may be affected so that the necessary preventive and protective measures can be identified and record the significant findings of the risk assessment – trivial risks do not need to be recorded;
▶ introduce preventive and protective measures to control the risks identified by the risk assessment;
▶ set up an effective health and safety management system to implement their health and safety policy to include organising, planning, monitoring, auditing and review, and keeping records;
▶ provide appropriate health surveillance of employees required by specific health and safety regulations, e.g. COSHH;
▶ appoint competent people to help devise and apply measures needed to comply with health and safety legislation;
▶ establish emergency procedures to be followed in the event of serious and imminent danger to persons at work;
▶ arrange necessary contacts with external services, e.g. first aid, emergency medical care and rescue work;
▶ give employees information about health and safety matters;
▶ work together with other employers who share the same workplace;
▶ give other employees and self-employed people working in that business information about health and safety matters;
▶ make sure that employees have adequate health and safety instruction and training and are capable enough at the job to avoid the risk;
▶ provide health and safety information to temporary workers to meet their special needs;
▶ ensure that young persons employed by him are protected at work from risks to their health and safety which are a consequence of their

lack of experience, absence of awareness and immaturity.

The Regulations also:

▶ place a duty on the employees to follow health and safety instructions and training in the use of equipment.

1.10 Provision and Use of Work Equipment Regulations 1998 (PUWER)

The Regulations require risks to people's health and safety from equipment they use at work to be prevented or controlled. Although power presses are included as work equipment, part IV of PUWER contains specific requirements for power presses and is dealt with in Chapter 17. In addition to the requirements of PUWER, lifting equipment is also subject to the requirements of the Lifting Operations and Lifting Equipment Regulations 1998 and is dealt with in Chapter 20.

Work equipment has wide meaning and, generally, any equipment which is used by an employee at work is covered:

▶ machines such as circular saws, drilling machines, photocopiers, mowing machines, tractors, dumper trucks and power presses;
▶ hand tools such as screwdrivers, hammers and hand saws;
▶ lifting equipment such as lift trucks, elevating work platforms, vehicle hoists and lifting slings;
▶ other equipment such as ladders and water pressure cleaners;
▶ an installation such as a series of machines connected together, an enclosure to provide sound insulation or scaffolding.

Similarly if you allow employees to provide their own equipment, it too will be covered and the employer will need to make sure it complies.

Examples of use of equipment covered by the Regulations include starting and stopping the equipment, programming, setting, repairing, modifying, maintaining, servicing, cleaning and transporting.

PUWER cannot be considered in isolation from other health and safety legislation but needs to be considered alongside other health and safety laws,

e.g. HSWA and the Workplace (Health, Safety and Welfare) Regulations 1992.

In general terms the Regulations require that equipment provided for use at work is:

▶ suitable for use, and for the purpose and conditions in which it is used;
▶ safe for use, maintained in a safe condition and, in certain circumstances, inspected to ensure this remains the case;
▶ used only by people who have received adequate information and training;
▶ accompanied by suitable safety measures, e.g. protective devices, markings and warnings.

The employer should also ensure that risks created by the use of equipment are eliminated where possible or controlled by:

▶ taking appropriate 'hardware measures', e.g. providing suitable guards, protection devices, markings and warning devices, system control devices (such as emergency stop buttons) and personal protective equipment;
▶ taking appropriate 'software measures' such as following safe systems of work (e.g. ensuring maintenance is only performed when equipment is shut down) and providing adequate information, instruction and training.

Working with machinery can be dangerous because moving machinery can cause injuries in many ways:

▶ workers can be hit and injured by moving parts of machinery or ejected material, and parts of the body can be drawn into or trapped between rollers, belts and pulley drives;
▶ sharp edges can cause cuts and severing injuries, sharp pointed parts can stab or puncture the skin and rough surfaces can cause friction or abrasion injuries;
▶ workers can be crushed both between parts moving together or towards a fixed part of the machine, wall or other object, and two parts moving past one another can cause shearing;
▶ parts of the machine, materials and emissions (such as steam or water) can be hot or cold enough to cause burns or scalds and electricity can cause electrical shock and burns;
▶ injuries can also occur due to machinery becoming unreliable and developing faults due to poor or lack of maintenance or when

1

machines are used improperly through inexperience or lack of training.

The specific requirements of PUWER include:

▶ *The suitability of work equipment* – equipment must be suitable by design and construction for the actual work it is provided to do and installed, located and used in such a way as to reduce the risk to users and other workers, e.g. ensure there is sufficient space between moving parts of work equipment and fixed and moving parts in its environment. Ensure that, where mobile work equipment with a combustion engine is in use, there is sufficient air of good quality.

▶ *Maintenance of work equipment in good repair* – from simple checks on hand tools such as loose hammer heads to specific checks on lifts and hoists. When maintenance work is carried out it should be done in safety and without risk to health.

▶ *Information and instruction on use of the work equipment* – including instruction sheets, manuals or warning labels from manufacturers or suppliers. Adequate training for the purposes of health and safety in the use of specific work equipment.

▶ *Dangerous parts of machinery* – guarding machinery to avoid the risks arising from mechanical hazards. The principal duty is to take effective measures to prevent contact with dangerous parts of machinery by providing:
  ▶ fixed enclosing guards;
  ▶ other guards (see Fig. 1.4) or protection devices;
  ▶ protection appliances (jigs, holders);

▶ information, instruction, training and supervision.

▶ *Protection against specified hazards*
  ▶ material falling from equipment;
  ▶ material ejected from a machine;
  ▶ parts of the equipment breaking off, e.g. grinding wheel bursting;
  ▶ parts of equipment collapsing, e.g. scaffolding;
  ▶ overheating or fire, e.g. bearing running hot, ignition by welding torch;
  ▶ explosion of equipment, e.g. failure of a pressure-relief device;
  ▶ explosion of substance in the equipment, e.g. ignition of dust.

▶ *High and very low temperature* – prevent the risk of injury from contact with hot (blast furnace, steam pipes) or very cold work equipment (cold store).

▶ *Controls and control systems* – starting work equipment should only be possible by using a control and it should not be possible for it to be accidentally or inadvertently operated nor 'operate itself' (by vibration or failure of a spring mechanism).

Stop controls should bring the equipment to a safe condition in a safe manner. Emergency stop controls are intended to effect a rapid response to potentially dangerous situations and should be easily reached and activated. Common types are mushroom-headed buttons (see Fig. 1.5), bars, levers, kick plates or pressure-sensitive cables.

It should be possible to identify easily what each control does. Both the controls and

Figure 1.4 Guard fitted to vertical milling machine

Figure 1.5 Mushroom-headed stop button

their markings should be clearly visible and factors such as colour, shape and position are important.

▶ *Isolation from source of energy* – to allow equipment to be made safe under particular circumstances, e.g. when maintenance is to be carried out or when an unsafe condition develops. Isolation means establishing a break in the energy supply in a secure manner, i.e. by ensuring that inadvertent reconnection is not possible. Isolation may be achieved by simply removing a plug from an electrical socket or by operating an isolating switch or valve.

Sources of energy may be electrical, pressure (hydraulic or pneumatic) or heat.

▶ *Stability* – there are many types of work equipment that might fall over, collapse or overturn unless they are fixed. Most machines used in a fixed position should be bolted down. Some types of work equipment such as mobile cranes may need counterbalance weights.

Ladders should be at the correct angle (a slope of four units up to each one out from the base), correct height (at least 1 metre above the landing place) and tied at the top or secured at the foot.

▶ *Lighting* – if the lighting in the workplace is insufficient for detailed tasks then additional lighting will need to be provided, e.g. local lighting on a machine (Fig. 1.6).

▶ *Markings* – there are many instances where marking of equipment is appropriate for health and safety reasons, e.g. start/stop controls, safe working load on cranes, types of fire extinguishers and pipework colour coded to indicate contents. Markings may use words, letters, numbers or symbols and the use of colour and shape may be significant. Markings should conform to published standards (see 1.20 The Health and Safety (Safety Signs and Signals) Regulations 1996).

▶ *Warnings* – normally in the form of a permanent printed notice or similar, e.g. 'head protection must be worn' (see page 24). Portable warnings are also necessary during temporary operations such as maintenance.

Warning devices can be used which may be audible, e.g. reversing alarms on heavy vehicles, or visible, e.g. lights on a control panel. They may indicate imminent danger, development of a fault or the continued presence of a potential hazard.
They must all be easy to see and understand, and they must be unambiguous.

1.11 Workplace (Health, Safety and Welfare) Regulations 1992 (as amended)

These Regulations cover a wide range of basic health, safety and welfare issues and apply to most workplaces. They aim to ensure that workplaces meet the health, safety and welfare needs of all members of the workforce, including people with disabilities. Where necessary, parts of the workplace, including in particular doors, passageways, stairs, showers, washbasins, lavatories and workstations, should be made accessible for disabled people.

A workplace in these Regulations applies to a very wide range of workplaces, not only factories, shops and offices, but also, for example, schools, hospitals, hotels and places of entertainment, and also includes private roads and paths on industrial estates and business parks.

The Regulations set general requirements in three broad areas, which are outlined here.

▶ Health
  ▶ Ventilation: enclosed workspaces should be sufficiently well ventilated.

Figure 1.6 Local lighting on a centre lathe

1

- ▶ Temperature: inside the workplace should provide reasonable comfort without the need for special clothing.
- ▶ Hot and cold environment: insulate or provide air cooling or as last resort provide protective clothing.
- ▶ Lighting: should be sufficient to enable people to work without experiencing eye strain and to safely move about. Provide artificial lighting if necessary.
- ▶ Cleanliness and waste materials: regularly clean workplace to ensure dirt and refuse is not allowed to accumulate. Spillages and deposits should be removed and cleaned as soon as possible.
- ▶ Room dimensions: work rooms should have enough free space to allow easy access and to move about within the room and not restrict the workers' movements while performing their work. This includes a adequate ceiling height.
- ▶ Workstations: should be arranged so that each task can be carried out safely and comfortably and allows adequate freedom of movement. Work surface height and seating should be arranged appropriate to the work and worker. Everything required for the work should be within easy reach without the need for undue bending or stretching.
- ▶ Safety
  - ▶ Maintenance: equipment that could fail and put workers at serious risk should be properly maintained and checked at regular intervals, as appropriate, by inspection, testing, adjustment, lubrication, repair and cleaning. Equipment which require a system of maintenance include emergency lighting, fencing, powered doors, fixed equipment used for window cleaning and anchorage points for safety harnesses.
  - ▶ Condition of floor: floors and traffic routes should be sound, not uneven or slippery and should be free of obstructions and substances which could cause a slip, trip or fall.
  - ▶ Falls: every vessel containing a dangerous substance should be adequately fenced or covered to prevent a person falling into it. A vessel includes sumps, silos, vats,

pits or tanks. Every open-sided staircase should be securely fenced. A secure and substantial handrail should be provided and maintained on at least one side of every staircase. Falls from height account for a significant number of workplace injuries and unnecessary deaths every year. Employers must ensure that all work at height is properly planned, supervised and carried out by competent people. Duties relating to falls from height are covered by Work at Height Regulations 2005 (WAHR).
  - ▶ Transparent surfaces: windows and transparent or translucent surfaces in doors, gates, walls and partitions should be made of a safety material and marked to make it apparent.
  - ▶ Windows: it should be possible to reach or operate skylights, operable windows and ventilators safely. Controls should be placed so that people are not likely to fall through or out of windows. All windows and skylights in a workplace should be able to be cleaned safely.
  - ▶ Organisation of traffic routes: there should be enough traffic routes of sufficient width and headroom to allow people on foot or vehicles to circulate safely and without difficulty. A traffic route is defined as a route for pedestrian traffic, vehicles or both, and include any stairs, staircases, fixed ladders, doorways, gateways, loading bays or ramps. Potential hazards on traffic routes used by vehicles and pedestrians should be indicated by suitable warning signs. Suitable road markings and signs should be used to alert drivers.
  - ▶ Doors and gates: should be suitably constructed and if necessary be fitted with safety devices. Doors and gates which swing in both directions should have a transparent panel. Sliding doors should have a stop to prevent the door coming off its track. Upward opening doors should have an effective device to prevent them from falling back.
- ▶ Welfare
  - ▶ Toilets: suitable and sufficient toilet facilities shall be provided which are convenient and allow everyone at work to

use them without unreasonable delay. The rooms containing them shall be adequately ventilated and lit and kept in a clean and orderly condition. Separate conveniences should be provided for men and women.

▶ Washing: washbasins should have hot, cold or warm running water. Showers or baths should be provided if required by the nature of the work, e.g. dirty or particularly strenuous.

▶ Drinking water: an adequate supply of wholesome drinking water, normally obtained directly from the mains supply, shall be provided for all persons in the workplace.

▶ Accommodation for clothing: provide accommodation for work clothing and workers' own personal clothing so it can be hung in a clean, warm, dry, well-ventilated place. Where special work clothing becomes dirty, damp or contaminated due to work activities, it should be accommodated separately from the workers' own clothing. In this case, separate changing rooms should be provided.

▶ Rest and eating meals; provide suitable and sufficient rest facilities equipped with an adequate number of tables and chairs. Where workers regularly eat meals at work, provide suitable and sufficient facilities which would include preparing or obtaining a hot drink. Good hygiene standards should be maintained in those parts of the rest facility used for preparing or eating food and drink.

1.12 Working at Height Regulations 2005 (WAHR)

Falls from height are one of the biggest causes of workplace fatalities and major injuries. Common causes are falls from ladders and through fragile roofs. The purpose of the Regulations is to prevent death and injury from a fall from height.

Work at height means work in any place where, if there were no precautions in place, a person could fall a distance liable to cause personal injury. You are working at height if you:

▶ work above ground/floor level;
▶ could fall from an edge, through an opening or fragile surface;
▶ could fall from ground level into an opening in a floor or a hole in the ground.

The Regulations apply to all work at height where there is a risk of a fall liable to cause personal injury. They place duties on employers and those who control any work at height activity.

As part of the Regulation the employer must ensure:

▶ all work at height is properly planned and organised;
▶ those involved in work at height are competent (have had instruction and have sufficient skills, knowledge and experience);
▶ the risks from work at height are assessed, and appropriate work equipment is selected and used;
▶ the risks from working on or near fragile surfaces are properly managed;
▶ the equipment used for work at height is properly inspected and maintained.

Before working at height the following simple steps should be considered:

▶ Avoid working at height where it is reasonably practicable to do so.
▶ Where work at height cannot be avoided, prevent falls using either an existing place of work that is already safe or the right type of equipment.
▶ Minimise the distance and consequences of a fall by using the right type of equipment where the risk cannot be eliminated.

The surface from which work at height can be carried out is referred to by the Regulations as a working platform. A working platform can be a roof, floor, plant and machinery with a fixed guardrail, platform on a scaffold, tower scaffolds, mobile elevating work platforms and the rungs and treads of ladders and stepladders. For the purpose of this book I shall restrict coverage to ladders and stepladders.

The use of ladders and stepladders are not banned under health and safety law. The law says that ladders can be used for work at height when a risk assessment has shown that using equipment offering a higher level of fall protection is not justified because of the low risk and short duration of use. For tasks of low risk and short duration, ladders and stepladders can be a sensible and practicable option. If the risk assessment determines it is correct to use a ladder, the risk can be further minimised by making sure the workers:

▶ use the right type of ladder for the job;
▶ are competent;
▶ use the equipment provided safely and follow a safe system of work;
▶ are fully aware of the risks and measures to help control them.

Ladders should only be used in situations where they can be used safely, e.g. where the ladder will be level and stable, and where it is reasonably practicable to do so, the ladder can be secured.

The employer needs to make sure that any ladder or stepladder is both suitable for the work task and is in a safe condition before use. Only use ladders or stepladders that:

▶ have no visible defects (pre-use check each working day);
▶ have an up-to-date record of regular detailed visual inspections by a competent person;
▶ are suitable for the intended use (strong and robust enough for the job);
▶ have been maintained and stored in accordance with the manufacturer's instructions.

Before starting a task you should always carry out a visual 'pre-use' check which should look for:

▶ twisted, bent or dented stiles;
▶ cracked, worn, bent or loose rungs or treads;
▶ missing or damaged tie rods;
▶ cracked or damaged welded joints, loose rivets or damaged stays and locking mechanisms;
▶ missing or damaged feet;
▶ split or buckled platform on stepladder.

1.12.1 Safe use of ladders/ stepladders

Once a 'pre-use' check has been carried out, there are simple precautions that can be taken to minimise the risk of a fall:

▶ Always position ladder on a firm level base which is clean and free from loose materials.
▶ Take all necessary precautions to avoid vehicles or people hitting the bottom of the

ladder (protect the area using suitable barriers or cones).
▶ Avoid positioning the ladder where it could be pushed over by other hazards such as doors and windows.
▶ Make sure the ladder is at 75° – use the one in four rule (i.e. one unit out for every four units up).
▶ Make sure the ladder is long enough for the task, at least 1 m (three rungs) above where you are working.
▶ Tie the ladder to a suitable point, making sure both stiles are tied.
▶ Never over-reach – be safe, get down and move it.
▶ Don't try to move or extend ladders while standing on the rungs.
▶ Always grip the ladder and face the rungs while climbing or descending.
▶ Don't work off the top three rungs (always maintain a handhold).
▶ Don't stand ladders on movable objects such as pallets, bricks, lift trucks, tower scaffolds and excavator buckets.
▶ Avoid holding items while climbing (consider using a tool belt).
▶ Maintain three points of contact while climbing and wherever possible at the work position (this means a hand and two feet).
▶ Don't work within 6 m horizontally of an overhead power line unless it has been made 'dead'.

1.13 Personal Protective Equipment at Work Regulations 1992 (as amended)

These Regulations require that every employer shall ensure that suitable personal protective equipment (PPE) is provided to his employees who may be exposed to the risk to their health and safety while at work.

PPE should always be relied upon as a last resort to protect against health and safety. Engineering controls and safe systems of work should always be considered first. Where the risks are not adequately controlled by other means, the employer has a duty to ensure that suitable PPE is provided, free of charge.

PPE will only be suitable if it is appropriate for the risks and the working conditions, takes account of the worker's needs and fits properly, gives adequate protection and is compatible with any other item of PPE worn.

PPE at Work Regulations do not apply where separate Regulations require the provision and use of PPE against specific hazards, e.g. gloves for use with dangerous chemicals, covered by Control of Substances Hazardous to Health Regulations 2002.

The employer also has duties to:

▶ assess the risks and PPE intended to be issued and that it is suitable;
▶ maintain, clean and replace PPE;
▶ provide storage for PPE when it is not being used;
▶ ensure that PPE is properly used;
▶ give training, information and instruction to employees on the use of PPE and how to look after it.

1.13.1 Personal protective equipment

PPE is defined as all equipment which is intended to be worn or held to protect against risk to health and safety. This includes most types of protective clothing and equipment such as: eye protection, safety helmets, safety footwear, gloves, high-visibility clothing and safety harness. It does not include hearing protection and most respiratory equipment, which are covered by separate existing regulations.

1.13.1.1 Eye protection

Serves as a guard against the hazards of impact, splashes from chemicals or molten metal, liquid droplets (chemical mists and sprays), dust, gases and welding arcs. Eye protectors include safety spectacles, eye-shields, goggles, visors, welding filters, face shields and hoods (Fig. 1.7). Make sure the eye protection has the right combination of impact/dust/splash/molten metal protection for the task and fits the user properly.

1.13.1.2 Head protection

Includes industrial safety helmets to protect against falling objects or impact with fixed

Figure 1.7 Eye protection

objects; industrial scalp protectors to protect against striking fixed obstacles, scalping or entanglement and caps and hairnets to protect against scalping and entanglement. Some safety helmets incorporate or can be fitted with specially designed eye or hearing protection and can include neck protection for use during welding. Do not use head protection if it is damaged – always have it replaced.

1.13.1.3 Feet and leg protection

Serve to guard against the hazards of electrostatic build-up, abrasion, wet, slipping, cuts and punctures, falling objects, chemical and metal splash. Options include the provision of safety boots and shoes with protective toe caps and penetration-resistant mid soles as well as gaiters, leggings and spats. Footwear can have a variety of sole patterns and materials to help prevent slips in difficult conditions including oil and chemical spillage. They can also be antistatic and electrically or thermally insulating. It is important to select footwear appropriate for the identified risks.

1.13.1.4 Hand and arm protection

Gloves, gauntlets, mitts, wrist cuffs and armlets provide protection against a range of hazards such as abrasion, temperature extremes, cuts and punctures, impact, chemicals, electric shock, skin infection, disease or contamination. Avoid using gloves when operating machines such as bench drills where gloves can get caught. Barrier creams may sometimes be used as an aid to skin hygiene but are no substitute for proper PPE. Wearing gloves for long periods can lead to skin problems but using separate cotton inner gloves may help prevent this. Some people may be allergic to materials used in gloves, e.g. latex.

1

1.13.1.5 Body protection

Types of clothing used for body protection include conventional and disposable overalls, boiler suits, aprons, high-visibility clothing and specialist protective clothing, e.g. chain mail aprons. They are used to protect against hazards such as temperature extremes, adverse weather, chemical and metal splash, spray from spray guns, impact or penetration, contaminated dust or excessive wear or entanglement of own clothing. The choice of materials includes flame retardant, antistatic, chain mail, chemically impermeable and high visibility.

1.14 Health and Safety (Display Screen Equipment) Regulations 1992 (as amended)

These Regulations only apply to employers whose workers regularly use display screen equipment (DSE) as a significant part of their work, e.g. daily, for continuous periods of an hour or more. The Regulations do not apply to workers who use DSE infrequently or for short periods of time. DSE are devices or equipment with a graphic display screen and includes computer display screens, laptops and touch screens.

Some workers may experience fatigue, eye strain, upper limb disorders and backache from overuse or improper use of DSE. Upper limb disorders include problems of the arms, hands, shoulders and neck linked to work activities. These problems can also be experienced from poorly designed workstations or working environment. The cause may not be obvious and can be due to a combination of factors.

Any employer who has DSE users must:

▶ analyse workstations to assess and reduce the risks;
▶ make sure controls are in place;
▶ provide information and training;
▶ provide eye and eyesight tests on request, and special spectacles if required;
▶ review the assessment when the user or DSE changes.

1.14.1 Workstation analysis and controls

In using DSE, as in any other type of work, ill health can result from poor equipment or furniture, work organisation, working environment, job design and posture from inappropriate working methods. Health problems associated with DSE can be prevented in the majority of cases by good ergonomic design of the equipment, workplace and job. Ergonomics is the science of making sure that work tasks, equipment, information and the working environment are suitable for every worker, so that work can be done safely and productively.

Risk assessment should first identify any hazard and then enable an evaluation of the risks. Risks to health may arise from a combination of factors and are particularly likely to occur when the work, workplace and work environment do not take account of the workers' needs.

An employer shall ensure that a workstation meets the requirements of DSE Regulations by ensuring, for example:

▶ adequate lighting;
▶ adequate contrast – no glare or distracting reflections;
▶ distracting noise minimised;
▶ leg room and clearances to allow postural changes;
▶ window covering if needed to minimise glare;
▶ software appropriate to task, adapted to user, providing feedback on system status, e.g. error message;
▶ screen – stable image, adjustable, readable, glare/reflection free;
▶ keyboard – usable, adjustable, detachable, legible;
▶ work surface with space for flexible arrangement of equipment and documents – glare free;
▶ chair – stable and adjustable;
▶ footrest if the user needs one.

Breaking up long spells of DSE work helps prevent fatigue, eye strain, upper limb disorders and back ache. The employer needs to plan so that users can interrupt prolonged use of DSE with changes of activity. Organised or scheduled rest breaks may sometimes be a solution. The following may help users:

- stretch and change position;
- look into the distance from time to time and blink often;
- change activity before user gets tired;
- short, frequent breaks are better than longer, infrequent ones.

1.14.2 Training

The employer must provide information, instruction and health and safety training to users to help them identify risks and safe working practices. When training users the following should be considered:

- the risks from DSE work and the controls put in place;
- how to adjust furniture, work and equipment positions;
- how to organise the workplace to avoid awkward or frequently repeated stretching movements;
- how to clean the screen, keyboard and mouse;
- who to contact for help and to report problems or symptoms.

1.14.3 Eye and eyesight effects

Medical evidence shows that using DSE is not associated with permanent damage to eyes or eyesight nor does it make existing defects worse. It may, however, make workers with pre-existing vision defects more aware of them. Some workers may experience temporary visual fatigue, leading to a range of symptoms such as blurred vision, red or sore eyes and headaches. Visual symptoms may be caused by:

- staying in the same position and concentrating for a long time;
- poor positioning of the DSE;
- poor legibility of the screen, keyboard or source documents;
- poor lighting, including glare and reflections;
- a drifting, flickering or jittering image on the screen.

If a user requests an eye test, the employer is required to provide one. The provision of eye and eyesight tests and special corrective appliances under the DSE Regulations is at the expense of the employer. If the test shows that the user needs spectacles specifically for display screen

work, then the employer must pay for a basic pair of frames and lenses. If the user's normal spectacles are suitable for display screen work, the employer does not have to pay for them.

1.14.4 Review

DSE assessments need to be reviewed when:

- major changes are made to the equipment, furniture, work environment or software;
- users change workstations;
- the nature of the work task changes considerably;
- it is thought that the controls in place may be causing other problems.

1.15 The Reporting of Injuries, Diseases and Dangerous Occurrences Regulations 2013 (RIDDOR)

These Regulations require injuries, diseases and occurrences in specific categories to be notified to the relevant enforcing authorities. In the case of a factory the enforcing authority is the Health and Safety Executive (HSE).

The employer must report incidents and keep appropriate records of:

- work-related accidents which cause death;
- work-related accidents which cause serious injuries (reportable injuries);
- diagnosed cases of certain industrial diseases;
- certain dangerous occurrences (incident with potential to cause harm).

For the purpose of RIDDOR an accident is a separate, identifiable, unintended incident that causes physical injury.

Not all accidents need to be reported but a RIDDOR report is required when:

- the accident is work related;
- the acident results in an injury of a type which is reportable.

Reportable injuries are:

- death of a worker arising from a work-related accident;
- specified injuries to workers in an accident in the workplace which include fractures, amputation, loss of sight, crush injuries,

serious burns and other injury which results in admittance to hospital for more than 24 hours;

▶ over seven-day injuries, where a worker is away from work or unable to perform their normal working duties for more than seven consecutive days;

▶ work-related accidents involving members of the public or people not at work injured and taken from the scene to hospital for treatment.

Reportable occupational diseases are:

▶ diagnosis of certain occupational diseases where these are likely to have been caused or made worse by their work. These include carpal tunnel syndrome, severe cramp of hand or forearm, occupational dermatitis, hand–arm vibration syndrome, occupational asthma, tendonitis and any occupational cancer.

Reportable dangerous occurrences are:

▶ certain specified 'near miss' events, where something happens that does not result in an injury, but could have done. These include:

  ▶ the collapse, overturning or failure of load-bearing parts of lifts and lifting equipment;

  ▶ plant or equipment coming into contact with overhead power lines;

  ▶ explosions or fires causing work to be stopped for more than 24 hours.

Injuries, fatalities and dangerous occurrences must be notified to the enforcing authority, in the case of a factory, the HSE, by the quickest practicable means without delay and a report sent within 10 days. Reports of diseases must be sent to the HSE without delay.

Reports are made online by completing the appropriate online report form. The form will be submitted to the RIDDOR database and a copy sent back to the employer for his records. All incidents can be done online but a telephone service to the incident control centre is available for reporting fatal and specified injuries only.

1.15.1 Recording

The employer must keep a record of:

▶ any accident, occupational disease or dangerous occurrence which requires reporting under RIDDOR;

▶ any other occupational accident causing injuries which result in the worker being away from work for more than three consecutive days. The employer does not have to report injuries longer than three days unless the period off work exceeds seven days.

1.16 Control of Substances Hazardous to Health (COSHH) Regulations 2002 (as amended)

Using chemicals and other hazardous substances at work can put peoples health at risk. So the law requires employers to control exposure to hazardous substances to prevent ill health. Effects from hazardous substances include:

▶ skin irritation or dermatitis as a result of skin contact;

▶ asthma as a result of developing allergy to substances used at work;

▶ losing consciousness as a result of being overcome by toxic fumes;

▶ cancer, which may appear long after exposure to the substances which caused it.

Hazardous substances include:

▶ metalworking fluids;

▶ substances used directly in work activities, e.g. adhesives, paints and cleaning agents;

▶ substances generated during work activities, e.g. fumes from soldering or welding;

▶ naturally occurring substances, e.g. wood dust.

COSHH sets out eight basic measures that employers and employees must take to comply with the Regulations.

1. Assess the risks to health arising from hazardous substances present in their workplace.
2. Decide the precautions needed to avoid exposing employees to hazardous substances.
3. Prevent or adequately control exposure of employees to hazardous substances.
4. Ensure that control measures are used and maintained.
5. Monitor the exposure of employees to hazardous substances if necessary.
6. Carry out appropriate health surveillance where COSHH sets specific requirements.

7. Prepare plans and procedures to deal with accidents, incidents and emergencies involving hazardous substances.

8. Ensure that employees are properly informed, trained and supervised.

Employees must make full and proper use of any control measure; personal protective equipment (PPE) or any facility provided and report any defects found in these. Follow all instructions and safety information provided and only use and dispose of substances in the recommended manner. You should know the warning symbols and pay particular attention to any container bearing any of the symbols shown in Fig. 1.8.

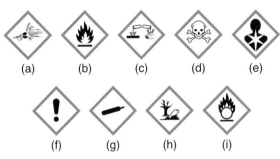

| (a) | (b) | (c) | (d) | (e) |

| (f) | (g) | (h) | (i) |

Figure 1.8 (a) Danger – explosive. (b) Danger – extremely flammable. (c) Warning – corrosive. (d) Danger – fatal if swallowed. (e) Danger – may cause cancer, allergy or asthma symptoms or breathing difficulties if inhaled. (f) Warning – causes skin irritation, harmful if swallowed. (g) Warning – contains gas under pressure; may explode if heated. (h) Warning – very toxic to aquatic life. (i) Danger – may cause or intensify fire; oxidiser.

European legislation on the classification, labelling and packaging (the CLP Regulations) of substances and mixtures came into force in 2009. From 1 June 2015, chemical suppliers must comply with the CLP Regulations.

Given the expanding international market in chemical substances and mixtures, to help protect people and the environment and to facilitate trade, the United Nations has developed a 'Globally Harmonised System' (GHS) on classification and labelling. The CLP Regulations adopts GHS across all European Union countries including the UK.

Most of the chemicals you might use are not dangerous but where a supplier concludes that

a chemical can cause harm they must provide information on the label. This information includes the use of symbols known as hazard pictograms (Fig. 1.8) and comprise a red diamond border with a black pictogram to indicate the hazard. One or more pictograms might appear on a single chemical. The CLP Regulations also introduces two signal words 'Danger' and 'Warning'. If the chemical has a more severe hazard, the label includes the signal word 'Danger' and in the case of less severe hazard, the signal word is 'Warning'.

1.17 Control of Noise at Work Regulations 2005

The Regulations are designed to protect against risks to both health and safety from exposure to noise at work – the health risk of hearing damage to those exposed and safety risks such as noise affecting the ability to hear instructions or warnings.

Sound and noise are an important part of everyday life. In moderation they are harmless but if they are too loud, they can permanently damage your hearing. The danger depends on how loud the noise is and how long the person is exposed to the noise. Once the damage is done there is no cure. Noise at work can cause hearing damage that is permanent and disabling. This can be hearing loss that is gradual because of exposure to noise over time but also damage caused by sudden, extremely loud noises. The damage is disabling as it can stop people being able to understand speech, keep up with conversations or use the telephone. Hearing loss is not the only problem as people can develop tinnitus (ringing, humming or buzzing in the ear), a distressing condition which can lead to disturbed sleep.

Noise at work can interfere with communication and make warnings harder to hear and can reduce people's awareness of their surroundings. These issues can lead to safety risks, putting people at risk of injury or death.

The Regulations require the employer to eliminate or reduce risks to health and safety from noise at work. To comply, the employer is required to:

▶ assess the risks to employees from noise at work;

- take action to eliminate or reduce the noise exposure that produces these risks;
- provide employees with personal hearing protection;
- make sure the legal limits on noise exposure are not exceeded;
- maintain and ensure the use of equipment provided to control noise risks;
- provide employees with information, instruction and training;
- carry out health surveillance (monitor the workers hearing ability).

Action will have to be taken about noise if any of the following apply:

- the noise is obtrusive, e.g. as noisy as a busy road or crowded restaurant or worse, for most of the working day;
- employees have to raise their voices to carry out a normal conversation when about 2 metres apart, for at least part of the day;
- employees use noisy power tools or machinery for more than half an hour each day;
- area is known to have noisy tasks, e.g. engineering, forging and stamping, foundries and press shops;
- there are noises due to impacts, e.g. hammering, drop forging, pneumatic impact tools.

Safety issues will have to be considered in relation to noise where:

- warning sounds are used to avoid or alert to dangerous situations;
- working practices rely on verbal communications;
- there is work around mobile machinery or traffic.

The employer then needs to carry out a risk assessment to decide what action needs to be taken. The findings of the risk assessment must be recorded together with any action taken or intended to be taken to comply with the law.

1.17.1 Noise exposure levels

Noise is measured in decibels (dB). The Regulations require the employer to take specific action at certain action values. The Regulations define 'exposure action values' – levels of noise exposure which, if exceeded, require the employer to take specific action. These relate to:

- the levels of exposure to noise that employees average over a working day or week;
- the maximum noise (peak sound pressure) to which employees are exposed in a working day.

Lower exposure action values:

- daily or weekly exposure of 80 dB;
- peak sound pressure of 135 dB.

Upper exposure action values:

- daily or weekly exposure of 85 dB;
- peak sound pressure of 137 dB.

There are also levels of noise exposure which must not be exceeded. These are called exposure limit values:

- daily or weekly exposure of 87 dB;
- peak sound pressure of 140 dB.

Examples of typical noise levels are shown in Table 1.1.

Table 1.1

Noise source	Noise level dB(A)
Normal conversation	50–60
Loud radio	65–75
Busy street	78–85
Electric drill	87
Sheet metal shop	93
Circular saw	99
Hand grinding metal	108
Chain saw	115–120
Jet aircraft taking off 25 m away	140

1.17.2 Noise control

Whenever there is noise at work, the employer should look for alternative processes, equipment and/or working methods which would make the work quieter or mean that people are exposed for shorter periods of time. The first step is to try to remove the source of noise altogether, e.g. housing the noisy machine where it cannot be heard by workers. If complete removal is not possible, the following should be investigated:

- use quieter equipment or a different, quieter process;

▶ introduce engineering controls – avoid metal to metal contact and add abrasion-resistant rubber and vibration dampening techniques;

▶ use screens, barriers, enclosures and absorbent materials to reduce the noise on its path to the people exposed;

▶ design and lay out the workplace to create quiet workstations;

▶ improve working techniques to reduce noise levels;

▶ limit the time people spend in noisy areas.

Any action taken should be 'reasonably practicable' in proportion to the risk. If the exposure is below the lower action value, the risk is low and it is likely no action is required. However, if there are simple inexpensive practical steps that would further reduce the risk, then implementation should be considered.

1.17.3 Hearing protection

Hearing protectors should be issued to employees:

▶ where extra protection is needed above what has been achieved using noise control;

▶ as a short-term measure while other methods of controlling noise are being developed.

The use of hearing protection should not be used as an alternative to controlling noise by technical and organisational means.

The Regulations require an employer to:

▶ provide employees with hearing protectors and ensure they are used fully and properly when their noise exposure exceeds the upper exposure action values;

▶ provide employees with hearing protectors if they ask for them and their noise exposure is between the lower and upper exposure action values;

▶ identify hearing protection zones – areas of the workplace where access is restricted, and where wearing hearing protection is compulsory.

Employees also have a legal duty to:

▶ co-operate with the employer to do what is needed to protect their hearing, use any noise control devices properly and follow any working methods that are put in place;

▶ properly wear any hearing protection they are given and wear it at all times;

▶ look after their hearing protection;

▶ attend any hearing checks that are part of employer's health surveillance;

▶ report any problems with noise control devices or hearing protection straight away.

As already stated, the use of hearing protection is the last line of defence against damage to hearing. The main types of hearing protector are:

▶ **Earmuffs** – which should totally cover your ears, fit tightly and have no gaps around the seals. Don't allow hair, jewellery, glasses or hats interfere with the seal. Keep the seals and insides clean and don't stretch the headband.

▶ **Earplugs** – go right in the ear canal. Practise fitting them and get help if you are having trouble. Clean your hands before you fit them and don't share them.

▶ **Semi-inserts/canal caps** – are held in or across the ear canal by a band usually plastic. Check for a good seal every time you put them on.

When selecting hearing protectors the employer should take account of the following:

▶ Protection should be sufficient to eliminate risks from noise but not so much protection that the wearers become isolated.

▶ Consider the work and working environment, e.g. physical activity, comfort and hygiene.

▶ Compatibility with other protective equipment, e.g. hard hats, masks and eye protection.

Employers have a duty to maintain hearing protection so that it works effectively such as headband tension and the condition of seals.

1.17.4 Information, instruction and training

Employees should be provided with training so that they understand the risks they may be exposed to, and their duties and responsibilities. This can include:

▶ their likely noise exposure and the risks to hearing this may create;

▶ what the employer is doing to control risks and exposure;

▶ where and how to obtain hearing protection;

▶ how to identify and report defects in noise control equipment and hearing protection;
▶ the employees' duties under the Regulations;
▶ what an employee should do to minimise the risk, such as the proper way to use noise control equipment and hearing protection;
▶ knowing the employer's health surveillance system.

1.17.5 Health surveillance

The employer must provide health surveillance for all employees who are likely to be exposed above the upper exposure action values, or are at risk for any reason, e.g. they already suffer from hearing loss or are particularly sensitive to damage.

Health surveillance usually means regular hearing checks, conducted annually for the first two years of being exposed and then at three-year intervals. This may need to be more frequent if a problem with hearing is detected or where the risk of hearing damage is high.

1.18 Control of Vibration at Work Regulations 2005

These Regulations are designed to protect persons from the risk to their health and safety of the effects of exposure to vibration. There are two types of vibration, hand–arm vibration (HAV) and whole-body vibration. Regular and frequent exposure to HAV can lead to permanent health effects. This is most likely when contact with a vibrating tool or work process is a regular part of a person's job. Too much exposure to HAV can cause hand–arm vibration syndrome (HAVS) and carpal tunnel syndrome (CTS). HAVS affects the nerves, blood vessels, muscles and joints of the hand, wrist and arm. Carpal tunnel syndrome is a nerve disorder which may involve pain, tingling, numbness and weakness in parts of the hand.

The symptoms of HAV include any combination of:

▶ tingling and numbness in the fingers;
▶ not being able to feel things properly;
▶ loss of strength in the hands;
▶ the fingers going white (blanching) and becoming red and painful on recovery (particularly in the cold and wet, and probably only in the tips at first).

A risk assessment must be carried out to identify the measures necessary to reduce vibration, taking into account the amount and duration of exposure. The vibration magnitude is measured in metres per second squared (m/s^2).

The exposure action value (EAV) is a daily amount of vibration exposure, above which employers are required to take action to control exposure. Daily exposure action value is $2.5\,m/s^2$.

The exposure limit value (ELV) is the maximum amount of vibration an employee may be exposed to on any single day. It represents a high risk above which employees should not be exposed. Daily ELV is $5\,m/s^2$.

Health and safety surveillance, information and training must be provided by the employers.

HAV is vibration transmitted from work processes into workers' hands and arms. It can be caused by operating hand-held power tools such as:

▶ hammer drills;
▶ hand-held grinders;
▶ impact wrenches;
▶ jigsaws;
▶ pedestal grinders;
▶ power hammers and chisels.

Whole-body vibration is not relevant to this book.

1.19 Electrical hazards

Electrical equipment of some kind is used in every factory. Electricity should be treated with respect – it cannot be seen or heard, but it can kill. Even if it is not fatal, serious disablement can result through shock and burns. Also, a great deal of damage to property and goods can be caused, usually through fire or explosion as a result of faulty wiring or faulty equipment.

The Electricity at Work Regulations 1989 came into force on 1 April 1990. The purpose of the Regulation is to require precautions to be taken against the risk of death or personal injury from electricity in work activities.

The British Standard BS 7671—'Requirements for Electrical Installations', also known as the IEE Wiring Regulations, relates to the design, selection, erection, inspection and testing of electrical installations. BS 7671 is a code of practice and although non-statutory, it is widely recognised and accepted in the UK and compliance with it is likely to achieve compliance with the relevant aspects of the Electricity at Work Regulations.

1.19.1 Major hazards

The major hazards arising from the use of electrical equipment are:

1.19.1.1 Electric shock

The body responds in a number of ways to electric current flowing through it, any one of which can be fatal. The chance of electric shock is increased in wet or damp conditions, or close to conductors such as working in a metal tank. Hot environments where sweat or humidity reduce the insulation protection offered by clothing increase the risk.

1.19.1.2 Electric burn

This is due to the heating effect caused by electric current passing through body tissue, most often the skin at the point of contact giving rise to the electric shock.

1.19.1.3 Fire

Caused by electricity in a number of ways including overheating of cables and electrical equipment due to overloading; leakage currents due to poor or inadequate insulation; overheating of flammable materials placed too close to electrical equipment; ignition of flammable materials by sparking of electrical equipment.

1.19.1.4 Arcing

Generates ultraviolet radiation causing a particular type of burn similar to severe sunburn. Molten metal resulting from arcing can penetrate, burn and lodge in the flesh. Ultraviolet radiation can also cause damage to sensitive skin and to eyes, e.g. arc eye in metal arc welding.

1.19.1.5 Explosion

These include the explosion of electrical equipment, e.g. switchgear or motors, or where electricity causes the ignition of flammable vapours, gases, liquids and dust by electric sparks or high temperature electrical equipment.

1.19.2 Electrical precautions

Where it is possible for electrical equipment to become dangerous if a fault should arise, then precautions must be taken to prevent injury. These precautions include:

1.19.2.1 Double insulation

The principle is that the live conductors are covered by two discrete layers of insulation. Each layer would provide adequate insulation in itself but together they ensure little likelihood of danger arising from insulation failure. This arrangement avoids the need for an earth wire. Double insulation is particularly suitable for portable equipment such as drills. However, safety depends on the insulation remaining in sound condition and the equipment must be properly constructed, used and maintained.

The most vulnerable item on any portable equipment is the cable, which can deteriorate from age or environmental effects, abuse or misuse, or fail because of repeated flexing or from mechanical damage through being struck or penetrated by objects.

1.19.2.2 Earthing

In the UK, the electricity supply is connected to earth. It is this system that enables earth faults on electrical equipment to be detected and the electrical supply to be cut off automatically. This automatic cut-off is performed by fuses or automatic circuit breakers: if a fault occurs the fuse will blow and break the circuit. Although they do not eliminate the risk of electric shock, danger may be reduced by the use of a residual current device (RCD) designed to operate rapidly at small leakage currents. RCDs should only be considered as providing a second line of defence. It is essential to regularly operate the test trip button to maintain their effectiveness.

1.19.2.3 Use of safe voltages

Reduced voltage systems (110 volts) are particularly suitable for portable electrical equipment in construction work and in high-conducting locations such as boilers, tunnels and tanks; where the risk to equipment and trailing cables is high; and where the body may be damp.

Where electrically powered tools are used, battery operated ones are the safest. Electrical risks can be eliminated by using air, hydraulic or hand-operated tools especially in harsh conditions.

1.19.3 The human body as part of a circuit

In order to minimise the risk of shock and fire, any metalwork other than the current-carrying conductor must be connected to earth. The neutral of the electrical supply is earthed at the source of distribution, i.e. the supply transformer, so that, if all appliances are also connected to earth, a return path for the current will be available through earth when a fault occurs (see Fig. 1.9).

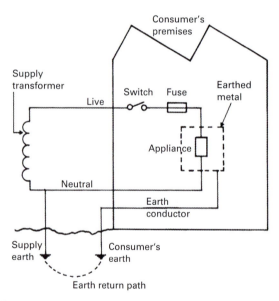

Figure 1.9 Electric circuit for premises

To be effective, this earth path must be of sufficiently low resistance to pass a relatively high current when a fault occurs. This higher current will in turn operate the safety device in the circuit, i.e. the fuse will blow.

Accidents happen when the body provides a direct connection between the live conductors – when the body or a tool touches equipment connected to the supply. More often, however, the connection of the human body is between one live

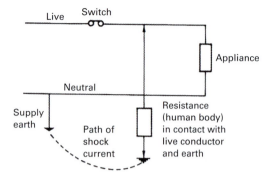

Figure 1.10   Human body as a resistance in electric circuit

conductor and earth, through the floor or adjacent metalwork (see Fig. 1.10). Metal pipes carrying water, gas or steam, concrete floors, radiators and machine structures all readily provide a conducting path of this kind.

Any article of clothing containing any metal parts increases the likelihood of accidental electrical contact. Metal fittings such as buttons, buckles, metal watch or wrist bracelets or dog tags, and even rings could result in shock or burns.

Wetness or moisture at surfaces increases the possibility of leakage of electricity, by lowering the resistance and thus increasing the current. Contact under these conditions therefore increases the risk of electric shock.

All metals are good electrical conductors and therefore all metallic tools are conductors. Any tool brought near a current-carrying conductor will bring about the possibility of a shock. Even tools with insulated handles do not guarantee that the user will not suffer shock or burns.

1.19.4 Electric shock and treatment

If the human body accidentally comes in contact with an electrical conductor which is connected to the supply, a current may, depending on the conditions, flow through the body. This current will at least produce violent muscular spasms which may cause the body to be flung across the room or fall off a ladder. In extreme cases the heart will stop beating.

Burns are caused by the current acting on the body tissue and internal heating can also take place leading to partial blockage of blood flow through the blood vessels.

In the event of someone suffering electric shock know what to do – it should form part of your training – do not put yourself in danger.

1. Shout for help – if the casualty is still in contact with electric current, switch off or remove the plug.
2. If the current cannot be switched off, take special care to stand on a dry non-conducting surface and pull or push the victim clear using a length of dry cloth, jacket or a broom. Remember: do not touch the casualty as you will complete the circuit and also receive a shock.
3. Once the casualty is free, immediately call for an ambulance and get help. Only those with the necessary training, knowledge and skill should carry out first aid.

Posters giving the detailed procedure to be followed in the event of a person suffering electric shock must be permanently displayed in your workplace. With this and your training you should be fully conversant with the procedures – remember it could save a life.

1.19.5 General electrical safety rules

▶ Ensure that a properly wired plug is used for all portable electrical equipment (see Fig. 1.11)

brown wire	**live** conductor
blue wire	**neutral** conductor
green/yellow wire	**earth** conductor.

▶ Never improvise by jamming wires in sockets with nails or matches.
▶ Moulded rubber plugs are preferable to the brittle plastic types, since they are less prone to damage.

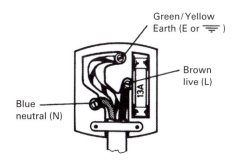

Figure 1.11 Correctly wired plug

▶ All electrical connections must be secure, loose wires or connections can arc.
▶ A fuse of the correct rating must be fitted – this is your safeguard if a fault develops – never use makeshift fuses such as pieces of wire.
▶ use an RCD where appropriate to provide additional safety.
▶ Any external metal parts must be earthed so that if a fault develops, the fuse will blow and interrupt the supply.
▶ Never run power tools from lamp sockets.
▶ Connection between the plug and equipment should be made with the correct cable suited to the current rating of the equipment.
▶ Old or damaged cable should never be used.
▶ Equipment should always be disconnected from the mains supply before making any adjustment, even when changing a lamp.
▶ Do not, under any circumstances, interfere with any electrical equipment or attempt to repair it yourself. All electrical work should be done by a qualified electrician. A little knowledge is often sufficient to make electrical equipment function but a much higher level of knowledge and expertise is usually needed to ensure safety.

1.20 The Health and Safety (Safety Signs and Signals) Regulations 1996

These Regulations cover various means of communicating health and safety information. These include the use of illuminated signs, hand and acoustic signals (e.g. fire alarms), spoken communication and the marking of pipework containing dangerous substances. These are in addition to traditional signboards such as prohibition and warning signs. Fire safety signs are also covered.

The Regulations require employers to provide specific safety signs where there is a significant risk to health and safety which has not been avoided or satisfactorily controlled by other means, e.g. by engineering controls and safe systems of work. Every employer must provide sufficient information, instruction and training in the meaning of safety signs and the measures to be taken in connection with safety signs.

1.21 Safety signs and colours

Colours play an essential safety role in giving information for use in the prevention of accidents, for warning of health hazards, to identify contents of gas cylinders, pipeline and services, the identification and safe use of cables and components in electronic and electrical installations as well as the correct use of fire-fighting equipment.

The purpose of a system of safety colours and safety signs is to draw attention to objects and situations which affect or could affect health and safety. The use of a system of safety colours and safety signs does not replace the need for appropriate accident prevention measures.

British Standard BS ISO 3864 'Graphical Symbols, Safety Colours and Safety Signs' is concerned with a system for giving safety information which does not, in general, require the use of words and covers safety signs, including fire safety signs.

The safety colours, their meaning, and examples of their use are shown in Table 1.2. Examples of the shape and colour of the signs are shown in Figs. 1.12–1.16.

Red   Smoking
             prohibited

Figure 1.12 Prohibition – indicates certain behaviour is prohibited

Red   Fire
             extinguisher

Figure 1.13 Fire equipment

Yellow   Caution
             toxic hazard

Figure 1.14 Hazard – indicates warning of possible danger

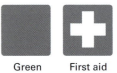

Green   First aid

Figure 1.15 Safe condition – conveys information about safe conditions

Blue   Head protection
             must be worn

Figure 1.16 Mandatory – indicates specific course of action to be taken

Table 1.2

Safety colour	Meaning	Examples of use
Red (white background colour with black symbols)	Stop Prohibition (**Don't** do)	Stop signs Emergency stops Prohibition signs
Red (white symbols and text)	Fire equipment	Position of fire equipment, alarms, hoses, extinguishers, etc.
Yellow (black symbols and text)	Hazard (risk of danger)	Indication of hazards (electrical, explosive, radiation, chemical, vehicle, etc.) Warning of threshold, low passages, obstacles
Green (white symbols and text)	Safe condition (the safe way)	Escape routes Emergency exits Emergency showers First-aid and rescue stations
Blue (white symbols and text)	Mandatory action (**MUST** do)	Obligation to wear personal safety equipment

1.21.1 Portable fire extinguishers

All fire extinguishers manufactured to European Standard BS EN 3-7 are coloured red with icons to indicate the type of fire to which they are suited and the means of operation.

The European Standard allows a small colour zone at the top half front of the extinguisher body relating to the old British Standard extinguisher colour coding system, i.e. red for water, cream for foam, blue for powder, black for CO_2 and canary yellow for wet chemical.

The colours should be visible through a horizontal arc of 180° when the extinguisher is properly mounted. The area of the colour zone should be up to 10% of the body area but not less than 3% of the body area.

1.22 Fire

Most fires can be prevented by taking responsibility for and adopting the correct behaviours and procedures.

Fires need three things to start:

1. Sources of ignition (heat)
2. Sources of fuel (something that burns)
3. Sources of oxygen.

For example:

▶ Sources of ignition include heaters, lighting, naked flames, hot metals, electrical equipment, smoking materials (cigarettes, matches, lighters, etc.), and anything else that can get very hot or cause sparks.
▶ Sources of fuel include wood, paper, plastics, rubber or foam, loose packaging materials, waste rubbish, oils and flammable liquids.
▶ Sources of oxygen including the air around us.

No one should underestimate the danger of fire. Many materials burn rapidly and the fumes and smoke produced, particularly from synthetic material, including plastics, may be deadly.

There are a number of reasons for fires starting:

▶ malicious ignition: i.e. deliberate fire raising;
▶ misuse or faulty electrical equipment: e.g. incorrect plugs and wiring, damaged cables, overloaded sockets and cables, sparking and

equipment such as soldering irons left on and unattended;
▶ cigarettes and matches: smoking in unauthorised areas, throwing away lighted cigarettes or matches;
▶ mechanical heat and sparks: e.g. faulty motors, overheated bearings, sparks produced by grinding and cutting operations;
▶ heating plant: flammable liquids/substances in contact with hot surfaces;
▶ rubbish burning: casual burning of waste and rubbish.

There are a number of reasons for the spread of fire including:

▶ delayed discovery;
▶ presence of large quantities of combustible materials;
▶ lack of fire-separating walls between production and storage areas;
▶ openings in floors and walls between departments;
▶ rapid burning of dust deposits;
▶ oils and fats flowing when burning;
▶ combustible construction of buildings;
▶ combustible linings of roofs, ceilings and walls.

1.22.1 Fire prevention

The best prevention is to stop a fire starting:

▶ where possible use materials which are less flammable;
▶ minimise the quantities of flammable materials kept in the workplace or store;
▶ store flammable material safely, well away from hazardous processes or materials, and where appropriate, from buildings;
▶ warn people of the fire risk by a conspicuous sign at each workplace, storage area and on each container;
▶ some items, like oil-soaked rags, may ignite spontaneously; keep them in a metal container away from other flammable material;
▶ before welding or similar work remove or insulate flammable material and have fire extinguishers to hand;
▶ control ignition sources, e.g. naked flames and sparks, and make sure that 'no smoking' rules are obeyed;

- do not leave goods or waste to obstruct gangways, exits, stairs, escape routes and fire points;
- make sure that vandals do not have access to flammable waste materials;
- comply with the specific precautions for highly flammable gas cylinders such as acetylene;
- after each spell of work, check the area for smouldering matter or fire;
- burn rubbish in a suitable container well away from buildings and have fire extinguishers to hand;
- never wedge open fire-resistant doors designed to stop the spread of fire and smoke;
- have enough fire extinguishers, of the right type and properly maintained, to deal promptly with small outbreaks of fire.

1.22.2 Regulatory Reform (Fire Safety) Order 2005

The Order requires:

A fire risk assessment to be carried out which addresses the following:

- measures to reduce the risk of fire and the risk of the spread of fire on the premises;
- measures in relation to the means of escape from the premises;
- measures for securing that, at all material times, the means of escape can be safely and effectively used;
- measures in relation to the means for fighting fires on the premises;
- measures in relation to the means for detecting fire on the premises and giving warning in the case of fire on the premises;
- measures in relation to the arrangements for action to be taken in the event of fire on the premises, including:
  - measures relating to the instruction and training of employees,
  - measures to mitigate the effects of the fire.

A responsible person be appointed to carry out the risk assessment and demonstrate that the fire safety precautions are adequate including adequate safety training provision.

1.22.3 Fire fighting

Every employee should know where the portable fire extinguishers, the hose reels and the controls for extinguishing are located and how to operate extinguishers in their working area. Training must include the use of extinguishers on simulated fires. Tackling a small fire with an extinguisher may make the difference between a small incident and a full-scale disaster. Portable fire extinguishers save lives and property by putting out small fires or containing them until the fire brigade arrives. They should only be used for fighting a fire in its very early stages.

It must be stressed that fire fighting should only be attempted if it is safe to do so and that an escape route must always be available.

It is also essential to emphasise the limits of first-aid fire fighting in order to show the need to attempt this safely and the importance of first raising the alarm. If in doubt, get out, call the fire brigade out and stay out.

As previously stated, a fire requires fuel, oxygen (air) and heat. This is shown by the 'fire triangle' in Fig. 1.17, where one side stands for fuel, another for heat and the third for air or oxygen. If any one side is removed the fire inside will go out.

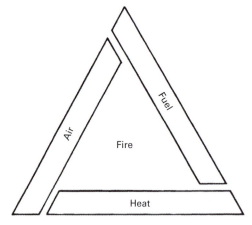

Figure 1.17 Fire triangle

The extinguishing of a fire is generally brought about by depriving the burning substances of oxygen and by cooling them to a temperature below which the reaction is not sustained.

By far the most important extinguishing agent, by reason of its availability and general effectiveness, is water. It is more effective than any other common substance in absorbing heat, thereby reducing the temperature of the burning mass. The steam produced also has a smothering action by lowering the oxygen content of the atmosphere near the fire.

For these reasons the use of a water hose reel in factories is common and is suitable for most fires except those involving flammable liquids or live electrical equipment.

Types of fire are classified as follows while the corresponding icons which have been established to provide easy identification regardless of language are as shown in Table 1.3.

▶ Class A fires – freely burning fires fuelled by ordinary combustible materials such as cloth, wood, paper and fabric.

▶ Class B fires – fires fuelled by flammable liquids such as oils, spirits and petrol.

▶ Class C fires – fires fuelled by flammable gases such as propane, butane and North Sea gas.

▶ Class D fires – fires involving flammable metals such as Magnesium, Lithium or Aluminium powders or swarf.

▶ Fires involving electrical hazards.

▶ Class F fires – fires fuelled by cooking oils and fats. Use of a wet chemical is the most effective way of extinguishing this type of fire.

1.22.4 Types of portable fire extinguishers (Fig. 1.18)

1.22.4.1 Water

Colour coded red – these are suitable for class A types of fires. Water is a fast, efficient means of extinguishing these materials and works by having a rapid cooling effect on the fire so that insufficient heat remains to sustain burning and continuous ignition ceases.

Figure 1.18 Portable fire extinguishers

1.22.4.2 Water with additives

These are suitable for class A types of fire. They contain special additives and are particularly effective for cooling and penetrating the fire and can be up to 300% more effective than the ordinary jet water extinguisher.

Table 1.3

Which extinguisher to use	Freely burning materials	Flammable liquids	Flammable gases	Flammable metals	Electrical hazards	Cooking oils and fat
Water	✓					
Water with additive	✓					
Spray foam	✓	✓				
Dry powder	✓	✓	✓		✓	
Dry powder special metal				✓		
CO$_2$ gas		✓				
Wet chemical	✓					

1

1.22.4.3 Spray foam

Colour coded red with cream colour zone – these are ideal in multi-risk situations where both class A and B type fires are likely. Spray foam has a blanketing effect which both smothers the flame and prevents re-ignition of flammable vapours by sealing the surface of the material.

1.22.4.4 Dry powder

Colour coded red with blue colour zone – these are suitable for class A, B and C fires and for vehicle protection. Because dry powder is non-conductive it is ideal for electrical hazards. Dry powder is a highly effective means of extinguishing fires as it interferes with the combustion process and provides rapid fire knockdown. A specialist range designed to tackle flammable metals is available.

1.22.4.5 Carbon dioxide (CO_2 gas)

Colour coded red with black colour zone – these are suitable for class B type fires and are also ideal for electrical hazards because CO_2 is non-conductive. CO_2 is an extremely fast fire control medium. These extinguishers deliver a powerful concentration of CO_2 gas under great pressure, which smothers the flames very rapidly by displacing air from the local area of the fire. CO_2 is a non-toxic, non-corrosive gas that is harmless to most delicate equipment and materials found in situations such as computer rooms.

1.22.4.6 Wet chemical

Colour coded red with canary yellow colour zone. These have been specifically developed to deal class F type fires. The specially formulated wet chemical, when applied to the burning liquid, cools and emulsifies the oil changing it into soap form, extinguishing the flame and sealing the surface to prevent re-ignition. It is also capable of fighting class A fires.

Table 1.3 shows the portable extinguishers most suited to each class of fire.

1.23 Dangerous Substances and Explosive Atmospheres Regulations (DSEAR) 2002

This set of Regulations is concerned with protecting against risks from fire and explosion arising from dangerous substances used or present in the workplace. Dangerous substances are any substances used or present at work that could, if not properly controlled, cause harm to people as a result of a fire or explosion. They include such things as solvents, paints, varnishes, flammable gases such as liquid petroleum gas (LPG) and dusts from machining. The Regulations require employers to control the risks to safety from fire and explosions.

Employers must:

▶ identify any dangerous substances in their workplace and the fire and explosion risks;
▶ put control measures in place to either remove those risks or, where this is not possible, control them;
▶ put controls in place to reduce the effects of any incidents involving dangerous substances;
▶ prepare plans and procedures to deal with accidents, incidents and emergencies involving dangerous substances;
▶ ensure employees are provided with information and training to enable them to control or deal with risks arising from dangerous substances;
▶ identify and classify areas of the workplace where explosive atmospheres may occur and avoid ignition sources (e.g. from unprotected equipment) in those areas.

1.24 First aid at work

People at work can suffer injuries or become ill. It doesn't matter whether the injury or illness is caused by the work they do or not, what is important is that they receive immediate attention and that an ambulance is called in serious cases.

The Health and Safety (First Aid) Regulations 1981 requires the employer to provide adequate and appropriate equipment, facilities and personnel to enable first aid to be given to employees if they are injured or become ill at work. It is important to remember that accidents can happen at any time and so first aid provision must be available at all times people are at work.

The minimum first aid provision for any workplace is:

▶ a suitably stocked first aid box;
▶ an appointed person to take charge of first aid arrangements;
▶ information for employees about first aid arrangements.

An appointed person is someone the employer chooses to:

▶ take charge when someone is injured or becomes ill, including calling an ambulance if required;
▶ look after the first aid equipment, e.g. restocking the first aid box.

Appointed persons do not need first aid training though emergency first aid courses are available.

Depending on the category of risk and the number of people employed, it may be necessary to appoint a first-aider.

A first-aider is someone who has undertaken training by a competent training provider, appropriate to the circumstances, and must hold a valid certificate of competence in one of the following:

▶ first aid at work (FAW);
▶ emergency first aid at work (EFAW);
▶ any other level of training or qualification that is appropriate to the circumstances.

The employer can use the findings of their first aid needs assessment to decide the appropriate level to which first aiders should be trained.

▶ EFAW training enables a first-aider to give emergency first aid to someone who is injured or becomes ill at work.
▶ FAW training includes the EFAW syllabus and also equips the first-aider to apply first aid to a range of specific injuries and illness.
▶ Additional training to deal with injuries caused by special hazards, e.g. chemicals.

To help keep their basic skills up to date, it is strongly recommended that first-aiders undertake annual refresher training.

Certificates for the purpose of first aid at work last for three years. Before their certificates expire, first-aiders will need to undertake a re-qualification course as appropriate to obtain another three-year certificate. Once a certificate has expired, the first-aider is no longer considered to be competent to act as a workplace first-aider.

The training organisation should only issue certificates to those students it has assessed as competent through demonstrating satisfactory knowledge, skills and understanding in all aspects of the training course.

1.25 Causes of accidents

Workplace accidents can be prevented – you only need commitment, common sense and to follow the safety rules set out for your workplace. Safety doesn't just happen – you have to make it happen.

Most accidents are caused by carelessness, through failure to think ahead or as a result of fatigue. Fatigue may be brought on by working long hours without sufficient periods of rest or even through doing a second job in the evening.

Taking medicines can affect people's ability to work safely, as can the effects of alcohol. Abuse of drugs or substances such as solvents can also cause accidents at work.

Serious injury and even death have resulted from horseplay, practical jokes or silly tricks. There is no place for this type of behaviour in the workplace.

Improper dress has led to serious injury: wearing trainers instead of safety footwear, and loose cuffs, torn overalls, floppy woollen jumpers, rings, chains, watch straps and long hair to get tangled up.

Don't forget, quite apart from the danger to your own health and safety, you are breaking the law if you fail to wear the appropriate personal protective equipment.

Unguarded or faulty machinery, and tools are other sources of accidents. Again within the health and safety law you must not use such equipment and furthermore it is your duty to report defective equipment immediately.

1

Accidents can occur as a result of the workplace environment, e.g. poor ventilation, temperature too high or too low, bad lighting, unsafe passages, doors, floors and dangers from falls and falling objects.

They can also occur if the workplace, equipment and facilities are not maintained, are not clean and rubbish and waste materials are not removed.

Many accidents befall new workers in an organisation, especially the young, and are the result of inexperience, lack of information, instruction, training or supervision all of which is the duty of the employer to provide.

1.26 General health and safety precautions

As already stated, you must adopt a positive attitude and approach to health and safety. Your training is an important way of achieving competence and helps to convert information into healthy and safe working practices.

Remember to observe the following precautions.

Horseplay

▶ work is not the place for horseplay, practical jokes, or silly tricks.

Hygiene

▶ always wash your hands using suitable hand cleaners and warm water before meals, before and after going to the toilet, and at the end of each shift;
▶ dry your hands carefully on the clean towels or driers provided – don't wipe them on old rags;
▶ paraffin, petrol or similar solvents should never be used for skin-cleaning purposes;
▶ use appropriate barrier cream to protect your skin;
▶ conditioning cream may be needed after washing to replace fatty matter and prevent dryness;
▶ take care working with metalworking fluids.

Housekeeping

▶ never throw rubbish on the floor;
▶ keep gangways and work area free of metal bars, components, etc.;
▶ if oil or grease is spilled, wipe it up immediately or someone might slip and fall;

▶ never put oily rags in overall or trouser pockets.

Moving about

▶ always walk – never run;
▶ keep to gangways – never take shortcuts;
▶ look out for and obey warning notices and safety signs;
▶ never ride on a vehicle not made to carry passengers, e.g. fork-lift trucks.

Personal protective equipment

▶ use all personal protective clothing and equipment, such as ear and eye protectors, dust masks, overalls, gloves, safety shoes and safety helmets;
▶ get replacements if damaged or worn.

Ladders

▶ refer to Section 1.12.1 on page 12.

Machinery

▶ ensure you know how to stop a machine before you set it in motion;
▶ keep your concentration while the machine is in motion;
▶ never leave your machine unattended while it is in motion;
▶ take care not to distract other machine operators;
▶ never clean a machine while it is in motion – always isolate it from the power supply first;
▶ never clean swarf away with your bare hands – always use a suitable rake;
▶ keep your hair short or under a cap or hairnet – it can become tangled in drills or rotating shafts;
▶ avoid loose clothing – wear a snug-fitting boiler suit, done up, and ensure that any neckwear is tucked in and secure;
▶ do not wear rings, chains or watches at work – they have caused serious injury when caught accidentally on projections;
▶ do not allow unguarded bar to protrude beyond the end of a machine, e.g. in a centre lathe;
▶ always ensure that all guards are correctly fitted and in position – remember, guards are fitted on machines to protect you and others from accidentally coming in contact with dangerous moving parts.

Harmful substances
▶ learn to recognise hazard warning signs and labels;
▶ follow all instructions;
▶ before you use a substance find out what to do if it spills onto your hand or clothes;
▶ never eat or drink in the near vicinity;
▶ do not take home any clothes which have become soaked or stained with harmful substances;
▶ do not put liquids or substances into unlabelled or wrongly labelled bottles or containers.

Electricity
▶ make sure you understand all instructions before using electrical equipment;
▶ do not use electrical equipment for any purpose other than, nor in the area other than the intended one;
▶ always switch off or isolate before connecting or disconnecting any electrical equipment.

Compressed air
▶ only use compressed air if allowed to do so;

▶ never use compressed air to clean a machine – it may blow in your face or someone else's and cause an injury.

Fire
▶ take care when using flammable substances;
▶ know the location of fire extinguishers;
▶ know the correct fire drill.

Smoking
▶ smoking in public places and the workplace is dealt with as a public health matter in Great Britain and not by the HSE. The laws create three specific offences:
  ▶ failing to display no-smoking signs;
  ▶ smoking in a smoke free-place;
  ▶ failing to prevent smoking in a smoke-free place.

First aid
▶ have first aid treatment for every injury however trivial;
▶ know the first aid arrangements for your workplace.

Review questions

1. State four major hazards which may arise from the use of electrical equipment.
2. What is the purpose of a system of safety colors and safety signs?
3. Show by means of the fire triangle, the three elements necessary to produce fire.
4. List four ways in which accidents may be caused.
5. State two aims of the HSWA.
6. State four precautions to be taken to avoid accidents when using machinery.
7. List three types of fire extinguishing medium and the types of fire for which each is best suited.
8. State two major responsibilities of an employee, while at work, under HSWA.
9. Name four of the effects from hazardous substances defined by COSHH.
10. Indicate four ways in which health and safety inspectors can take enforcement action on an organisation.
11. State two major responsibilities of employers under HSWA.
12. What do the letters PPE stand for?
13. Name four types of incident which an employer must report as required by RIDDOR.
14. What are the levels of noise exposure which must not be exceeded under the Control of Noise at Work Regulations?
15. Name two symptoms of injury which could result from exposure to hand–arm vibration.

CHAPTER 2

Hand processes

Hand tools are used to remove small amounts of material, usually from small areas of the workpiece. This may be done because no machine is available, the workpiece is too large to go on a machine, the shape is too intricate or simply that it would be too expensive to set up a machine to do the work.

Since the use of hand tools is physically tiring, it is important that the amount of material to be removed by hand is kept to an absolute minimum and that the correct tool is chosen for the task. Wherever possible, use should be made of the available powered hand tools, not only to reduce fatigue but also to increase the speed of the operation and so reduce the cost.

2.1 Engineer's files

Files are used to perform a wide variety of tasks, from simple removal of sharp edges to producing intricate shapes where the use of a machine is impracticable. They can be obtained in a variety of shapes and in lengths from 150 mm to 350 mm. When a file has a single series of teeth cut across its face it is known as *single-cut* file, and with two sets of teeth cut across its face it is known as *double-cut* file, Fig. 2.1.

The grade of cut of a file refers to the spacing of the teeth and determines the coarseness or

Figure 2.1   Single-cut and double-cut files

smoothness of the file. Three standard grades of cut in common use, from coarsest to smoothest, are *bastard, second cut* and *smooth*. In general, the bastard cut is used for rough filing to remove the most material in the shortest time, the second cut to bring the work close to finished size and the smooth cut to give a good finish to the surface while removing the smallest amount of material.

2.1.1 File indentification

Files are identified either by their general shape – i.e. hand, flat or pillar – or by their cross-section – i.e. square, three-square, round, half-round or knife – Fig. 2.2.

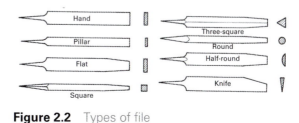

Figure 2.2   Types of file

2

2.1.1.1 Hand file

The hand file is for general use, typically on flat surfaces. It is rectangular in cross-section, parallel in width along its length, but tapers slightly in thickness for approximately the last third of its length towards the point. It is double-cut on both faces, single-cut on one edge and is plain on the second edge. The plain edge with no teeth is known as the 'safe' edge and is designed to file up to the edge of a surface without damaging it. The taper in thickness enables the file to enter a slot slightly less than its full thickness.

2.1.1.2 Pillar file

This file has the same section as a hand file but of a thinner section. It is used for narrow slots and keyways.

2.1.1.3 Flat file

The flat file is also for general use, typically on flat surfaces. It is rectangular in cross-section and tapers in both width and thickness for approximately the last third of its length towards the point. Both faces are double-cut and both edges single-cut. The tapers in width and thickness enable this file to be used in slots which are narrower than its full width and thickness and which require filing on length and width.

2.1.1.4 Square file

The square file is of square cross-section, parallel for approximately two-thirds of its length, then tapering towards the point. It is double-cut on all sides. This file is used for filing keyways, slots and the smaller square or rectangular holes with 90° sides.

2.1.1.5 Three-square file

The three-square or triangular file has a 60° triangle cross-section, parallel for approximately two-thirds of its length, then tapering towards the point. The three faces are double-cut and the edges sharp. This file is used for surfaces which meet at less than 90°, angular holes and recesses.

2.1.1.6 Round file

The round file is of circular cross-section, parallel for approximately two-thirds of its length and then tapering towards the point. Second-cut and smooth files are single-cut, while the bastard is double-cut. This file is used for enlarging round holes, elongating slots and finishing internal round corners.

2.1.1.7 Half-round file

The half-round file has one flat and one curved side. It is parallel for approximately two-thirds of its length, then tapers in width and thickness towards the point. The flat side is double-cut and the curved side is single-cut on second-cut and smooth files. This is an extremely useful double-purpose file for flat surfaces and for curved surfaces too large for the round file.

2.1.1.8 Knife file

The knife file has a wedge-shaped cross-section, the thin edge being straight while the thick edge tapers to the point in approximately the last third of its length. The sides are double-cut. This file is used in filing acute angles.

2.1.1.9 Dreadnought files

When soft material is being filed, the material is more readily removed and the teeth of an engineer's file quickly become clogged. When this happens, the file no longer cuts but skids over the surface. This results in constant stoppages to clear the file so that it again cuts properly. To overcome the problem of clogging, files have been developed which have deep curved teeth milled on their faces and these are known as *dreadnought* files, Fig. 2.3.

Figure 2.3   Dreadnought file

These files are designed to remove material faster and with less effort, since the deep curved teeth produce small spiral filings which clear themselves from the tooth and so prevent clogging. Their principal use is in filing soft materials such as aluminium, lead, white metal, copper, bronze and brass. They can also be used on large areas of steel, as well as on non-metallic materials such as plastics, wood, fibre and slate.

This type of file is available as hand, flat, half-round and square, from 150 mm to 400 mm long. The available cuts are broad, medium, standard, fine and extra fine.

2.1.1.10 Needle files

Needle files are used for very fine work in tool making and fitting, where very small amounts of material have to be removed in intricate shapes or in a confined space. This type of file is available from 120 mm to 180 mm long, of which approximately half is file-shaped and cut, the remainder forming a slender circular handle, Fig. 2.4.

Figure 2.4   Needle file

2.1.2 Filing

One of the greatest difficulties facing the beginner is to produce a filed surface which is flat. By carefully observing a few basic principles and carrying out a few exercises, the beginner should be able to produce a flat surface.

Filing is a two-handed operation, and the first stage is to grip the file correctly. The handle is gripped in the palm of the right hand with the thumb on top and the palm of the left hand resting at the point of the file. Having gripped the file correctly, the second stage is to stand correctly at the vice. The left foot is placed well forward to take the weight of the body on the forward stroke. The right foot is placed well back to enable the body to be pushed forward.

Remember that the file cuts on the forward stroke and therefore the pressure is applied by the left hand during the forward movement and is released coming back. Do not lift the file from the work on the back stroke, as the dragging action helps clear the filings from the teeth and also prevents the 'see-saw' action which results in a surface which is curved rather than flat. Above all, take your time – long steady strokes using the length of the file will remove metal faster and produce a flatter surface than short rapid strokes.

As already stated, a smooth-cut file is used to give a good finish to the surface while removing small amounts of material. An even finer finish to the surface can be achieved by a method known as drawfiling. With this method, the file, rather than being pushed across, is drawn back and forth along the surface at right angles to its normal cutting direction.

An even finer finish can be obtained using abrasive cloth supported by the file to keep the surface flat. Abrasive cloth is available on rolls 25 mm wide, in a variety of grit sizes from coarse to fine. By supporting the cloth strip on the underside of the file and using a traditional filing stroke, extremely fine surface finishes can be obtained while removing very small amounts of material. This process is more of a polishing operation.

2.1.3 Care of files

A file which cuts well saves you extra work. It is important, therefore, that all the teeth are cutting. Never throw files on top of each other in a drawer, as the teeth may be chipped. Never knock the file on its edge to get rid of filings in the teeth – use a file brush. A file brush should be used regularly to remove filings from the teeth, as failure to do so will cause scratching of the work surface and inefficient removal of metal. Always clean the file on completion of the job before putting it away. Do not exert too much pressure when using a new file, or some of the teeth may break off due to their sharpness – work lightly until the fine tooth points are worn slightly. For the same reason, avoid using a new file on rough surfaces of castings, welds or hard scale.

Always use a properly fitted handle of the correct size – on no account should a file be used without a handle or with a handle which is split; remember, one slip and the tang could pierce your hand.

2.2 The hacksaw

The hacksaw is used to cut metal. Where large amounts of waste metal have to be removed, this is more easily done by hacksawing away the surplus rather than by filing. If the workpiece is left slightly too large, a file can then be used to obtain the final size and surface.

The hacksaw blade fits into a hacksaw frame on two holding pins, one of which is adjustable in order to tension the blade. The hacksaw frame should be rigid, hold the blade in correct alignment, tension the blade easily and have a comfortable grip.

The blade is fitted to the frame with the teeth pointing away from the handle, Fig. 2.5, and is correctly tensioned by turning the wing nut to take up the slack and then applying a further three turns only. A loose blade will twist or buckle and not cut straight, while an overtightened blade could pull out the ends of the blade.

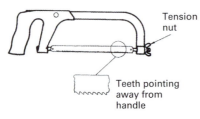

Figure 2.5 Hacksaw

The standard hacksaw blade is 300 mm long × 13 mm wide × 0.65 mm thick and is available with 14, 18, 24 and 32 teeth per 25 mm, i.e. for every 25 mm length of blade there are 14 teeth, 18 teeth and so on.

A hacksaw blade should be chosen to suit the type of material being cut, whether hard or soft, and the nature of the cut, whether thick section or thin. Two important factors in the choice of a blade are the pitch, or distance between each tooth and the material from which the blade is made.

When cutting soft metals, more material will be cut on each stroke and this material must have somewhere to go. The only place the material can go is between the teeth, and therefore if the teeth are further apart there is more space for the metal being cut. The largest space is in the blade having the least number of teeth, i.e. 14 teeth per 25 mm. The opposite is true when cutting harder metals. Less material will be removed on each stroke, which will require less space between each tooth. If less space is required, more teeth can be put in the blade,

more teeth are cutting and the time and effort in cutting will be less.

When cutting thin sections such as plate, at least three consecutive teeth must always be in contact with the metal or the teeth will straddle the thin section. The teeth will therefore have to be closer together, which means more teeth in the blade, i.e. 32 teeth per 25 mm.

Like a file, the hacksaw cuts on the forward stroke, which is when pressure should be applied. Pressure should be released on the return stroke. Do not rush but use long steady strokes (around 70 strokes per minute when using high-speed-steel blades). The same balanced stance should be used as for filing.

Table 2.1 gives recommendations for the number of teeth per 25 mm on blades used for hard and soft materials of varying thickness.

Table 2.1 Selection of hacksaw blades

Material thickness (mm)	No. of teeth per 25 mm	
	Hard materials	Soft materials
Up to 3	32	32
3 to 6	24	24
6 to 13	24	18
13 to 25	18	14

Three types of hacksaw blade are available: all-hard, flexible and bimetal.

▶ *All hard* – this type is made from hardened high-speed steel. Due to their all-through hardness, these blades have a long blade life but are also very brittle and are easily broken if twisted during sawing. For this reason they are best suited to the skilled user.

▶ *Flexible* – this type of blade is also made from high-speed steel, but with only the teeth hardened. This results in a flexible blade with hard teeth which is virtually unbreakable and can therefore be used by the less experienced user or when sawing in an awkward position. The blade life is reduced due to the problem of fully hardening the teeth only.

▶ *Bimetallic* – this type of blade consists of a narrow cutting-edge strip of hardened high-speed steel joined to a tough alloy-steel back by electron beam welding. This blade combines the qualities of hardness of the all-hard blade and the unbreakable qualities of the flexible blade, resulting in a shatterproof blade with long life and fast-cutting properties.

2.3 Cold chisels

Cold chisels are used for cutting metal. They are made from high-carbon steel, hardened and tempered at the cutting end. The opposite end, which is struck by the hammer, is not hardened but is left to withstand the hammer blows without chipping.

2.3.1 Cold chisel classification

Cold chisels are classified as 'flat' or 'cross-cut', according to the shape of the point.

2.3.1.1 Flat

This chisel has a broad flat point and is used to cut thin sheet metal, remove rivet heads or split corroded nuts. The cutting edge is ground to an angle of approximately 60°, Fig. 2.6.

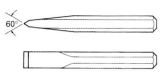

Figure 2.6 'Flat' cold chisel

2.3.1.2 Cross-cut

This chisel has a narrower point than the flat chisel and is used to cut keyways, narrow grooves, square corners and holes in sheet metal too small for the flat chisel, Fig. 2.7.

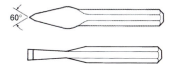

Figure 2.7 'Cross-cut' cold chisel

2.3.2 Using the chisel

When using a cold chisel on sheet-material, great care must be taken not to distort the metal. To prevent distortion, the sheet must be properly supported. A small sheet is best held in a vice,

Fig. 2.8. A large sheet can be supported by using two metal bars securely clamped, Fig. 2.9.

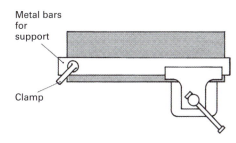

Figure 2.8 Sheet metal in vice

Figure 2.9 Sheet metal in support bars

To remove a section from the centre of a plate, the plate can be supported on soft metal. It is best to mark out the shape required, drill a series of holes in the waste material, and use the chisel to break through between the holes, Fig. 2.10.

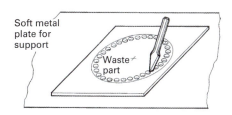

Figure 2.10 Chisel cutting a hole supported on soft metal plate

The chisel should be held firmly but not too tight, and the head should be struck with sharp blows from the hammer, keeping your eye on the cutting edge, not the chisel head. Hold the chisel at approximately 40°, Fig. 2.11. Do not hold the chisel at too steep an angle, otherwise it will tend to dig into the metal. Too shallow an angle will cause the chisel to skid and prevent it cutting. Use a hammer large enough to do the job, grasping it

2

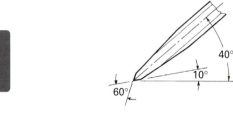

Figure 2.11 Correct angle of chisel

well back at the end of the handle, not at the end nearest the head. Never allow a large 'mushroom' head to form on the head of a chisel, as a glancing blow from the hammer can dislodge a chip which could fly off and damage your face or hand. Always grind off any sign of a mushroom head as it develops, Fig. 2.12.

Figure 2.12 Correct chisel

Cold chisels can be sharpened by regrinding the edge on an off-hand grinder. When resharpening, do not allow the chisel edge to become too hot, otherwise it will be tempered, lose its hardness and be unable to cut metal.

2.4 Scrapers

Scraping, unlike filing or chiselling, is not done to remove a great deal of material. The material is removed selectively in small amounts, usually to give a flat or a good bearing surface. A surface produced by machining or filing may not be good enough as a bearing where two surfaces are sliding or rotating. The purpose of scraping is therefore to remove high spots to make the surface flat or circular, and at the same time to create small pockets in which lubricant can be held between the two surfaces. Surface plates and surface tables are examples of scraping being used when flatness is of prime importance. Examples where both flatness and lubricating properties are required can be seen on the sliding surfaces of centre lathes and milling, shaping and grinding machines.

The flat scraper, for use on flat surfaces, resembles a hand file thinned down at the point, but it does not have any teeth cut on it, Fig. 2.13. The point is slightly curved, and the cutting edges are kept sharp by means of an oilstone. The scraper cuts on the forward stroke, the high spots being removed one at a time by short forward rocking strokes. The flatness is checked with reference to a surface plate. A light film of engineer's blue is smeared evenly on the surface plate, and the surface being scraped is placed on top and moved slightly from side to side. Any high spots show up as blue spots, and these are reduced by scraping. The surface is again checked, rescraped and the process is repeated until the desired flatness is obtained. Flatness of the surface is indicated when the whole area being scraped is evenly covered by blue from the surface plate.

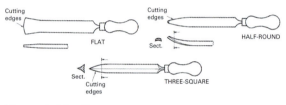

Figure 2.13 Scrapers

The same procedure is used on internal curved surfaces, using a half-round scraper slightly hollow on the underside, to prevent digging in, and with a cutting edge on each side, Fig. 2.13. The reference surface in this case is the shaft which is to run in the curved surface and which is smeared with engineer's blue. Entry of the shaft in the bearing indicates the high spots, which are removed by scraping, and this process is repeated until the desired surface is produced.

The three-square or triangular scraper, Fig. 2.13 is commonly used to remove the sharp edges from curved surfaces and holes. It is not suited to scraping internal curved surfaces, due to the steeper angle of the cutting edges tending to dig into the surface. However, the sharp point is useful where a curved surface is required up to a sharp corner.

2.5 Engineer's hammers

The engineer's hammer consists of a hardened and tempered steel head, varying in mass from

0.1 kg to about 1 kg, firmly fixed on a tough wooden handle, usually hickory or ash.

The flat striking surface is known as the face, and the opposite end is called the pein. The most commonly used is the ball-pein, Fig. 2.14, which has a hemispherical end and is used for riveting over the ends of pins and rivets.

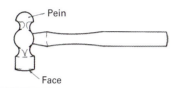

Figure 2.14   Ball-pein hammer

For use with soft metal such as aluminium or with finished components where the workpiece could be damaged if struck by an engineer's hammer, a range of hammers is available with soft faces, usually hide, copper or a tough plastic such as nylon. The soft faces are usually in the form of replaceable inserts screwed into the head or into a recess in the face, Fig. 2.15.

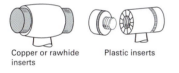

Copper or rawhide   Plastic inserts
inserts

Figure 2.15   Soft-faced hammers

Always use a hammer which is heavy enough to deliver the required force but not too heavy to be tiring in use. The small masses, 0.1 kg to 0.2 kg, are used for centre punching, while the 1 kg ones are used with large chisels or when driving large keys or collars on shafts. The length of the handle is designed for the appropriate head mass, and the hammer should be gripped near the end of the handle to deliver the required blow. To be effective, a solid sharp blow should be delivered and this cannot be done if the handle is held too near the hammer head.

Always ensure that the hammer handle is sound and that the head is securely fixed.

2.6 Screwdrivers

The screwdriver is one of the most common tools, and is also the one most misused. Screwdrivers

should be used only to tighten or loosen screws. They should never be used to chisel, open tins, scrape off paint or lever off tight parts such as collars on shafts. Once a screwdriver blade, which is made from toughened alloy steel, has been bent, it is very difficult to keep it in the screw head.

There are a number of different head drives. The four most common are slotted, Phillips, Pozidriv and Torx as shown in Fig. 2.16. Always select the screwdriver to suit the size and type of head drive. Use of the incorrect size or type results in damage to both the screwdriver and the screw head, leading to a screw that is very difficult to loosen or tighten.

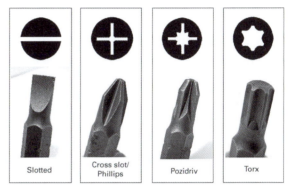

Slotted   Cross slot/   Pozidriv   Torx
                 Phillips

Figure 2.16   Screw head drives

Phillips and Pozidriv are numbered with 1, 2 and 3 the most common. Torx cover a range numbered T5 to T55 from smallest to the largest.

Straight slots in screws are machined with parallel sides. It is essential that any screwdriver used in such a slot has the sides of the blade parallel to slightly tapered up to about 10°, Fig. 2.17(a). A screwdriver sharpened to a point like a chisel will not locate correctly and will require great force to keep it in the slot, Fig. 2.17(b). Various blade lengths are available with corresponding width and thickness to suit the screw size.

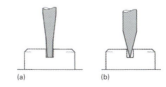

(a)   (b)

Figure 2.17   Screwdriver point: (a) right (b) wrong

In the interests of personal safety, never hold the work in your hand while tightening or loosening a screw – the blade may slip and cause a nasty injury. Always hold the work securely in a vice or clamped to a solid surface.

2.7 Taps

Tapping is the operation of cutting an internal thread by means of a cutting tool known as a tap. When tapping by hand, straight-flute hand taps are used. These are made from hardened high-speed steel and are supplied in sets of three. The three taps differ in the length of chamfer at the point, known as the lead. The one with the longest lead is referred to as the taper or first tap, the next as the second or intermediate tap and the third, which has a very short lead, as the bottoming or plug tap, Fig. 2.18. A square is provided at one end so that the tap can be easily rotated by holding it in a tap wrench, Fig. 2.19. The chuck type of wrench is used for the smaller tap sizes.

The first stage in tapping is to drill a hole of the correct size. This is known as the tapping size and is normally slightly larger than the root diameter of the thread. Table 2.2 shows the tapping sizes for ISO metric threads which have replaced most threads previously used in Great Britain.

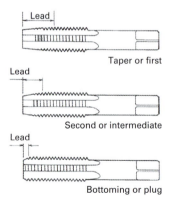

Figure 2.18 Set of taps

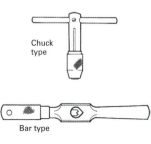

Figure 2.19 Tap wrenches

Table 2.2 Tapping sizes for ISO metric threads

Thread diameter and pitch (mm)	Drill diameter for tapping (mm)
1.6 × 0.35	1.25
2 × 0.4	1.6
2.5 × 0.45	2.05
3 × 0.5	2.5
4 × 0.7	3.3
5 × 0.8	4.2
6 × 1.0	5.0
8 × 1.25	6.8
10 × 1.5	8.5
12 × 1.75	10.2

Tapping is then started using the taper or first tap securely held in a tap wrench. The long lead enables it to follow the drilled hole and keep square. The tap is rotated, applying downward pressure until cutting starts. No further pressure is required, since the tap will then screw itself into the hole. The tap should be turned back quite often, to help clear chips from the flutes.

If the hole being tapped passes through the component, it is only necessary to repeat the operation using the second or intermediate tap. Where the hole does not pass through – known as a blind hole – it is necessary to use the plug or bottoming tap. This tap has a short lead and therefore forms threads very close to the bottom of the hole. When tapping a blind hole, great care should be taken not to break the tap. The tap should be occasionally withdrawn completely and any chips be removed before proceeding to the final depth.

For easier cutting and the production of good-quality threads, a proprietary tapping compound should be used.

2.8 Dies

Dies are used to cut external threads and are available in sizes up to approximately 36 mm thread diameter. The common type, for use by hand, is the circular split die, made from high-speed steel hardened and tempered and split at one side to enable small adjustments of size to be made, Fig. 2.20.

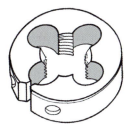

Figure 2.20 Circular split die

The die is held in a holder known as a die stock, which has a central screw for adjusting the size and two side locking screws which lock in dimples in the outside diameter of the die, Fig. 2.21. The die is inserted in the holder with the split lined up with the central screw. The central screw is then tightened so that the die is expanded, and the two side locking screws are tightened to hold the die in position.

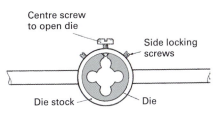

Figure 2.21 Die holder

Dies have a lead on the first two or three threads, to help start cutting, but it is usual also to have a chamfer on the end of the component. The die is placed squarely on the end of the bar and is rotated, applying downward pressure until cutting starts, ensuring that the stock is horizontal. No further pressure is required, since the die then screws itself forward as cutting proceeds. The die should be rotated backwards every two or three turns, to break up and clear the chips. The thread can now be checked with a nut. If it is found to be tight, the central screw is slackened, the side locking screws are tightened equally, and the die is run down the thread again. This can be repeated until the final size is reached.

As with tapping, easier cutting and better threads are produced when a proprietary cutting compound is used.

Dienuts are also available and are generally used for reclaiming or cleaning up existing threads, not for cutting a thread from solid.

2.9 Hand reamer

Where a more accurate-size hole with better surface finish, than can be achieved by drilling, is required, the hole can be finished using a reamer. The hole is drilled undersize (see Table 8.2) and the reamer, which has a square at the end of the shank, to fit a tap wrench, is carefully 'wound' into the hole, removing the excess material. Hand reamers have a long lead to assist with cutting and alignment. A suitable lubricant should be used to prevent wear on the tool, improve surface finish and prevent scratching, usually a light oil or proprietary tapping compound. Always withdraw the reamer to prevent the flutes becoming blocked with swarf.

2.10 Powered hand tools

The main advantages of powered hand tools are the reduction of manual effort and the speeding up of the operation. The operator, being less fatigued, is able to carry out the task more efficiently, and the speeding up of the operation results in lower production costs. Being portable, a powered hand tool can be taken to the work, which can also lead to a reduction in production costs. Accuracy of metal-removal operations is not as good with powered hand tools, since it is difficult to remove metal from small areas selectively. A comparison of hand and powered hand tools is shown in Table 2.3.

Table 2.3 Comparison of hand and powered hand tools

	Speed of production	Cost of tool	Accuracy	Fatigue
Hand tools	Low	Low	High	High
Powered hand tools	High	High	Low	Low

Powered hand tools can be electric or air-operated. In general, electric tools are heavier than the equivalent air tool, due to their built-in motor, e.g. electric screwdrivers weigh 2 kg while an equivalent air-operated screwdriver weighs 0.9 kg. The cost of powered tools is much greater than the equivalent hand tools and must be taken into account when a choice has to be made.

2

Air-operated tools can be safely used in most work conditions, while electrical tools should not be used in conditions which are wet or damp or where there is a risk of fire or explosion, such as in flammable or dusty atmospheres. A selection of air-operated tools is shown in Fig. 2.22.

2.10.1 Hand drills

Electric and air-operated drills are available with a maximum drilling capacity in steel of about 30 mm diameter for electric and about 10 mm diameter for air models. Air-operated tools are more ideally suited to the rapid drilling of the smaller diameter holes, Fig. 2.22(a).

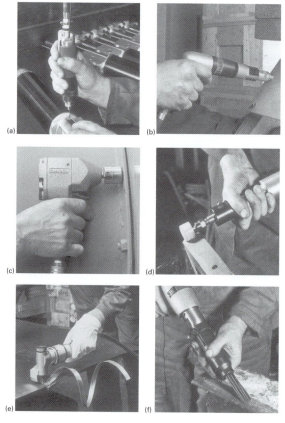

Figure 2.22   Air-operated tools: (a) hand drill (b) screwdriver (c) impact wrench (d) grinder (e) metal shears (f) hammer

2.10.2 Screwdriver

Used for inserting screws of all types, including machine, self-tapping, self-drilling and tapping

and wood screws. Some models are reversible and can be used with equal ease to remove screws. The tool bits are interchangeable to suit the different screw-head types, such as slotted, 'supadriv', 'pozidriv', hexagon-socket or hexagon-headed. Electric and air-operated screwdrivers are available with a maximum capacity of about 8 mm diameter thread with a variety of torque settings to prevent the screw being overtightened or sheared off, Fig. 2.22(b).

2.10.3 Impact wrench

Used for tightening and also, with the reversal mechanism, for loosening hexagon-headed nuts and screws. Air-powered models are available with a maximum capacity of 32 mm diameter threads and with torque settings to suit a range of thread sizes. They have the advantage of being able to tighten all nuts or screws to the same predetermined load, Fig. 2.22(c).

2.10.4 Grinder

Used to remove metal from the rough surfaces of forgings, castings and welds usually when the metal is too hard or the amount to be removed is too great for a file or a chisel. Electric and air-operated grinders are available with straight grinding wheels up to 230 mm diameter or with small mounted points of various shapes and sizes, Fig. 2.22(d).

2.10.5 Metal shears

Used to cut metal, particularly where the sheet cannot be taken to a guillotine or where profiles have to be cut. Electric or air-operated shears are available capable of cutting steel sheet up to 2 mm thick by means of the scissor-like action of a reciprocating blade, Fig. 2.22(e).

2.10.6 Hammer

Can be fitted with a wide range of attachments for riveting, shearing off rivet heads, removing scale or panel cutting. Air-operated models are available which deliver between 3000 and 4000 blows per minute, Fig. 2.22(f).

Review questions

1. Why is it essential to use the correct size screwdriver?
2. Describe how a circular split die may be adjusted to give the correct size thread.
3. What are the main advantages of using power tools?
4. Explain the importance of the number of teeth on a hacksaw blade when cutting different types of material.
5. Why is it necessary to drill the correct tapping size hole when producing an internal thread?
6. Name eight types of file.
7. Describe where soft-faced hammers are used and name three materials used for the soft faces.
8. Thread-cutting taps are supplied in sets of three. Name each tap and state where each would be used.
9. Why is it necessary to keep the cutting edge cool when regrinding a cold chisel?
10. When would a scraper be used rather than filing or chiselling?

CHAPTER 3

Marking out

Marking out is the scratching of lines on the surface of a workpiece, known as scribing, and is usually carried out only on a single workpiece or a small number of workpieces. The two main purposes of marking out are:

▶ to indicate the workpiece outline or the position of holes, slots, etc. If the excess material will have to be removed, a guide is given for the extent to which hacksawing or filing can be carried out;
▶ to provide a guide to setting up the workpiece on a machine. The workpiece is set up relative to the marking out and is then machined. This is especially important when a datum has to be established when castings and forgings are to be machined.

It is important to note that the scribed lines are only a guide, and any accurate dimension must be finally checked by measuring.

3.1 Datum

The function of a datum is to establish a reference position from which all dimensions are taken and hence all measurements are made. The datum may be a point, an edge or a centre line, depending on the shape of the workpiece. For any plane surface, two datums are required to position a point and these are usually at right angles to each other.

Fig. 3.1 shows a workpiece where the datum is a point; Fig. 3.2 shows a workpiece where both datums are edges; Fig. 3.3 shows a workpiece where both datums are centre lines and Fig. 3.4 shows a workpiece where one datum is an edge and the other is a centre line.

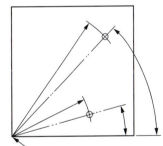

Datum point

Figure 3.1 Datum point

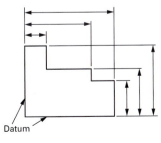

Datum

Figure 3.2 Datum edges

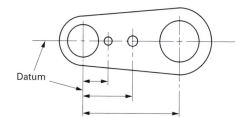

Figure 3.3 Datum centre lines

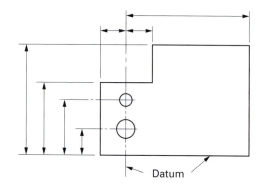

Figure 3.4 Datum edge and centre line

The datums are established by the draughtsman when the drawing is being dimensioned and, since marking out is merely transferring drawing dimensions to the workpiece, the same datums are used.

3.2 Co-ordinates

The draughtsman can dimension drawings in one of two ways.

▶ Cartesian or *rectangular co-ordinates* – where the dimensions are taken relative to the datums at right angles to each other, i.e. the general pattern is rectangular. This is the method shown in Figs 3.2 and 3.4.

▶ *Polar co-ordinates* – where the dimension is measured along a radial line from the datum. This is shown in Fig. 3.1. Marking out polar co-ordinates requires not only accuracy of the dimension along the radial line but accuracy of the angle itself. As the polar distance increases, any slight angular error will effectively increase the inaccuracy of the final position.

The possibility of error is less with rectangular co-ordinates, and the polar co-ordinate dimensions shown in Fig. 3.1 could be redrawn as rectangular co-ordinates as shown in Fig. 3.5.

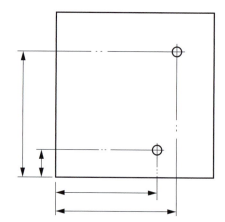

Figure 3.5 Rectangular co-ordinates

3.3 Marking out equipment

3.3.1 Surface table and surface plate

In order to establish a datum from which all measurements are made a reference surface is required. This reference surface takes the form of a large flat surface called a surface table (Fig. 3.6) upon which the measuring equipment is used.

Surface plates (Fig. 3.7) are smaller reference surfaces and are placed on a bench for use with smaller workpieces. For general use, both surface tables and surface plates are made from cast iron machined to various grades of accuracy. For

Figure 3.6 Surface table

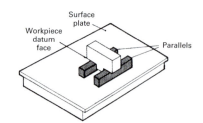

Figure 3.7 Surface plate and parallels

high-accuracy inspection work and for use in standards rooms, surface tables and plates made from granite are available.

3.3.2 Parallels (Fig. 3.7)

The workpiece can be set on parallels to raise it off the reference surface and still maintain parallelism. Parallels are made in pairs to precisely the same dimensions, from hardened steel, finish ground, with their opposite faces parallel and adjacent faces square. A variety of sizes should be available for use when marking out.

3.3.3 Jacks and wedges

When a forging or casting has to be marked out, which has an uneven surface or is awkward in shape, it is still essential to maintain the datum relative to the reference surface. Uneven surfaces can be prevented from rocking and kept on a parallel plane by slipping in thin steel or wooden wedges (Fig. 3.8) at appropriate positions. Awkward shapes can be kept in the correct position by support from adjustable jacks (see Fig. 3.9).

Figure 3.8 Wedge

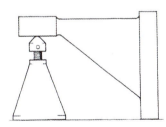

Figure 3.9 Jack used to support

3.3.4 Angle plate

When the workpiece has to be positioned at 90° to the reference surface, it can be clamped to an angle plate (Fig. 3.10). Angle plates are usually made from cast iron and the edges and faces are accurately machined flat, square and parallel. Slots are provided in the faces for easy clamping of the workpiece. Angle plates may be plain or adjustable.

Figure 3.10 Angle plate and surface gauge

3.3.5 Vee blocks (Fig. 3.11)

Holding circular work for marking out or machining can be simplified by using a vee block. The larger sizes are made from cast iron, the smaller sizes from steel hardened and ground, and provided with a clamp. They are supplied in pairs marked for identification. The faces are machined to a high degree of accuracy of flatness, squareness, and parallelism, and the 90° vee is central with respect to the side faces and parallel to the base and side faces.

Figure 3.11 Vee block in use

3.3.6 Engineer's square (Fig. 3.12)

An engineer's square is used when setting the workpiece square to the reference surface (see Fig. 3.13) or when scribing lines square to the datum edge (Fig. 3.14). The square consists of a stock and blade made from hardened steel and ground on all faces and edges to give a high degree of accuracy in straightness, parallelism and squareness. It is available in a variety of blade lengths.

3.3.7 Combination set (Fig. 3.15)

The combination set consists of a graduated hardened steel rule on which any of three separate heads – protractor, square or centre head – can be mounted. The rule has a slot in which each head slides and can be locked at any position along its length.

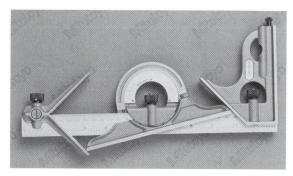

Figure 3.15 Combination set

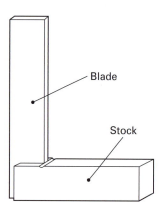

Figure 3.12 Engineer's square

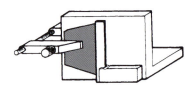

Figure 3.13 Setting workpiece square to reference surface

Figure 3.14 Scribing line square to datum

3.3.7.1 Protractor head (Fig. 3.16)

This head is graduated from 0° to 180°, is adjustable through this range, and is used when scribing lines at an angle to a workpiece datum.

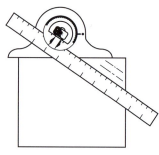

Figure 3.16 Protractor head

3.3.7.2 Square head (Fig. 3.17(a))

This head is used in the same way as an engineer's square, but, because the rule is adjustable, it is not as accurate. A second face is provided at 45° (Fig. 3.17(b)). A spirit level is incorporated which is useful when setting workpieces such as castings level with the reference surface. Turned on end, this head can also be used as a depth gauge (see Fig. 3.17(c)).

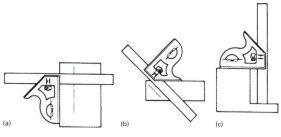

(a)   (b)   (c)

Figure 3.17 Squarehead

3.3.7.3 *Centre head (Fig. 3.18)*

With this head the blade passes through the centre of the vee and is used to mark out the centre of a circular workpiece or round bar.

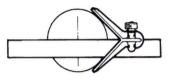

Figure 3.18   Centre head

3.3.8 Marking dye

On surfaces of metal other than bright metals, scribed lines may not be clearly visible. In such cases the surface can be brushed or sprayed with a quick drying coloured dye before marking out. This provides a good contrast, making the scribed lines easy to see.

3.3.9 Scriber (Fig. 3.19)

The scriber is used to scribe all lines on a metal surface and is made from hardened and tempered steel, ground to a fine point which should always be kept sharp to give well-defined lines. Another type is shown in Fig. 3.14.

Figure 3.19   Scriber

3.3.10 Surface gauge (Fig. 3.10)

The surface gauge, also known as a scribing block, is used in conjunction with a scriber to mark out lines on the workpiece parallel with the reference surface. The height of the scriber is adjustable and is set in conjunction with a steel rule. The expected accuracy from this set-up will be around 0.3 mm but with care this can be improved.

3.3.11 Vernier height gauge (Fig. 3.20)

Where greater accuracy is required than can be achieved using a surface gauge, marking out can be done using a vernier height gauge. The vernier scale carries a jaw upon which various attachments can be clamped. When marking

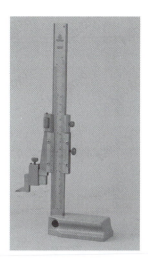

Figure 3.20   Vernier height gauge

out, a chisel pointed scribing blade is fitted. Care should be taken to allow for the thickness of the jaw, depending on whether the scribing blade is clamped on top or under the jaw. The precise thickness of the jaw is marked on each instrument. These instruments can be read to an accuracy of 0.02 mm and are available in a range of capacities reading from 0 to 1000 mm.

3.3.12 Dividers and trammels

Dividers are used to scribe circles or arcs and to mark off a series of lengths such as hole centres. They are of spring bow construction, each of the two pointed steel legs being hardened and ground to a fine point and capable of scribing a maximum circle of around 150 mm diameter (Fig. 3.21). Larger circles can be scribed using trammels, where the scribing points are adjustable along the length of a beam (Fig. 3.22).

Dividers and trammels are both set in conjunction with a steel rule by placing one point in a

Figure 3.21   Dividers

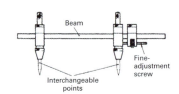

Figure 3.22   Trammels

convenient graduation line and adjusting the other to coincide with the graduation line the correct distance away.

3.3.13 Hermaphrodite calipers (Fig. 3.23)

These combine a straight pointed divider leg with a caliper or stepped leg and are used to scribe a line parallel to the edge of a workpiece. They are more commonly known as 'odd-legs' or 'jennies'.

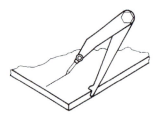

Figure 3.23   Hermaphrodite calipers

3.3.14 Precision steel rule

These are made from hardened and tempered stainless steel, photo-etched for extreme accuracy and have a non-glare satin chrome finish. Rules are available in lengths of 150 mm and 300 mm and graduations may be along each edge of both faces usually in millimetres and half millimetres.

Accuracy of measurement depends on the quality of the rule and the skill of the operator. The width of the lines on a high-quality rule are quite fine and accuracies of around 0.15 mm can be achieved but an accuracy of double this can more realistically be achieved.

3.3.15 Centre punch (Fig. 3.24)

The centre punch is used to provide a centre location for dividers and trammels when scribing circles or arcs, or to show permanently the position of a scribed line by a row of centre dots. The centre dot is also used as a start for small diameter drills.

Figure 3.24   Centre punch

Centre punches are made from high carbon steel, hardened and tempered with the point ground at 30° when used to provide a centre location for dividers and at 90° for other purposes.

Care should be taken in the use of centre dots on surfaces which are to remain after machining, since, depending upon the depth, they may prove difficult to remove.

3.3.16 Clamps

Clamps are used when the workpiece has to be securely fixed to another piece of equipment, e.g. to the face of an angle plate (Fig. 3.10).

The type most used are toolmaker's clamps (Fig. 3.25), which are adjustable within a range of about 100 mm but will only clamp parallel surfaces. Greater thicknesses can be clamped using 'G' clamps, so named because of their shape (Fig. 3.26). Due to the swivel pad on the end of the clamping screw, the 'G' clamp is also capable of clamping surfaces which are not parallel.

Care should be taken to avoid damage to the surfaces by the clamp.

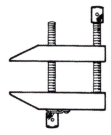

Figure 3.25   Toolmaker's clamp

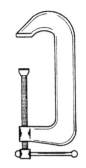

Figure 3.26   'G' clamp

3.4 Examples of marking out

We will now see how to mark out a number of components.

3.4.1 Example 3.1: component shown in Fig. 3.27

The plate shown at step 1 has been filed to length and width with the edges square and requires the position of the steps to be marked out.

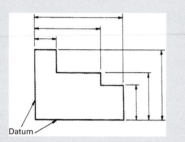

Figure 3.27 Component used in Example 3.1

3.4.1.1 Step 1

Use a square on one datum edge and measure the distance from the other datum edge using a precision steel rule. Scribe lines.

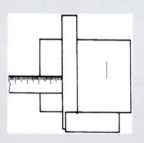

3.4.1.2 Step 2

Repeat with the square on the second datum edge and scribe lines to intersect.

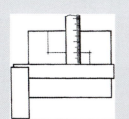

3.4.2 Example 3.2: component shown in Fig. 3.28

The plate shown at step 1 has been cut out 2 mm oversize on length and width and has not been filed. All four sides have sawn edges.

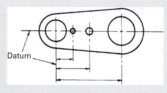

Figure 3.28 Component used in Example 3.2

3.4.2.1 Step 1

Measure from each long edge and find the centre using a precision steel rule. Scribe the centre line using the edge of the rule as a guide. Find the centre of the small radius by measuring from one end the size of the radius plus 1 mm (this allows for the extra left on the end). Centre dot where the lines intersect.

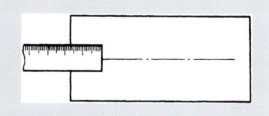

3.4.2.2 Step 2

Using dividers, set the distance from the centre of the small radius to the centre of the first small hole. Scribe an arc. Repeat for the second small hole and the large radius. Centre dot at the intersection of the centre lines. The dividers are set using the graduations of a precision steel rule.

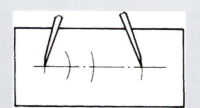

3.4.2.3 Step 3

Set dividers to the small radius. Locate on the centre dot and scribe the radius. Repeat for the large radius, and if necessary, the two holes.

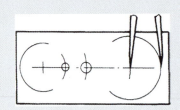

3.4.2.4 Step 4

Complete the profile by scribing a line tangential to the two radii using the edge of a precision steel rule as a guide.

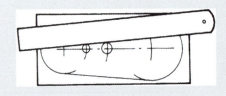

3.4.3 Example 3.3: component shown in Fig. 3.29

The plate shown at step 1 has been roughly cut to size and requires complete marking out of the profile and holes.

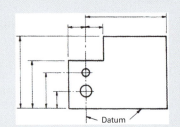

Figure 3.29 Component used in Example 3.3

3.4.3.1 Step 1

Clamp the plate to the face of an angle plate, ensuring that the clamps will not interfere with marking out. Use a scriber in a surface gauge and set the heights in conjunction with a precision steel rule. Scribe the datum line.

Scribe each horizontal line the correct distance from the datum.

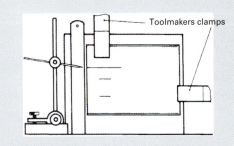

3.4.3.2 Step 2

Without unclamping the plate, swing the angle plate on to its side (note the importance of clamp positions at step 1). This ensures that the lines about to be scribed are at right angles to those scribed in step 1, owing to the accuracy of the angle plate. Scribe the datum centre line. Scribe each horizontal line the correct distance from the datum to intersect the vertical lines.

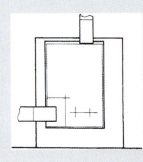

3.4.4 Example 3.4: component shown in Fig. 3.30

The plate shown at step 1 is to be produced from the correct width (W) bright rolled strip

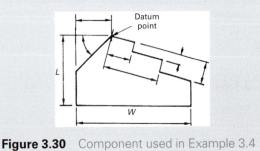

Figure 3.30 Component used in Example 3.4

3

and has been sawn 2 mm oversize on length (L). The base edge has been filed square to the ends and it is required to mark out the angled faces.

3.4.4.1 Step 1

Using a precision steel rule measure from two adjacent edges to determine the datum point. Centre dot the datum point.

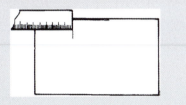

3.4.4.2 Step 2

Set protractor at required angle and scribe line through datum point.

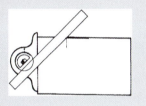

3.4.4.3 Step 3

Reset protractor at second angle and scribe one line through datum point. Scribe the remaining two lines parallel to and the correct distance from the first line using the protractor at the same setting.

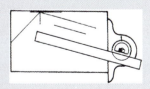

3.4.4.4 Step 4

Set dividers at correct distances, locate in datum centre dot and mark positions along scribed line.

3.4.4.5 Step 5

Reset protractor and scribe lines through marked positions.

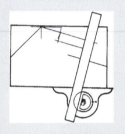

3.4.5 Example 3.5: shaft shown in Fig. 3.31

The shaft shown is to have a keyway cut along its centre line for a required length. Accurate machining is made possible by setting up the shaft relative to the marked-out position.

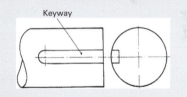

Figure 3.31 Component used in Example 3.5

3.4.5.1 Step 1

Scribe line on the end face through the centre of the shaft using the centre head of a combination set.

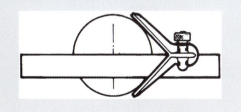

3.4.5.2 Step 2

Clamp shaft in a vee block ensuring that the line marked at step 1 is lying horizontal. This can be checked using a scriber in a surface gauge. Transfer the centre line along the required length of shaft. Scribe two further lines to indicate the width of slot.

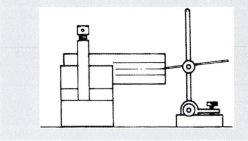

3.4.5.3 Step 3

The required length of slot can be marked without removing it simply by turning the vee block on its end and scribing a horizontal line at the correct distance from the end of the shaft.

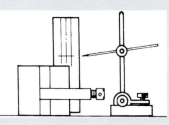

Review questions

1. State two uses of a centre punch when marking out.
2. What are trammels used for?
3. Name the three heads which comprise a combination set and describe the use of each one.
4. When would a vee block be used during a marking out operation?
5. Where would it be necessary to use a marking dye?
6. What are the two main purposes of marking out engineering components?
7. Why are surface tables and plates used when marking out?
8. Why is it necessary to create a datum when marking out?
9. Describe where jacks and wedges would be used.
10. State the difference between rectangular and polar co-ordinates.

CHAPTER 4

Sheet-metal operations

Many engineering components are produced from a flat sheet of metal which is cut to shape and then folded to form the finished article. The edges are then secured by a variety of methods such as welding, brazing, soldering and riveting. The accuracy of the size and shape of the finished article depends upon the accuracy of drawing the shape on the flat sheet, known as the development. Allowance is made at this stage for folding or bending, the amount varying with the radius of the bend and the metal thickness.

The thickness of metal sheet is identified by a series of numbers known as standard wire gauge, or SWG. Table 4.1 lists the most frequently used gauges and gives their thickness in millimetres.

Table 4.1 Most frequently used standard wire gauges

SWG	Thickness (mm)
10	3.2
12	2.6
14	2.0
16	1.6
18	1.2
19	1.0
20	0.9
22	0.7
24	0.6

Cutting is carried out using simple snips for thin-gauge steel up to around 20 SWG, treadle-operated guillotines capable of cutting 14 SWG steel, and hand-lever shears for sheet up to 1.5 mm thick.

Holes and apertures can be cut using a simple hand-operated punch or a punch and die fitted in a fly press.

For simple bends in thin material, bending can be carried out in a vice. For bends in thicker material with a specific bend radius, folding machines give greater accuracy with less effort.

4.1 Cutting and bending sheet metal

Light-gauge metal can be easily cut using snips. These may have straight or curved blades, Fig. 4.1, the latter being used to cut around a curved profile. Lengths of handle vary from 200 mm to 300 mm, the longer handle giving greater leverage for cutting heavier gauge material. For cutting thicker metals, up to 1.5 mm, hand-lever shears are available, usually bench-mounted, Fig. 4.2. The length of the lever and

Figure 4.1   Straight- and curved-blade snips

4

Figure 4.2 Hand-lever shears

the linkage to the moving shear blade ensure adequate leverage to cut the thicker metals.

Where larger sheets are required to be cut with straight edges, the guillotine is used. Sheet widths of 600 mm × 2 mm thick and up to 1200 mm × 1.6 mm thick can be accommodated in treadle-operated guillotines, Fig. 4.3. These have a moving top blade, which is operated by a foot treadle, and a spring which returns the blade to the top of its stroke. The table is provided with guides, to maintain the cut edges square, and adjustable stops to provide a constant size over a number of components. When the treadle is operated, a clamp descends to hold the work in

position while cutting takes place, and this also acts as a guard to prevent injury. *These machines can be extremely dangerous if not used correctly, so take great care.*

When holes are to be cut in sheet metal, up to 16 SWG, this can be done simply and effectively using a 'Q-Max' sheet metal punch as shown in Fig. 4.4. A pilot hole is drilled in the correct position, the screw is inserted with the punch and die on either side of the sheet, and the screw is tightened. The metal is sheared giving a correct size and shape of hole in the required position.

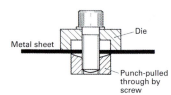

Figure 4.4 'Q-Max' sheet metal punch

Where a number of components require the same size hole in the same position, it may be economical to manufacture a punch and die for the operation. The operation is carried out on a fly press, Fig. 4.5, with the punch, which is the size and shape of the hole required, fitted in the moving part of the press. The die, which contains a hole the same shape as the punch, but slightly larger to give clearance, is clamped to the table directly in line with the punch. When the handle of the fly press is rotated, the punch descends and a sheet of metal inserted between the punch and die will have a piece removed the same shape as the punch, Fig. 4.6.

Figure 4.3 Treadle guillotine

Figure 4.5 Fly press

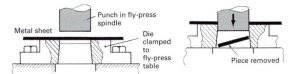

Figure 4.6 Punch and die in fly press

With the use of simple tools, the fly press can also be used for bending small components, Fig. 4.7. The top tool is fixed to the moving part and the bottom tool, correctly positioned under the top tool, is fixed to the table of the press. Metal bent in this way will spring back slightly, and to allow for this the angle of the tool is made less than 90°. In the case of mild steel, an angle of 88° is sufficient for the component to spring back to 90°.

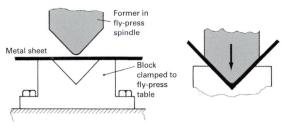

Figure 4.7 Bending tool in fly press

The simplest bends can be produced by holding the component in a vice and bending it over using a soft hammer. If the component is wider than the vice jaws, it can be clamped between metal bars. Unless a radius is put on one of the bars, this method produces a sharp inside corner, which may not always be desirable.

Folding machines, Fig. 4.8, are used with larger work of thicker gauges and for folding box sections. The top clamping beam is adjustable

Figure 4.8 Folding machine

to allow for various thicknesses of material and can be made up in sections known as fingers to accommodate a previous fold. Slots between the fingers allow a previous fold not to interfere with further folds, as in the case of a box section where four sides have to be folded. The front folding beam, pivoted at each end, is operated by a handle which folds the metal past the clamping blade, Fig. 4.9.

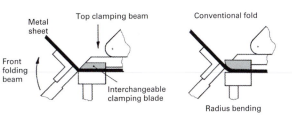

Figure 4.9 Folding operations

4.2 Development

The development of sheet-metal components ranges from the simple to the extremely complex. Let us consider three simple shapes: a cylinder, a cone and a rectangular tray.

If a cylinder is unfolded, like unrolling a carpet, the length of the development is equal to the circumference, Fig. 4.10.

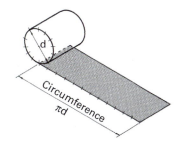

Figure 4.10 Development of cylinder

If a cone is unfolded while pivoting about the apex O, the development is a segment of a circle of radius Oa whose arc ab is equal in length to the circumference of the base, Fig. 4.11. To find the length of arc ab, the base diameter is equally divided into 12 parts. The 12 small arcs 1–2, 2–3, etc. are transferred to the arc with point 12 giving the position of point b. A part cone (frustum) is developed in exactly the same way, with the arc representing the small diameter having a radius Oc, Fig. 4.12.

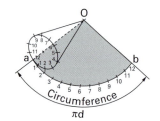

Figure 4.11 Development of cone

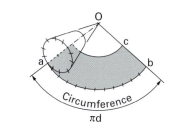

Figure 4.12 Development of part cone

In practice, the circumference must take account of the material thickness. Any metal which is bent will stretch on the outside of the bend and be compressed on the inside. Unless the metal is of very light gauge, an allowance must be made for this. The allowance is calculated on the assumption that, since the outside of the bend stretches and the inside is compressed, the length at a distance half way between the inside and outside diameters, i.e. the mean diameter, will remain unchanged.

4.2.1 Example 4.1

The cylinder shown in Fig. 4.13 has an outside diameter of 150 mm and is made from 19 SWG (1 mm thick) sheet. Since the outside diameter is 150 mm and the thickness 1 mm

mean diameter = 150 − 1 = 149 mm

giving

mean circumference = $\pi \times 149 = 468$ mm

The circumference at the outside of the cylinder is

$\pi \times 150 = 471$ mm

Thus a blank cut to a length of 468 mm will stretch to a length of 471 mm at the outside and give a component of true 150 mm diameter.

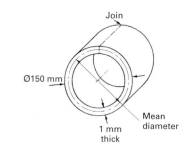

Figure 4.13 Cylinder

4.2.2 Example 4.2

The development of a rectangular tray is simply the article with the sides and ends folded down, Fig. 4.14(a). The development would be as shown in Fig. 4.14(b), the dotted lines indicating the position of the bend. If sharp inside corners are permissible, the bend lines are the inside dimensions of the tray. If the tray is dimensioned to the outside, you must remember to deduct twice the metal thickness for length and for width.

A tray which is to be joined by spot welding or soldering requires a tab, and in this case the tab bend line must allow for the metal thickness so that the tab fits against the inside face of the tray, Fig. 4.14(c). Sharp inside corners for bends are not always possible or desirable, and as a general rule an inside radius is made, equal to twice the thickness of the metal used.

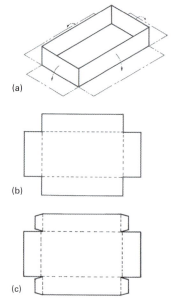

Figure 4.14 Development of rectangular tray

To find the development length on the flat sheet, it is necessary to find the length of the mean line by calculating the lengths of the flat portions and the bends separately. The stretched-out length of the bend is called the bend allowance and for a 90° bend is found by multiplying the mean radius by 1.57 (i.e. π/2).

4.2.3 Example 4.3

Figure 4.15 shows a right-angled bracket made from 1 mm thick material. To obtain the development, first find

length ab = 60 – inside radius – metal thickness

$= 60 - 2 - 1$

$= 57$ mm

next

length cd = 80 – 2 – 1

$= 77$ mm

finally

length bc = mean radius × 1.57

$= $ (inside radius $+ \frac{1}{2}$ metal thickness) $\times 1.57$

$= (2 + 0.5) \times 1.57$

$= 2.5 \times 1.57$

$= 3.9$, say 4 mm

∴ total length of development $= 57 + 77 + 4$

$= 138$ mm

It can be seen that the development of the bracket shown in Fig. 4.15 is made up of a 57 mm straight length plus a bend allowance of 4 mm plus a further straight length of 77 mm, as shown in Fig. 4.16. The bend is half way across the bend allowance, and therefore the bend line must be $57 + 2 = 59$ mm from one edge.

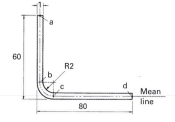

Figure 4.15   Right-angled bracket

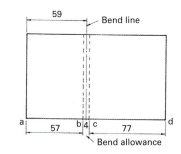

Figure 4.16   Development of right-angled bracket

Review questions

1. Why is it necessary to overbend material during a bending operation?
2. Under what circumstances would it be more appropriate to use a folding machine rather than a vice when bending sheet metal?
3. Why is it necessary to calculate the developed length of sheet metal components using the mean line?
4. How is the thickness of sheet metal identified?
5. Describe two methods of producing a hole in sheet metal.
6. Calculate the cut length of blank required to produce a plane cylinder of 180 mm finished diameter from 14 SWG (2 mm) sheet (ans. 559.2 mm).
7. A right angle bracket has leg lengths of 90 and 50 mm with a 3 mm corner radius and is made from 14 SWG (2 mm) sheet. Calculate the developed length (ans. 138.3).
8. State the restrictions in using hand snips to cut sheet metal.

CHAPTER 5

Standards, measurement and gauging

There shall be but one measure throughout the land. This famous excerpt from the Magna Carta expresses clearly the role of government in the regulation of fair trade. Central to this task was the construction and maintenance of national standards of length, weight and capacity to serve as the ultimate reference from which trading weights and measures throughout the land were to be derived.

Standards are necessary in industry in order to ensure uniformity and to establish minimum requirements of quality and accuracy. The adoption of standards eliminates the national waste of time and material involved in the production of an unnecessary variety of patterns and sizes of articles for one and the same purpose. Standards are desirable not only in the manufacture of articles but also for the instruments used to ensure the accuracy of these articles.

The word 'standard' can refer to a physical standard such as length or to a standard specification such as the paper standard. The National Measurement Office (NMO) is responsible for all aspects of the National Measuring System (NMS). In the UK all our measurement issues are looked after by the NMS, which is the nation's infrastructure of measurement laboratories delivering world-class measurement, science and technology. It provides traceable and increasingly accurate standards of measurement for use in trade, industry, academia and government.

The National Physical Laboratory (NPL) is the UK's National Measurement Institute which sits at the heart of the NMS. For more than a century NPL has developed and maintained the UK's primary measurement standards. These standards underpin an infrastructure of traceability throughout the UK and the world which ensures accuracy and consistency in measurement. The national primary standards, which constitute the basis of measurement in the UK, are maintained in strict accordance with internationally agreed recommendations. They are based on the International System of Units (SI).

Within the UK, the national primary standards are used to calibrate secondary standards and measuring equipment manufactured and used in industry. This calibration service is provided by NPL or through the approved laboratories of the United Kingdom Accreditation Service (UKAS).

UKAS, a national service which, for particular measurements, specially approves laboratories which are then authorised to issue official certificates for such measurements. These laboratories are located in industry, educational institutions and government establishments.

A UKAS certificate issued by an approved laboratory indicates that measurements are traceable to national standards and provides a high degree of assurance of the correctness of the calibration of the equipment or instrument identified on the certificate.

The preparation of standard specifications in the UK is the responsibility of the British Standards Institution (BSI), whose main function is to draw up and promote the adoption of voluntary standards and codes of good practice by agreement among all interested parties, i.e. manufacturers, users, etc.

BSI plays a large and active part in European and international standards.

In the UK, British Standards are issued prefixed with the letters BS and are used mainly in the UK.

The prefix EN means the standard is a European standard and is used throughout Europe.

The prefix ISO means the standard is an international standard and may be used throughout the world.

5.1 Length

The universal standard of length is the metre, and the definition of this in terms of wavelength of light was agreed by all countries in 1960. The metre was defined at this time as 1 650 763.73 wavelengths of the orange radiation of the krypton-86 isotope in vacuo. At the same time the yard was defined as 0.9144 m, which gives an exact conversion of 1 inch = 25.4 mm.

It was also in 1960 that the first laser was constructed and by the mid-1970s lasers were being used as length standards. In 1983 the krypton-86 definition was replaced and the metre was defined as 'the length of the path travelled by light in a vacuum during a time interval of 1/299 792 458 of a second' and this is done at NTL by an iodine-stabilised helium–neon laser with an uncertainty of 3 parts in 10 (to the power 11). This is the equivalent to measuring the Earth's mean circumference to about 1 mm. Lasers have a reproducibility better then ±3 parts in a hundred thousand million.

The great advantage of using lasers as length standards is that it is constant, unlike material length standards where small changes in the length of metal bars can occur over periods of time. Also these laser standards can be directly transferred to a material standard in the form of an end gauge, e.g. gauge blocks, and to a line standard, i.e. a measuring scale such as a vernier scale, to a high degree of accuracy.

End standards of length used in the workshop are of two types: gauge blocks and length bars. These are calibrated and intended for use at 20 °C.

5.1.1 Gauge blocks

Gauge blocks are made from a wear-resisting material – hardened and stabilised high-grade steel, tungsten carbide and ceramic and their dimensions and accuracy are covered by BS EN ISO 3650: 1999.

Steel gauge blocks have proven their reliability over many years and are the most commonly accepted for length standards. They provide high resistance to wear and wring well to other gauge blocks but need protection against corrosion. Provided they are correctly used and handled they will remain reliable for many years.

Tungsten carbide gauge blocks are ten times more resistant to wear than steel gauge blocks and are intended for frequent use.

Ceramic gauge blocks are extremely resistant to wear and scratches and wring perfectly to steel and carbide blocks. Due to the properties of this material, any minor damage (no burr is produced) is unlikely to affect the wringing characteristic of the surfaces. Being corrosion resistant, these gauge blocks are unaffected by moisture or sweaty hands. Although ceramic is a brittle material, the thinnest block resists breakage due to normal handling forces. They are, however, roughly double the cost of steel gauge blocks.

Two uses for gauge blocks are recognised: (i) general use for precise measurement where accurate work sizes are required; (ii) as standards of length used with high-magnification comparators to establish the sizes of gauge blocks in general use.

Each gauge block is of a rectangular section with the measuring faces finished by precision lapping to the required distance apart – known as the gauge length, Fig. 5.1, within the tolerances of length, flatness and parallelism set out in the standard. The measuring faces are of such a high degree of surface finish and flatness that gauge blocks will readily wring to each other, i.e. they will adhere when pressed and slid together. Thus a set of selected-size gauge blocks can be combined to give a very wide range of sizes, usually in steps of 0.001 mm.

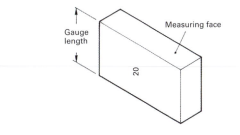

Figure 5.1 Gauge block

Table 5.1

Size (mm)	Increment (mm)	Number of pieces
1.0005	–	1
1.001 to 1.009	0.001	9
1.01 to 1.49	0.01	49
0.5 to 9.5	0.5	19
10 to 100	10	10
		Total 88 pieces

Table 5.2

Size (mm)	Increment (mm)	Number of pieces
1.0005	–	1
2.001 to 2.009	0.001	9
2.01 to 2.49	0.01	49
0.5 to 9.5	0.5	19
10 to 100	10	10
		Total 88 pieces

Standard sets are available with differing numbers of pieces. A typical set is shown in Fig. 5.2. These sets are identified by a number indicating the number of pieces prefixed by the letter M, to indicate metric sizes, and followed by the number 1 or 2. This latter number refers to a 1 mm- or 2 mm-based series, this being the base gauge length of the smaller blocks. For example, an 88 piece set to a 1 mm base is designated M88/1 and contains the sizes shown in Table 5.1.

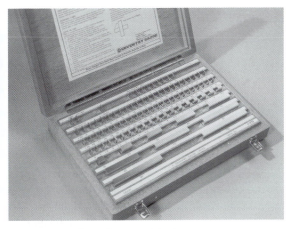

Figure 5.2 Set of gauge blocks

An 88 piece set to a 2 mm base is designated M88/2 and contains the sizes shown in Table 5.2.

The 2 mm-based series are recommended as they are less likely to suffer deterioration in flatness than the thinner 1 mm blocks.

Four grades of accuracy are provided for in the British Standard: grades 0, 1, 2 with 0 the finest, and calibration grade (K).

The choice of grade is solely dependent on the application.

Calibration-grade gauges should not be used for general inspection work: they are intended for calibrating other gauge blocks. This means that the actual gauge length is known, and this is obtained by referring to a calibration chart of the set, i.e. a chart showing the actual size of each block in the set. For this reason, relatively large tolerances on gauge length can be allowed, but calibration-grade gauges are required to have a high degree of accuracy of flatness and parallelism.

To build up a size combination, the smallest possible number of gauge blocks should be used. This can be done by taking care of the micrometres (0.001 mm) first, followed by hundredths (0.01 mm), tenths (0.1 mm) and whole millimetres; e.g. to determine the gauge blocks required for a size of 78.748 mm using the M88/2 set previously listed:

```
        78.748
Subtract 2.008 .... 1st gauge block
        76.740
Subtract 2.24 .... 2nd gauge block
        74.50
Subtract 4.50 .... 3rd gauge block
        70.00 .... 4th gauge block
```

The 2.24 mm second gauge block is used as it conveniently leaves a 0.5 increment for the third gauge block.

In some instances protector gauge blocks, usually of 2 mm gauge length, are supplied with a set, while in other instances they have to be ordered separately. These are available in pairs, are marked with the letter P, and are placed one at each end of a build-up to prevent wear on the gauge blocks in the set. If wear takes place on the protector gauge blocks, then only these need be replaced. Allowance for these, if used, must be made in calculating the build-up.

Having established the sizes of gauge blocks required, the size combination can be built up. Select the required gauge blocks from the case and close the lid. It is important that the lid is kept closed at all times when not in actual use, to protect the gauge blocks from dust, dirt and moisture.

Clean the measuring faces of each gauge block with a clean chamois leather or a soft linen cloth and examine them for damage. Never attempt to use a gauge block which is damaged, as this will lead to damage of other gauge blocks. Damage is likely to occur on the edges through the gauge block being knocked or dropped, and, in the event of damage, it is preferable to return the gauge block to the manufacturer for the surface to be restored.

When two gauge blocks are pressed and slid together on their measuring surfaces, they will adhere and are said to 'wring' together. They will only do this if the measuring surfaces are clean, flat and free from damage.

Wring two gauge blocks by placing one on the top of the other as shown in Fig. 5.3 and sliding them into position with a rotary movement. Repeat this for all gauge blocks in the size combination, starting with the largest gauge blocks. Never wring gauge blocks together while holding them

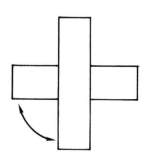

Figure 5.3 Position for wringing gauge blocks

above an open case, as they could be accidentally dropped and cause damage.

Do not finger the measuring faces, as this leads to the risk of tarnishing, and avoid unnecessary handling as this can lead to an increase in temperature and hence dimension.

Immediately after use, slide the gauge blocks apart, clean each one carefully, replace in the case and close the lid. Do not break the wringing joint but slide the blocks apart, and never leave the gauge blocks wrung together for any length of time.

Gauge blocks, separately or in combination, may be used for direct measurement as shown in Fig. 5.4(a) or for comparative measurement using a knife-edged straightedge or a dial indicator, Figs 5.4(b) and (c). The precise size of the gauge-block combination is usually arrived at by trial and error. The gauge-block combination is built up

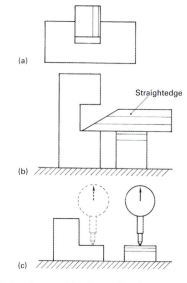

Figure 5.4 Gauge-block applications

until the top of the work and the gauge blocks are the same height, this being so when no light is visible under the knife edge of the straightedge when it is placed simultaneously on both surfaces, Fig. 5.4(b), and viewed against a well-illuminated background. The same principle is employed using a dial indicator, the work and the gauge-block combination being the same height when the same reading is obtained on the dial indicator from each surface, Fig. 5.4(c).

Gauge blocks are also widely used in conjunction with sine bars and with calibrated steel balls and rollers, as well as for checking straightness and squareness. They are also used in conjunction with a range of accessories.

Gauge-block combinations are suitable for providing end standards up to around 150 mm, but above this size can be difficult to handle. Where longer end standards are required, long series steel gauge blocks (sometimes referred to as 'long blocks') are available with standard cross-section and lengths ranging from 125 mm to 1000 mm. Each block has clamping holes towards each end for attaching connecting clamps to improve stack stability when two or more are wrung together, but it is always better to use a single block if possible as it is much easier to handle no matter how good the connection method. Gauge blocks are available singly or as a set. An eight piece set comprises 125, 150, 175, 200, 250, 300, 400 and 500 mm. Standard gauge blocks can then be wrung to the long blocks to give any desired length.

5.1.1.1 Gauge-block accessories

The use of gauge blocks for measuring can be extended by the use of a range of accessories. These are covered by BS 4311-2: 2009. A typical set of gauge-block accessories is shown in Fig. 5.5 and consists of

(i) two pairs of jaws, type A and type B, which when combined with gauge blocks form an external or internal caliper, Figs 5.6(a) and (b);
(ii) a centre point and a scriber, used in combination with gauge blocks to scribe arcs of precise radius, Fig. 5.6(c);
(iii) a robust base for converting a gauge-block combination together with the scriber into a height gauge, Fig. 5.6(d);

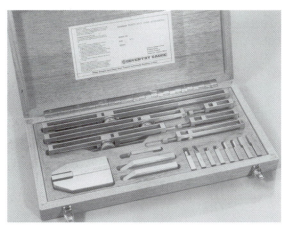
Figure 5.5 Set of gauge-block accessories

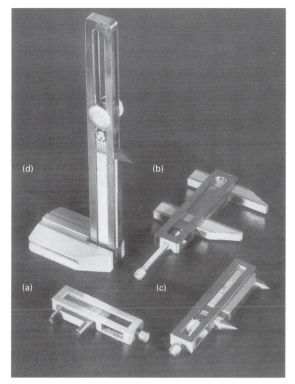
Figure 5.6 Assembled gauge blocks and accessories

(iv) a knife-edged straightedge;
(v) holders for supporting the various combinations when in use.

The accessories other than the holders are made from high-quality steel, hardened and stabilised, with their wringing faces precision-lapped to give flatness, parallelism and surface finish to the same degree as gauge blocks, to which they are wrung to assemble any combination. Accessories can be purchased in sets as shown or as individual items.

5.1.2 Length bars

Length bars can be used for greater stability when the use of longer end standards is required. Length bars are of round section, 22 mm diameter and are made of the same high-quality steel as gauge blocks. The gauge faces are hardened and stabilised and finished by precision lapping to the required length, flatness and parallelism set out in BS 5317: 1976(2012), 'Metric length bars and their accessories'. Length bars are available as individual gauges in lengths from 10 mm to 1200 mm or as sets supplied in a wooden case. Fig. 5.7 shows a set of length bars and length bar accessories. This set contains nine bars of lengths 10, 20, 40, 60, 80, 100, 200, 300 and 400 mm.

Four grades are provided for in BS 5317: Reference, Calibration, grade 1 and grade 2.

Reference grade bars are intended for use as reference standards and are of the highest accuracy.

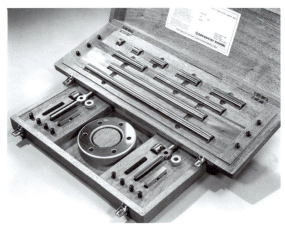

Figure 5.7 Set of length bars and length bar accessories

Calibration grade bars are intended for use in the calibration of length measuring standards and equipment. Both reference and calibration grades should only be used in a 'standards room' controlled at 20°C. Both have completely plain end faces and have UKAS calibration certificates.

Grade 1 bars are intended for use in inspection departments and toolrooms.

Grade 2 bars are intended as workshop standards for precision measurement of gauges, jigs, workpieces etc.

Grades 1 and 2 have internally threaded ends, Fig. 5.8, so that each bar can be used in combination with another by means of a freely fitting connecting screw to make secure lengths. These thread connections are assembled hand-tight only. The screws also allow the use of a range of accessories. In order to obtain specific lengths, gauge blocks can be wrung to the end faces of the length bar.

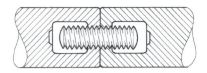

Figure 5.8 Internally threaded ends of length bars

Length bars intended for use in a horizontal position should be supported at two symmetrically placed points a calculated distance apart, known as Airy points. If a bar is supported at any two points, the sag caused by its own weight can affect the length between the end faces. In 1922, the then Astronomer Royal, Sir George Airy, established a formula for the distance separating two points such that the upper and lower surfaces at the ends of the bar would lie in a horizontal plane and so give a true length between the end faces. This distance is 0.577 of the length of the bar, Fig. 5.9. All bars 150 mm in length and over have the Airy points indicated by means of symmetrically spaced lines engraved around the diameter of the bar. For any combination of bars, the Airy points are calculated and the lines on the individual bars ignored.

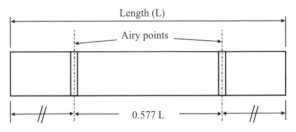

Figure 5.9 Position of Airy points

5.1.2.1 *Length-bar accessories*

The application of length bars can be extended by the use of a range of accessories, the specification for which is also contained in BS 5317: 1976 (2012).

A typical set of length bar accessories is shown in Fig. 5.7 and shown assembled for various applications in Fig. 5.10. These are for use with the grade 1 and 2 bars having internally threaded ends. This set comprises:

▶ knurled nuts and connecting screws for assembly of length bars and accessories.
▶ a base 25 mm thick, with opposite faces finished by precision lapping to a high degree of surface finish, flatness, parallelism and size. Length bars can be assembled to the base using a connecting screw in one of the threaded holes to give good stability when length bars are used vertically, Fig. 5.10(a).
▶ a pair of large radiused jaws used in conjunction with a length bar combination to form an internal or external caliper, Fig. 5.10(b). One face is finished flat and the other is radiused by precision lapping to a width of 25 mm. The jaws have a plain bore and are assembled using a connecting screw and a knurled nut. One jaw can be used in conjunction with a length bar combination and the base to form a height gauge.

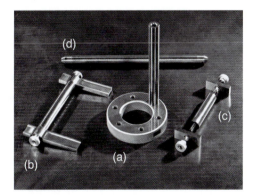

Figure 5.10 Assembled length bars and accessories

▶ a pair of small plain-faced jaws used in conjunction with a length bar combination to assemble a precise setting gauge, Fig. 5.10(c). The two opposite wringing faces are precision lapped to a high degree of surface finish, flatness, parallelism and thickness. These jaws also have a plain bore and are assembled using a connecting screw and knurled nut.
▶ a pair of spherical end pieces, 25 mm in length, used in conjunction with a length bar combination to assemble a precise setting rod or internal measuring pin, Fig. 5.10(d). The bore is threaded internally and is assembled using a connecting screw.

The accessories other than the nuts and connecting screws are made from high-grade steel, hardened and stabilised.

To obtain a specific size combination, gauge blocks can be inserted between the end of the length bar and the accessory. Two combinations of gauge blocks are required with any assembly, to fit between the accessory at either side of the connecting screw.

5.2 Angle

The measurement of angles can be simply carried out using a protractor, but the level of accuracy obtained is, at best, 5 minutes using a vernier instrument. Greater accuracy can be obtained by using angle gauge blocks or by using a sine bar in conjunction with gauge blocks.

5.2.1 Angle gauge blocks

Angle standards are available in the form of angle gauge blocks. They can be used to set and calibrate angles, tapers, dividing heads, rotary tables, etc. as well as to check squareness. They are available in three grades of accuracy: reference with 1 second accuracy, calibration with 2 second accuracy and workshop grade with 5 second accuracy. A typical set comprising 16 blocks is shown in Fig. 5.11. Working grade angle blocks are designed for use in inspection departments, toolrooms and workshops for accurately setting up and measuring of workpieces, jigs, tools and fixtures. A typical application is shown in Fig. 5.12 used to set up an adjustable angle plate by means of angle gauge blocks in conjunction with a dial test indicator to enable the subsequent machining of a workpiece at a precise angle.

The gauge blocks are made from high-grade steel, hardened and stabilised. Each working face is

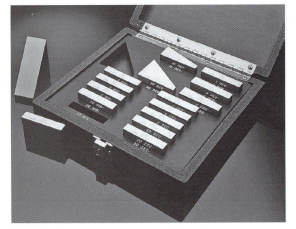

Figure 5.11 Set of angle gauge blocks

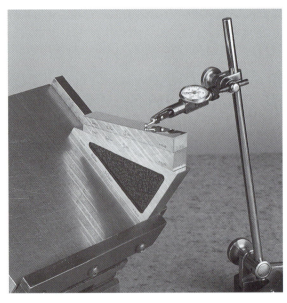

Figure 5.12 Set-up using angle gauge blocks

lapped to a high degree of accuracy of angle and flatness enabling blocks to be wrung together to build up the desired angle.

The set shown in Fig. 5.11 comprises:

▶ six blocks in degrees – 1, 3, 5, 15, 30 and 45;
▶ five blocks in minutes – 1, 3, 5, 20 and 30;
▶ five blocks in seconds – 1, 3, 5, 20 and 30.

The angle blocks in this set can be wrung together in combination by adding or subtracting to form any angle between 0° and 99° in 1 second steps. For example, wringing 30° and 3° angle gauges, as shown in Fig. 5.13(a) with both 'plus' ends together, will result in a total angle of 33°. If the

3° angle gauge is wrung in the opposite direction, with the 'minus' end of the 3° over the 'plus' end of the 30°, this has the effect of subtracting the 3° and results in a total angle of 27° as shown in Fig. 5.13(b).

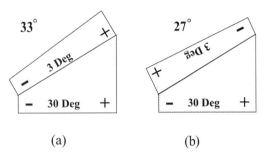

(a)   (b)

Figure 5.13 Angle gauge block combination

5.2.2 The sine bar

A number of different designs of sine bar are available and are covered by BS 3064: 1978, which also specifies three lengths – 100, 200 and 300 mm – these being the distances between the roller axes. The more common type of sine bar is shown in Fig. 5.14 the body and rollers being made of high-quality steel, hardened and stabilised, with all surfaces finished by lapping or fine grinding to the tolerances specified in the British Standard.

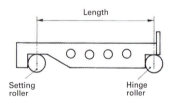

Figure 5.14 Sine bar

The main requirements are:

(a) the mean diameters of the rollers shall be equal to each other within 0.0025 mm;

(b) the upper working surface shall be parallel to the plane tangential to the lower surface of the rollers within 0.002 mm;

(c) the distance between the roller axes shall be accurate to within

0.0025 mm for 100 mm bars,
0.005 mm for 200 mm bars,
0.008 mm for 300 mm bars.

In use, gauge blocks are placed under the setting roller with the hinge roller resting on a datum surface, e.g. a surface plate. The angle of inclination is then calculated from the length of the sine bar (L) and the height of the gauge-block combination (h), Fig. 5.15. Since the roller diameters are the same, we can consider triangle ABC, where AB = L and BC = h; thus

$$\sin \theta = \frac{BC}{AB} = \frac{h}{L}$$

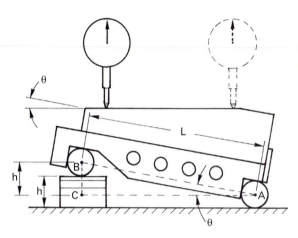

Figure 5.15 Sine-bar set-up

If a known angle of inclination is required then the necessary gauge-block combination can be calculated from $h = L \sin \theta$. However, it is more usual to need to measure the angle of inclination, and in this case the object to be measured is placed on the upper working surface of the sine bar and a gauge-block combination is made up by trial and error. This is done in conjunction with a dial indicator mounted in a surface gauge, the correct height of gauge blocks being established when the dial indicator gives the same reading at each end of the object being measured. The angle is then calculated from $\sin \theta = h/L$.

The degree of accuracy obtainable using a sine bar depends, among other factors, on the size of angle being checked. For example, using a 200 mm sine bar, the following gauge-block heights are required:

for an angle of 10° = 34.73 mm
for 10° 1′ = 34.79 mm
for 60° = 173.20 mm
for 60° 1′ = 173.23 mm

Thus for a 1′ difference at 10° a variation in gauge-block height of 0.06 mm is required, while the same 1′ difference at 60° requires a 0.03 mm variation.

This means, in theory at least, that if the smallest increment of gauge block available is 0.001 mm then, using a 200 mm sine bar, this represents an angular difference of 1 second at small angles and an angular difference of 2 seconds at around 60°. In practice, however, these accuracies would not be possible under workshop conditions, due to inaccuracies in gauge blocks, the sine bar, setting up, and using the dial indicator; but it can be appreciated that this simple piece of workshop equipment is extremely accurate and capable of measuring angles within 1 minute.

It can be seen from the above that the larger angles cannot be measured to the same degree of accuracy as smaller angles. Because of this and the fact that the larger angles require a large gauge-block combination which makes the complete set-up unstable, it is recommended that for larger angles the complement of the angle be measured, i.e. 90° minus the required angle.

Other equipment based on the same principle is available – sine tables with a larger working surface and inclinable about a single axis for checking larger work, compound sine tables inclinable about two axes for checking compound angles, and sine centres which have centres in each end and are the most convenient for checking taper diameters.

5.3 Dimensional deviation

Engineering workpieces cannot be consistently produced to an exact size. This is due to a number of reasons such as wear on cutting tools, errors in setting up, operator faults, temperature differences or variations in machine performance. Whatever the reason, allowance must be made for some error. The amount of error which can be tolerated – known as the tolerance – depends on the manufacturing method and on the functional requirements of the workpiece. For example, a workpiece finished by grinding can be consistently made to finer tolerances than one produced on a centre lathe. In a similar way, a workpiece required for agricultural equipment would not require the same fine tolerance required for a wrist-watch part. In fact it would be expensive and pointless to produce parts to a greater accuracy than was necessary for the part to function. Besides taking account of manufacturing methods, the need for limits and fits for machined workpieces was brought about mainly by the requirements for interchangeability between mass produced parts.

Establishing a tolerance for a dimension has the effect of creating two extremes of size, or limits – a maximum limit of size and a minimum limit of size within which the dimension must be maintained.

British Standard BS EN ISO 286-1: 2010 provides a comprehensive standardised system of limits and fits for engineering purposes. This British Standard relates to tolerances and limits of size for workpieces, and to the fit obtained when two workpieces are to be assembled.

BS EN ISO 286-1 is based on a series of tolerances graded to suit all classes of work covering a range of workpiece sizes up to 3150 mm. A series of qualities of tolerance – called tolerance grades – is provided, covering fine tolerances at one end to coarse tolerances at the other. The standard provides 20 tolerance grades, designated IT01, IT0, etc. (IT stands for ISO series of tolerances.) The numerical values of these standard tolerances for nominal work sizes up to 500 mm are shown in Table 5.3. You can see from the table that for a given tolerance grade the magnitude of the tolerance is related to the nominal size, e.g. for tolerance grade IT6 the tolerance for a nominal size of 3 mm is 0.006 mm while the tolerance for a 500 mm nominal size is 0.04 mm. This reflects both the practicalities of manufacture and of measuring.

5.3.1 Terminology

Limits of size – the maximum and minimum sizes permitted for a feature.
Maximum limit of size – the greater of the two limits of size.
Minimum limit of size – the smaller of the two limits of size.
Nominal size – the size to which the two limits of size are fixed. The basic size is the same for both members of a fit and can be referred to as Nominal size.
Upper deviation – the algebraic difference between the maximum limit of size and the corresponding basic size.
Lower deviation – the algebraic difference between the minimum limit of size and the corresponding basic size.
Tolerance – the difference between the maximum limit of size and the minimum limit of size (or, in other words, the algebraic difference between the upper deviation and lower deviation).
Mean size – the dimension which lies mid-way between the maximum limit of size and the minimum limit of size.
Maximum material condition – where most material remains on the workpiece, i.e. upper limit of a shaft or lower limit of a hole.
Minimum material condition – where least material remains on the workpiece, i.e. lower limit of a shaft or upper limit of a hole.

5.4 Gauging

Establishing a tolerance on a manufactured workpiece results in two extremes of size for each dimension – a maximum limit of size and a minimum limit of size within which the final workpiece dimension must be maintained.

This leads to the simpler and less expensive inspection technique of gauging during production, eliminating the need for more lengthy and expensive measuring techniques. Gauges used to check these maximum and minimum limits are known as limit gauges.

The simplest forms of limit gauges are those used to check plain parallel holes and shafts with other types available for checking tapered and threaded holes and shafts.

Limit gauges are arranged so that the 'GO' portion of the gauge checks the maximum material condition (i.e. the upper limit of the shaft or lower limit of the hole) while the 'NOT GO' portion of the gauge checks the minimum material condition (i.e. the lower limit of the shaft or upper limit of the hole).

In practice this means that if a workpiece is within its required limits of size, the 'GO' end should go

Table 5.3 Standard tolerances

Nominal size mm		Standard tolerance grades																			
		IT01	IT0	IT1	IT2	IT3	IT4	IT5	IT6	IT7	IT8	IT9	IT10	IT11	IT12	IT13	IT14	IT15	IT16	IT 17	IT18
		Standard tolerance values																			
Above	Up to and including	μm													mm						
–	3	0.3	0.5	0.8	1.2	2	3	4	6	10	14	25	40	60	0.1	0.14	0.25	0.4	0.6	1	1.4
3	6	0.4	0.6	1	1.5	2.5	4	5	8	12	18	30	48	75	0.12	0.18	0.3	0.48	0.75	1.2	1.8
6	10	0.4	0.6	1	1.5	2.5	4	6	9	15	22	36	58	90	0.15	0.22	0.36	0.58	0.9	1.5	2.2
10	18	0.5	0.8	1.2	2	3	5	8	11	18	27	43	70	110	0.18	0.27	0.43	0.7	1.1	1.8	2.7
18	30	0.6	1	1.5	2.5	4	6	9	13	21	33	52	84	130	0.21	0.33	0.52	0.84	1.3	2.1	3.3
30	50	0.6	1	1.5	2.5	4	7	11	16	25	39	62	100	160	0.25	0.39	0.62	1	1.6	2.5	3.9
50	80	0.8	1.2	2	3	5	8	13	19	30	46	74	120	190	0.3	0.46	0.74	1.2	1.9	3	4.6
80	120	1	1.5	2.5	4	6	10	15	22	35	54	87	140	220	0.35	0.54	0.87	1.4	2.2	3.5	5.4
120	180	1.2	2	3.5	5	8	12	18	25	40	63	100	160	250	0.4	0.63	1	1.6	2.5	4	6.3
180	250	2	3	4.5	7	10	14	20	29	46	72	115	185	290	0.46	0.72	1.15	1.85	2.9	4.6	7.2
250	315	2.5	4	6	8	12	16	23	32	52	81	130	210	320	0.52	0.81	1.3	2.1	3.2	5.2	8.1
315	400	3	5	7	9	13	18	25	36	57	89	140	230	360	0.57	0.89	1.4	2.3	3.6	5.7	8.9
400	500	4	6	8	10	15	20	27	40	63	97	155	250	400	0.63	0.97	1.55	2.5	4	6.3	9.7

5

into a hole or over a shaft while the 'NOT GO' end should not.

5.4.1 Plain plug gauges

These are used to check holes and are usually renewable-end types. The gauging member and the handle are manufactured separately, so that only the gauging member need be replaced when worn or damaged or when the workpiece limits are modified. The handle is made of a suitable plastics material which reduces weight and cost and avoids the risk of heat transference. A drift slot or hole is provided near one end of the handle to enable the gauging members to be removed when replacement is necessary.

The GO and NOT GO gauges may be in the form of separate 'single-ended' gauges or may be combined on one handle to form a 'double-ended' gauge, Fig. 5.16.

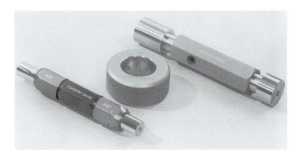

Figure 5.16   Plug and ring gauges

Since the GO gauging member must enter the hole being checked, it is made longer than the NOT GO gauging member, which of course should never enter. Large gauges which are heavy and difficult to handle do not have a full diameter but are cut-away and are known as segmental cylindrical gauges, Fig. 5.17.

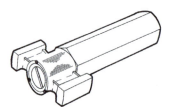

Figure 5.17   Segmental cylindrical gauge

When checking a hole, the GO cylindrical plug gauge should enter the hole being inspected when applied by hand without the use of excessive force, and the hole should be checked throughout its length. A GO segmental gauge should be applied in at least two positions equally spaced around the circumference.

NOT GO plug gauges should not enter the hole when applied by hand without using excessive force.

5.4.2 Ring and gap gauges

These are used to check shafts. Plain ring gauges are ordinarily used only as GO gauges, Fig. 5.16, the use of gap gauges being recommended for the NOT GO gauge. The use of NOT GO gauges is confined to setting pneumatic comparators, and these gauges are identified by a groove around the outside diameter.

Plain gap gauges are produced from flat steel plate, suitably hardened, and may be made with a single gap or with both the GO and the NOT GO gaps combined in the one gauge, Fig. 5.18.

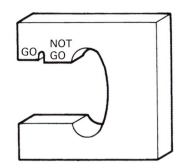

Figure 5.18   Solid gap gauge

Adjustable gap gauges, Fig. 5.19, consist of a horseshoe frame fitted with plain anvils the spacing of which can be adjusted to any particular limits required, within the range of the gauge. Setting to within 0.005 mm of a desired size is possible with well-made adjustable gauges.

When checking a shaft, the GO gap gauge should pass over the shaft, under its own weight when the axis of the shaft is horizontal or without the use of excessive force when the axis of the shaft is vertical. A cylindrical ring GO gauge should pass over the complete length of the shaft without the use of excessive force.

Figure 5.19   Adjustable gap gauges

NOT GO gap gauges should not pass over the shaft and this check should be made at not less than four positions around and along the shaft.

5.4.3 Screw thread gauges

Limit gauges for checking internal and external threads are available in a similar form to those for plain holes and shafts but with threaded rather than plain diameters.

BS 3643-2: 2007 specifies four fundamental deviations and tolerances for internal threads – 4H, 5H, 6H and 7H – and three for external threads – 4h, 6g and 8g – resulting in a variety of different types of fit, with the most common 6H/6g for general engineering work, regarded as a medium fit.

For checking internal threads, e.g. nuts, a double-ended screw plug gauge, Fig. 5.20, may be used. This gauge has a 'GO' end which checks that the major and effective diameters are not too small

Figure 5.20   Screw plug gauge

(i.e. maximum material condition). It also checks for pitch and flank errors in the thread.

The 'NOT GO' end checks only that the effective diameter is not too large (i.e. minimum material condition).

Figure 5.21   Screw ring gauge

For checking external threads 'GO' and 'NOT GO' ring gauges, Fig. 5.21, or caliper gauges having 'GO' and 'NOT GO' anvils may be used. Wherever practical, however, external threads should be checked using ring gauges as these provide a full functional check of all thread features, i.e. pitch, angle, thread form and size.

5.4.4 Taper gauges

Taper gauges are used to check the male and female features of tapered workpieces. A taper ring gauge is used to check the male workpiece feature, i.e. a shaft; while a taper plug gauge is used to check the female workpiece feature, i.e. a hole – Fig. 5.22.

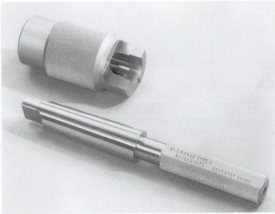

Figure 5.22   Taper ring and plug gauges

A taper gauge is used to check the correctness of the angle of taper of the workpiece and also to check the diameter at some point on the taper, usually the large diameter of a female feature and the small diameter of a male feature.

The correctness of the angle of taper is determined by rocking the gauge, any error being indicated by the amount of 'rock' felt by the person doing the checking. The taper is correct if no 'rock' can be detected. A positive indication of correctness of angle of taper can be achieved by smearing the taper portion of the gauge with engineer's blue, offering the gauge to the workpiece, and rotating the gauge slightly. Complete transference of engineer's blue from the gauge to the entire length of the workpiece surface indicates a perfect match of gauge and workpiece and therefore a correct angle of taper.

Checking the diameter at some point on the taper, usually at one end, is achieved by providing a step at the end of the gauge, giving two faces, equivalent to the GO and NOT GO limits of the diameter being checked. A correct workpiece is indicated by the diameter being checked passing the gauge face equivalent to the GO limit and not reaching the gauge face equivalent to the NOT GO limit, i.e. the workpiece surface must lie somewhere between the two gauge faces, Fig. 5.23. (This may require scratching the workpiece surface with a fingernail to determine whether or not it protrudes.)

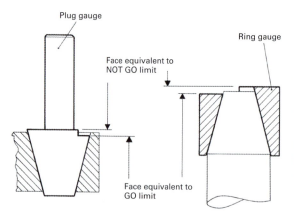

Plug gauge

Ring gauge

Face equivalent to NOT GO limit

Face equivalent to GO limit

Figure 5.23 Use of taper gauges

5.5 Straightness

The workshop standard against which the straightness of a line on a surface is compared is the straightedge. An error in straightness of a feature may be stated as the distance separating two parallel straight lines between which the surface of the feature, in that position, will just lie. Three types of straightedge are available: toolmaker's straightedges, cast-iron straightedges and steel or granite straightedges of rectangular section.

Toolmaker's straightedges, covered by BS 852: 1939(2012), are of short length up to 300 mm and are intended for very accurate work. They are made from high-quality steel, hardened and suitably stabilised, and have the working edge ground and lapped to a 'knife edge' as shown in the typical cross-section in Fig. 5.24. Above 25 mm length, one end is finished at an angle. This type of straightedge is used by placing the knife edge on the work and viewing against a well-illuminated background. If the work is perfectly straight at that position, then no white light should be visible at any point along the length of the straightedge. It is claimed that this type of test is sensitive to within $1 \mu m$.

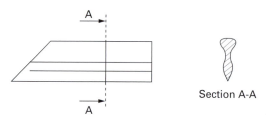

A

A

Section A-A

Figure 5.24 Toolmarker's straightedge

Cast-iron straightedges, of bow-shaped (Fig. 5.25(a)) and I-section design (Fig. 5.25(b)), are covered by BS 5204-1: 1975. Two grades of accuracy are provided for each type – grade A and grade B – with grade A the more accurate. The straightedges are made from close-grained plain or alloy cast iron, sound and free from blowholes and porosity. The working surfaces of grade A are finished by scraping, and those of grade B by scraping or by smooth machining. The recommended lengths for the bow-shaped type are 300, 500, 1000, 2000, 4000, 6000 and

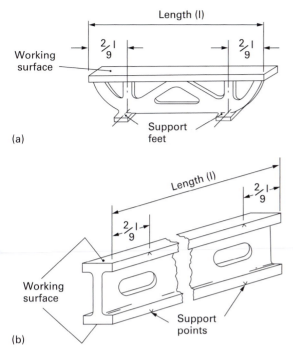

(a)

(b)

Figure 5.25 Cast-iron straightedges

8000 mm and for the I-section type 300, 500, 1000, 2000, 3000, 4000 and 5000 mm.

Steel and granite straightedges of rectangular section are covered by BS 5204-2: 1977. Two grades of accuracy are provided: grade A and grade B. Grade-A steel straightedges are made from high-quality steel with the working faces hardened. Grade-B steel straightedges may be supplied hardened or unhardened. Grade-A and grade-B straightedges may be made of close-grained granite of uniform texture, free from defects. The working faces are finished by grinding or lapping. The recommended lengths for rectangular-section straightedges are 300, 500, 1000, 1500 and 2000 mm.

When a straightedge is used on edge, it is likely to deflect under its own weight. The amount of deflection depends on the number and position of supports along the length of the straightedge. For minimum deflection, a straightedge must be supported at two points located two-ninths of its length from each end, Fig. 5.25 – i.e. at supports symmetrically spaced five-ninths (0.555) of its length apart. For this reason, rectangular and I-section straightedges have arrows together with the word 'support' engraved on their side faces to indicate the points at which the straightedge

should be supported for minimum deflection under its own weight.

If a rectangular or I-section straightedge is used on edge, it should not be placed directly on the surface being checked – it should be supported off the surface on two equal-size gauge blocks placed under the arrows marked 'support'. The straightness of the surface being checked is then established by determining the width of gap under the working face of the straightedge at various points along its length, using gauge blocks. Alternatively a dial indicator held in a surface gauge can be traversed along the surface being checked while in contact with the straightedge. Since the straightedge is straight, any deviation shown on the dial indicator will show the error of straightness of the surface.

Straightedges – especially the bow-shaped type – are used extensively to check the straightness of machine-tool slides and slideways. This is done by smearing a thin even layer of engineer's blue on the working surface of the straightedge, placing the straightedge on the surface to be checked, and sliding it slightly backwards and forwards a few times. Engineer's blue from the straightedge is transferred to the surface, giving an indication of straightness by the amount of blue present. Due to the width of the working face of the straightedge, an indication of flatness is also given.

5.6 Flatness

The workshop standard against which the flatness of a surface is compared is the surface plate or table. The error in flatness of a feature may be stated as the distance separating two parallel planes between which the surface of the feature will just lie. Thus flatness is concerned with the complete area of a surface, whereas straightness is concerned with a line at a position on a surface; e.g. lines AB, BC, CD and DA in Fig. 5.26 may all be straight but the surface is not flat, it is twisted.

For high-precision work, such as precision tooling and gauge work, toolmaker's flats and high-precision surface plates are available and are covered by BS 869: 1978. This standard recommends four sizes of toolmaker's flat – 63, 100, 160 and 200 mm diameter – made from high-quality steel, hardened and stabilised, or

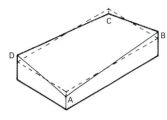

Figure 5.26   Error in flatness of a surface

from close-grained granite of uniform texture, free from defects. Two sizes of high-precision flat are recommended – 250 and 400 mm diameter – made from plain or alloy cast iron or from granite. The working surface of flats and plates are finished by high-grade lapping and must be free from noticeable scratches and flat within 0.5 μm for flats up to 200 mm diameter, 0.8 μm for 250 mm diameter plates and 1.0 μm for 400 mm diameter plates.

Surface plates and tables are covered by BS 817: 2008 which specifies the requirements for rectangular and square surface plates ranging from 160 mm × 100 mm to 2500 mm × 1600 mm in four grades of accuracy – 0, 1, 2, 3 – with grade 0 the most accurate. The accuracy relates to the degree of flatness of the working surface.

The highest accuracy grade 0 plates and tables are used for inspection purposes, grade 1 for general purpose use, grade 2 for marking out while grade 3 for low-grade marking out and as a general support plate.

The plates may be made from good-quality close-grained plain or alloy cast iron, sound and free from blowholes and porosity, and must have adequate ribbing on the underside to resist deflection. Alternatively, plates may be made of close-grained granite of uniform texture, free from defects and of sufficient thickness to resist deflection. The working surface must have a smooth finish.

The smaller sizes of plates may be used on a bench; the larger ones are usually mounted on a stand and are then known as surface tables.

The simplest method of checking the flatness of a surface is to compare it with a surface of known accuracy, i.e. a surface plate. This is done by smearing a thin even layer of engineer's blue on one surface, placing the surface to be checked on the surface plate, and moving it slightly from side to side a few times. Engineer's blue will be transferred from one surface to the other, the amount of blue present and its position giving an indication of the degree of flatness.

The main use of surface plates and tables is as a reference or datum surface upon which inspection and marking-out equipment are used.

5.7 Squareness

Two surfaces are square when they are at right angles to each other. Thus the determination of squareness is one of angular measurement. There is no absolute standard for angular measurement in the same way as there is for linear measurement, since the requirement is simply to divide a circle into a number of equal parts. The checking of right angles is a common requirement, and the workshop standard against which they are compared is the engineer's square, of which there are a number of types. BS 939: 2007 specifies the requirements for engineer's try-squares (Fig. 5.27(a)), cylindrical squares. (Fig. 5.27(b)), and

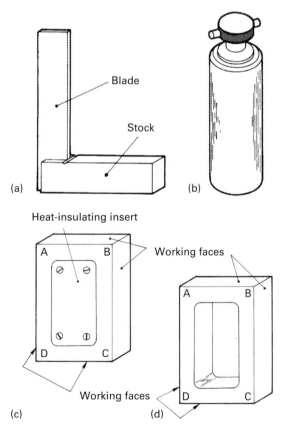

Figure 5.27   Types of square

block squares of solid (Fig. 5.27(c)) or open form (Fig. 5.27(d)).

Engineer's try-squares consist of a stock and a blade and are designated by a size which is the length from the tip of the blade to the inner working face of the stock. Recommended sizes are 50, 75, 100, 150, 200, 300, 450, 600, 800 and 1000 mm. Three grades of accuracy are specified – AA, A and B – with grade AA the most accurate. Try-squares are made of good-quality steel with the working surfaces of grades AA and A hardened and stabilised.

Grade AA try-squares have the inner and outer edges of the blade bevelled. All working surfaces of the blade and stock are lapped, finely ground or polished to the accuracy specified for each grade.

Cylindrical squares, of circular section, are designated by their length. Recommended lengths are 75, 150, 220, 300, 450, 600 and 750 mm. One grade of accuracy is specified: grade AA. Cylindrical squares are made of high-quality steel, hardened and stabilised; close-grained plain or alloy cast iron, sound and free from blowholes and porosity; or close-grained granite of uniform texture, free from defects. Granite is particularly suitable for the larger sizes, as its mass is approximately half that of the equivalent size in steel or cast iron. In order to reduce weight, it is recommended that steel or cast-iron cylindrical squares 300 mm long and above are of hollow section. All external surfaces are finished by lapping or fine grinding.

Solid-form block squares are designated by their length and width and are available in sizes from 50 mm × 40 mm up to and including 1000 mm × 1000 mm. Two grades of accuracy are specified: AA and A. Solid-form block gauges are made of high-quality steel, cast iron, or granite – the same as cylindrical squares. Again, granite is recommended due to its lower mass. The front and back surfaces of each solid-form steel block square are recessed and fitted with a heat-insulating material, to avoid heat transfer and hence expansion when handled. The working faces of the solid-form steel block square are finished by lapping, and those made of cast iron or granite are finished by lapping or fine grinding.

Open-form block squares are designated by their length and width and are available in sizes from 150 mm × 100 mm up to and including 600 mm × 400 mm. Two grades of accuracy are specified – A and B – grade A being the more accurate. They are made of close-grained plain or alloy cast iron, sound and free from blowholes and porosity, and may be hardened or unhardened. All external surfaces are finished by lapping or fine grinding.

5.7.1 Use of squares

Grade AA engineer's try-squares have the inner and outer edges of the blade bevelled to produce a 'knife edge'. This increases the sensitivity of the square in use. If, however, the square is used with its blade slightly out of normal to the surface being checked, an incorrect result may be obtained. For this reason, try-squares with bevelled-edge blades are unsuitable for checking cylindrical surfaces. For this purpose a try-square with a square edge or a block square should be used, since the cylindrical surface itself will provide the necessary sensitivity by means of line contact.

Cylindrical squares are ideal for checking the squareness of try-squares and block squares and for work with flat faces, since line contact by the cylindrical surface gives greater sensitivity.

To check the squareness of two surfaces of a workpiece using a try-square, the stock is placed on one face and the edge of the blade is rested on the other. Any error in squareness can be seen by the amount of light between the surface and the underside of the blade. This type of check only tells whether or not the surfaces are square to each other, however, and it is difficult to judge the magnitude of any error.

When accurate results are required, the workpiece and square may be placed on a surface plate and the square slid gently into contact with the surface to be checked. The point of contact can then be viewed against a well-illuminated background. If a tapering slit of light can be seen, the magnitude of the error present can be checked using gauge blocks at the top and bottom of the surface, Fig. 5.28, the difference between the two gauge blocks being the total error of squareness.

5

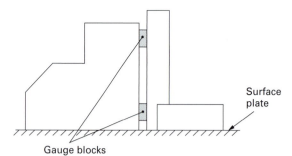

Figure 5.28 Checking squareness with square and gauge blocks

5.8 Roundness

There are a number of workshop tests which are used to determine the roundness of a part; however, not all of these give a precisely true indication. A part is round when all points on its circumference are equidistant from its axis but, as a result of different methods and of machine tools used in the production of cylindrical parts, errors in roundness can occur.

The simplest check for roundness is to measure directly at a number of diametrically opposite points around the circumference of a part, using a measuring instrument such as a micrometer or vernier caliper. Any difference in reading will give an indication of the out-of-roundness of the part. However, it is possible, when using this method, for errors in roundness to go undetected. For example, an incorrect set-up on a centreless grinding machine can produce a tri-lobed shape such as is shown exaggerated in Fig. 5.29. Measuring at diametrically opposite points will give identical readings, but the part is not round.

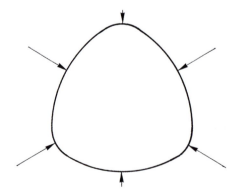

Figure 5.29 Tri-lobed shape

Alternatively, a part which contains centres at each end may be checked for roundness by mounting between bench centres and rotating the part under a dial indicator, Fig. 5.30. An error reading on the dial indicator would show the part to be not round. However, this method can also be misleading, since the centres in the part themselves may not be round or may not be the central axis of the part and it may be these errors which are represented on the dial indicator and not the error in roundness of the part.

Figure 5.30 Checking work between centres

A part with a plain bore may be loaded on to a mandrel before being placed between the bench centres. In this case, since a mandrel is accurately ground between true centres, it can be assumed that the centres are on the true central axis. A constant reading on the dial indicator would show the part to be round. It is also an indication that both the bore and the outside diameter lie on the same axis – a condition known as concentricity. An error reading on the dial indicator could therefore be an error in concentricity and the part be perfectly round.

The ideal workshop test which overcomes the problems already outlined is to rotate the part under a dial indicator with the part supported in a vee block. Because the points of support in the vee block are not diametrically opposite the plunger of the dial indicator, errors in roundness will be identified. For example, the tri-lobed condition undetected by direct measurement would be detected by this method, as shown in Fig. 5.31.

5.9 Surface roughness

No manufactured surface, however it may appear to the naked eye, is absolutely perfect. The

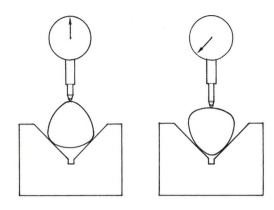

Figure 5.31 Checking work in a vee block

Table 5.4 Preferred surface-roughness values

Nominal R_a values (μm)
50
25
12.5
6.3
3.2
1.6
0.8
0.4
0.2
0.1
0.05
0.025
0.0125

degree of smoothness or roughness of a surface depends on the height and width of a series of peaks and valleys which give the surface a certain texture. This surface texture is characteristic of the method used to produce it. For example, surfaces produced by cutting tools have tool marks in well-defined directions controlled by the method of cutting, and equally spaced according to the feed rates used.

The control of surface texture is necessary to obtain a surface of known type and roughness value which experience has shown to be the most suitable to give long life, fatigue resistance, maximum efficiency and interchangeability at the lowest cost for a particular application. This is not necessarily achieved by the finest surface. For example, two surfaces required to slide over each other would not function if finished to the same high degree as the surface of a gauge block – they would not slide but simply wring together. At the opposite extreme, the same two sliding surfaces with a very poor texture would wear quickly. The cost of producing these extremes of surface would also vary greatly. It is essential that sliding surfaces have sufficient roughness to retain oil molecules for lubrication.

The measurement of roughness is taken as the average of the peaks and valleys over a given length. This roughness average is denoted by R_a and the values expressed in micrometers (μm).

The R_a values specified should be selected from a range of preferred values contained in BS 1134: 2010 shown in Table 5.4. The values are given as 'preferred' in order to discourage

unnecessary variation of the values expressed on drawings. Where a single value is stated on a drawing it is understood that work with any value from zero to the stated value is acceptable.

The method of assessing surface roughness is either by stylus type instruments or by using surface roughness comparison specimens (Fig. 5.32).

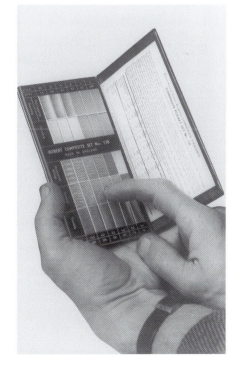

Figure 5.32 Surface roughness comparison specimens

79

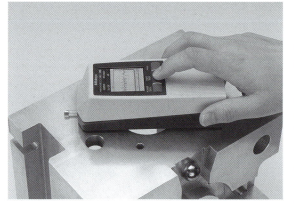

Figure 5.33 Portable surface roughness tester

Small hand-held portable surface roughness testers (Fig. 5.33) are available which give a measurement of the average roughness of a surface The measurements are taken by a traversing drive unit carrying a diamond stylus moving across a selected length (known as the sampling length) of the surface being tested at right angles to the direction of the machining marks commonly referred to as 'across the lay'. A clear representation of the measurement values are shown in the instrument display screen. It is possible with some models to save and print out the data either through an attached unit or through a computer.

Comparison specimens shown in Fig. 5.32 are used to give draughtsmen/women and machine operators an idea of the relation of the feel and appearance of common machined surfaces to their numerical roughness value. By visual examination and by scratching the surface with a fingernail, a comparison can be made between the specimen and the workpiece surface being considered. Some average surface-roughness values obtainable from common production processes are shown in Table 5.5.

Table 5.5 Average surface-roughness values obtained by various manufacturing processes

Process	Roughness values ($\mu m\ R_a$)												
	0.0125	0.025	0.05	0.1	0.2	0.4	0.8	1.6	3.2	6.3	12.5	25	50
Superfinishing			█	█									
Lapping			█	█	█								
Diamond turning				█	█								
Honing				█	█	█							
Grinding				█	█	█							
Turning							█	█	█	█			
Boring							█	█	█	█			
Die-casting								█					
Broaching								█	█				
Reaming								█	█				
Milling								█	█				
Investment casting									█				
Drilling									█	█			
Shaping									█	█	█		
Shell moulding										█			
Sawing										█	█	█	
Sand casting											█	█	

Review questions

1. Name three materials from which straightedges are made.
2. What is meant by tolerance?
3. A 200 mm sine bar is used to check an angle of 30° 12'. Calculate the size of gauge blocks necessary. (ans. 100.604 mm).
4. By means of a sketch show the GO and NOT GO arrangement of a taper plug gauge.
5. What is the name of the calibration service within the UK?
6. British standard numbers are often prefixed BS EN ISO. What do these prefixes mean.
7. State two recognised uses of gauge blocks.
8. Name three types of square used in industry.
9. State the reason why standards are necessary in industry.
10. At what temperature are gauge blocks calibrated and intended for use?

5

CHAPTER 6

Measuring equipment

Some form of precise measurement is necessary if parts are to fit together as intended no matter whether the parts were made by the same person, in the same factory, or in factories a long way apart. Spare parts can then be obtained with the knowledge that they will fit a part which was perhaps produced years before.

To achieve any degree of precision, the measuring equipment used must be precisely manufactured with reference to the same standard of length. That standard is the metre, which is now defined using lasers. Having produced the measuring equipment to a high degree of accuracy, it must be used correctly. You must be able to assess the correctness of size of the workpiece by adopting a sensitive touch or 'feel' between the instrument and workpiece. This 'feel' can be developed only from experience of using the instrument, although some instruments do have an aid such as the ratchet stop on some micrometers. Having the correct equipment and having developed a 'feel', you must be capable of reading the instrument to determine the workpiece size. It is here that the two main standard types of length-measuring instrument differ: the micrometer indicates the linear movement of a rotating precision screw thread, while the vernier instruments compare two scales which have a small difference in length between their respective divisions.

In the late nineteenth century, Dr Ernst Abbe and Dr Carl Zeiss worked together to create one of the world's foremost precision optics companies. The Abbe principle resulted in observations about measurement errors.

The Abbe principle states that the reading axis and the measuring axis of a measuring instrument must be co-axial for maximum accuracy. Any separation of the two may lead to an error.

The micrometer is one instrument which complies with the Abbe principle (Fig. 6.1) since the graduations are located along the same axis as the measurement being taken. The vernier caliper does not comply with this principle (Fig. 6.2). To ensure greatest accuracy of measurement, the workpiece being measured should be moved as close to the main beam as possible. The vernier depth gauge, however, does comply with Abbe's principle.

Reading and measuring axis

Figure 6.1   Abbe's principle – micrometer

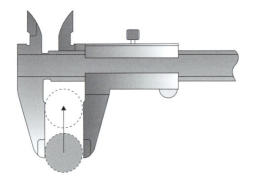

Figure 6.2 Abbe's principle – vernier caliper

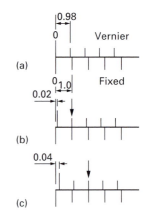

Figure 6.4 Vernier scale readings

Many of the electronic measuring instruments outlined in this chapter, e.g. vernier calipers, micrometers and dial test indicators, can be equipped to output data either through available data cables or by wireless transmitters/receivers. This data can be printed out or sent to a computer and collected and used by quality control to provide statistical information or for the generation of inspection certificates.

6.1 Vernier instruments

All instruments employing a vernier consist of two scales: one moving and one fixed. The fixed scale is graduated in millimetres, every 10 divisions equalling 10 mm, and is numbered 0, 1, 2, 3, 4 up to the capacity of the instrument. The moving or vernier scale is divided into 50 equal parts which occupy the same length as 49 divisions or 49 mm on the fixed scale (see Fig. 6.3). This means that the distance between each graduation on the vernier scale is $\frac{49}{50}$ mm = 0.98 mm, or 0.02 mm less than each division on the fixed scale (see Fig. 6.4(a)).

If the two scales initially have their zeros in line and the vernier scale is then moved so that its first graduation is lined up with a graduation on the fixed scale, the zero on the vernier scale will have moved 0.02 mm (Fig. 6.4(b)). If the second

graduation is lined up, the zero on the vernier scale will have moved 0.04 mm (Fig. 6.4(c)) and so on. If graduation 50 is lined up, the zero will have moved 50 × 0.02 = 1 mm.

Since each division on the vernier scale represents 0.02 mm, five divisions represent 5 × 0.02 = 0.1 mm. Every fifth division on this scale is marked 1 representing 0.1 mm, 2 representing 0.2 mm and so on (Fig. 6.3).

To take a reading, note how many millimetres the zero on the vernier scale is from zero on the fixed scale. Then note the number of divisions on the vernier scale from zero to a line which exactly coincides with a line on the fixed scale.

In the reading shown in Fig. 6.5(a) the vernier scale has moved 40 mm to the right. The eleventh line coincides with a line on the fixed scale, therefore 11 × 0.02 = 0.22 mm is added to the reading on the fixed scale, giving a total reading of 40.22 mm.

Similarly, in Fig. 6.5(b) the vernier scale has moved 120 mm to the right plus 3 mm and the sixth line coincides, therefore, 6 × 0.02 = 0.12 mm is added to 123 mm, giving a total of 123.12 mm.

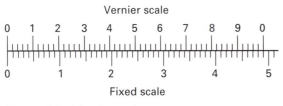

Figure 6.3 Vernier scale

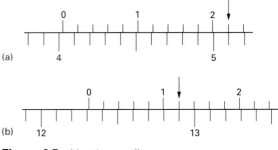

Figure 6.5 Vernier readings

It follows that if one part of a measuring instrument is attached to the fixed scale and another part to the moving scale, we have an instrument capable of measuring to 0.02 mm.

6.1.1 Vernier caliper

The most common instrument using the above principle is the vernier caliper (see Fig. 6.6). Because of its capability of external, internal, step and depth measurements (Fig. 6.7), its ease of operation and wide measuring range, the vernier caliper is possibly the best general purpose measuring instrument. They are available in a range of measuring capacities from 0–150 mm to 0–1000 mm.

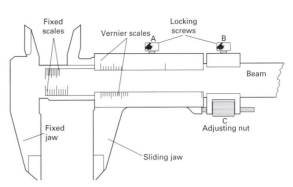

Figure 6.8   Vernier caliper adjustment

Figure 6.9   Dial caliper

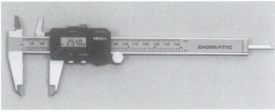

Figure 6.10   Digital caliper

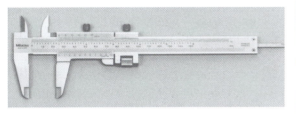

Figure 6.6   Vernier caliper

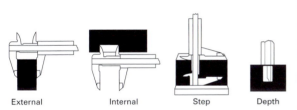

External   Internal   Step   Depth

Figure 6.7   External, internal, step and depth measurement

To take a measurement, slacken both locking screws A and B (Fig. 6.8). Move the sliding jaw along the beam until it contacts the surface of the work being measured. Tighten locking screw B. Adjust the nut C until the correct 'feel' is obtained, then tighten locking screw A. Re-check 'feel' to ensure that nothing has moved. When you are satisfied, take the reading on the instrument. Jaws may be carbide tipped for greater wear resistance.

Dial calipers (Fig. 6.9) are a form of vernier caliper where readings of 1 mm steps are taken from the vernier beam and subdivisions of this are read direct on a dial graduated in 0.02 mm divisions.

Conventional digital calipers (Fig. 6.10) make use of a basic binary system having a series of

light and dark bands under the slider and count these as they move along the track. There is no way the system can tell where the slider is, it purely depends on storing the number of bands passed over. Because of this, the jaws must first be closed and the display zeroed to reset the binary system before it starts counting. Digital calipers are now available which can read the slider location at any position without the need to reset zero. This type of caliper makes use of three sensors within the slider and three corresponding precision tracks embedded in the main beam. As the slider moves it reads the position of the tracks under the sensors and calculates its current absolute position. This eliminates the need for having to reset the caliper to zero before use.

An inch/metric conversion is built into the microprocessor so that a measurement in the desired units can be made. An additional benefit is that the zero point can be set anywhere within the instrument range. With this ability the instrument can be used as a comparator to determine if the workpiece is above or below zero and by how much. The resolution of these instruments is 0.01 mm/0.0005".

Although the vernier caliper is extremely versatile in its ability to carry out external, internal, step and depth measurements, calipers with special jaws are available for special purpose applications, a few of which are shown in Fig. 6.11.

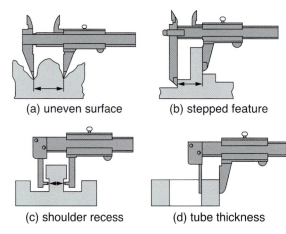

(a) uneven surface   (b) stepped feature

(c) shoulder recess   (d) tube thickness

Figure 6.11 Caliper special jaws

a. Point jaw type – for uneven surface measurement.
b. Offset jaw type – for stepped feature measurement.
c. Neck type – for outside diameter measurement such as behind a shrouded recess.
d. Tube thickness type – for tube or pipe wall thickness measurement.

6.1.1.1 Notes on using a vernier calliper

1. Before use, clean the measuring surfaces by gripping a piece of clean paper (copier paper is ideal) between the jaws and slowly pull it out. Ensure the instrument reads zero before taking a measurement. Always ensure digital models are zeroed.
2. Wipe sliding surfaces before use.
3. Look straight at the vernier graduations when making a reading. If viewed from an angle, an error of reading can be made due to parallax effect (Fig. 6.12).

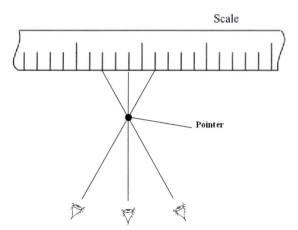

Figure 6.12 Parallax error

4. Ensure the surface to be measured is clean and free of swarf or grit before taking a measurement.
5. Do not use excessive force when taking a measurement.
6. Do not measure a workpiece at a position near the outer end of the jaws. Move the workpiece as close to the beam as possible to ensure greatest accuracy of measurement (see Abbe's principle Fig. 6.2).
7. Keep jaws square/parallel with surface being measured to ensure greatest accuracy.
8. For internal measurement, ensure the inside jaws are inserted as deeply as possible before taking a measurement.
9. Be careful not to drop or bang the caliper such as would cause damage to the instrument.
10. Caliper jaws are very sharp so handle with care to avoid personal injury.
11. Always store the instrument in a clean place when not in use, preferably in a case or box.
12. Do not leave the caliper jaws clamped together when not in use. Always leave a gap between the measuring faces (say 1 mm).

6.1.2 Vernier height gauge

The above principles apply to the vernier height gauge, Fig. 6.13. In this case the beam, carrying the fixed scale, is attached to a heavy base. The vernier scale carries a jaw upon which various attachments can be clamped. It is most widely used with a chisel-pointed scribing blade for accurate marking out, as well as for checking the height of steps in components. Care should

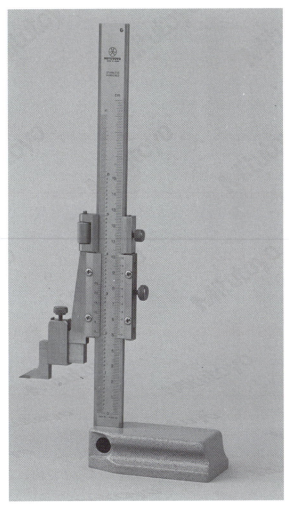

Figure 6.13 Vernier height guage

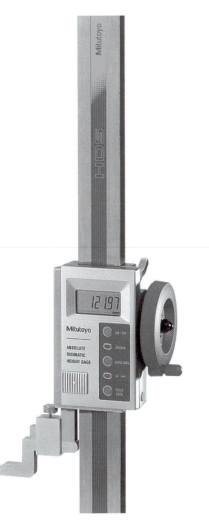

Figure 6.14 Digital height gauge

6

be taken to allow for the thickness of the jaw, depending on whether the attachment is clamped on top of or under the jaw. The thickness of the jaw is marked on each instrument. Height gauges are available in a range of capacities reading from zero up to 1000mm. A digital version is shown in Fig. 6.14.

6.1.3 Vernier depth gauge

Accurate depths can be measured using the vernier depth gauge (Fig. 6.15) again employing the same principles. The fixed scale is similar to a narrow rule. The moving scale is tee-shaped to provide a substantial base and datum from which readings are taken. The instrument reading is the amount which the rule sticks out beyond the base. Depth gauges are available in a range of capacities from 150mm to 300mm.

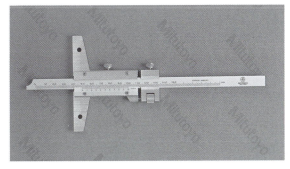

Figure 6.15 Vernier depth gauge

These are also available as an easy-to-read dial depth gauge (Fig. 6.16) and a digital model with an LCD readout (Fig. 6.17) operating in the same way as the caliper models.

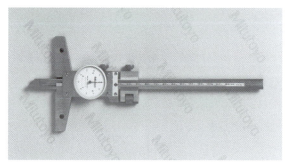

Figure 6.16   Dial depth gauge

Figure 6.17   Digital depth gauge

6.1.4 Vernier bevel protractor

As well as linear measurement, vernier scales can equally well be used to determine angular measurement. The vernier bevel protractor (Fig. 6.18) again uses the principle of two scales, one moving and one fixed. The fixed scale is graduated in degrees, every 10 degrees being numbered 0,10, 20, 30, etc. The moving or vernier scale is divided into 12 equal parts which occupy the same space as 23 degrees on the fixed scale (Fig. 6.19). This means that each division on the

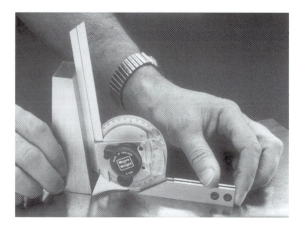

Figure 6.18   Vernier bevel protractor

Figure 6.19   Vernier protractor scale

vernier scale is $\frac{23}{12}$ degrees $= 1\frac{11}{12}$ degrees or 1 degree 55 minutes. This is 5 minutes less than two divisions on the fixed scale (Fig. 6.20(a)).

If the two scales initially have their zeros in line and the vernier scale is then moved so that its first graduation lines up with the 2 degree graduation on the fixed scale, the zero on the vernier scale will have moved 5 minutes (Fig. 6.20(b)). Likewise, the second graduation of the vernier lined up with the 4 degree graduation will result in the vernier scale zero moving 10 minutes (Fig. 6.20(c)) and so on until when the twelfth graduation lines up the zero will have moved $12 \times 5 = 60$ minutes = 1 degree. Since each division on the vernier scale represents 5 minutes, the sixth graduation is numbered to represent 30 minutes and the twelfth to represent 60 minutes.

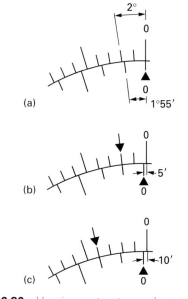

Figure 6.20   Vernier protractor scale readings

The stock of the vernier protractor carries the fixed scale. The removable blade is attached to the moving or vernier scale, which has a central screw to lock the scale at any desired position and give angular measurement to an accuracy of 5 minutes.

Since the vernier scale can be rotated in both directions, the fixed scale is graduated from 0–90, 90–0, 0–90, 90–0 through 360°. This requires a vernier scale for each, and therefore the vernier scale is also numbered 0–60 in each direction.

To take a reading, note how many degrees the zero on the vernier scale is from the zero on the fixed scale. Then counting in the same direction, note the number of divisions on the vernier scale from zero to a line which exactly coincides with a line on the fixed scale.

In the reading shown in Fig. 6.21(a) the vernier scale has moved to the left 45 degrees. Counting along the vernier scale in the same direction, i.e. to the left, the seventh line coincides with a line on the fixed scale. Thus $7 \times 5 = 35$ minutes is added, to give a total reading of 45 degrees 35 minutes.

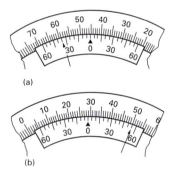

(a)

(b)

Figure 6.21   Vernier protractor readings

In the reading shown in Fig. 6.21(b), the vernier scale has moved to the right 28 degrees. Again counting along the vernier scale in the same direction, i.e. to the right, the eleventh line coincides with a line on the fixed scale, giving a total reading of 28 degrees 55 minutes.

Digital universal protractors are available with 150 mm and 300 mm blades and a measuring range of 360° (Fig. 6.22). The resolution is 1′ and an accuracy of +/− 2′. These instruments can be used as a protractor in the usual manner and can also be attached to a vernier height gauge giving greater versatility. They have an output function to enable transfer of data through a connecting cable.

Figure 6.22   Digital protractor

6.2 Micrometers

The micrometer relies for its measuring accuracy on the accuracy of the spindle screw thread. The spindle is rotated in a fixed nut by means of the thimble, which opens and closes the distance between the ends of the spindle and anvil (see Fig. 6.23). The pitch of the spindle thread, i.e. the distance between two consecutive thread forms, is 0.5 mm. This means that, for one revolution, the spindle and the thimble attached to it will move a longitudinal distance of 0.5 mm.

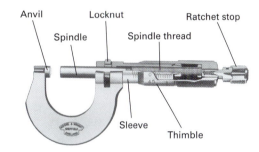

Figure 6.23   External micrometer

On a 0–25 mm micrometer, the sleeve around which the thimble rotates has a longitudinal line graduated in mm from 0 to 25 mm on one side of the line and subdivided in 0.5 mm intervals on the other side of the line.

The edge of the thimble is graduated in 50 divisions numbered 0, 5, 10, up to 45, then 0. Since one revolution of the thimble advances the spindle 0.5 mm, one graduation on the thimble must equal $0.5 \div 50$ mm = 0.01 mm. A reading is therefore the number of 1 mm and 0.5 mm divisions on the sleeve uncovered by the thimble plus the hundredths of a millimetre indicated by the line on the thimble coinciding with the longitudinal line on the sleeve.

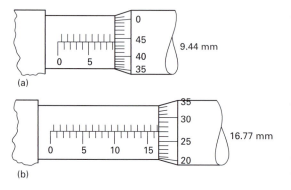

(a)

9.44 mm

(b)

16.77 mm

Figure 6.24 Micrometer readings

In the reading shown in Fig. 6.24(a), the thimble has uncovered 9 mm on the sleeve. The thimble graduation lined up with the longitudinal line on the sleeve is 44 = 44 × 0.01 = 0.44. The total reading is therefore 9.44 mm.

Similarly, in Fig. 6.24(b) the thimble has uncovered 16 mm and 0.5 mm and the thimble is lined up with graduation 27 = 27 × 0.01 = 0.27 mm, giving a total reading of 16.77 mm.

Greater accuracy can be obtained with external micrometers by providing a vernier scale on the sleeve. The vernier consists of five divisions on the sleeve, numbered 0, 2, 4, 6, 8, 0, these occupying the same space as nine divisions on the thimble (Fig. 6.25(a)). Each division on the vernier is therefore equal to 0.09 ÷ 5 = 0.018 mm. This is 0.002 mm less than two divisions on the thimble.

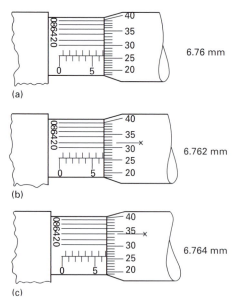

(a)

6.76 mm

(b)

6.762 mm

(c)

6.764 mm

Figure 6.25 Vernier micrometer readings

To take a reading from a vernier micrometer, note the number of 1 mm and 0.5 mm divisions uncovered on the sleeve and the hundredths of a millimetre on the thimble as with an ordinary micrometer. You may find that the graduation on the thimble does not exactly coincide with the longitudinal line on the sleeve, and this difference is obtained from the vernier. Look at the vernier and see which graduation coincides with a graduation on the thimble. If it is the graduation marked 2, then add 0.002 mm to your reading (Fig. 6.25(b)); if it is the graduation marked 4, then add 0.004 mm (Fig. 6.25(c)) and so on.

External micrometers with fixed anvils are available with capacities ranging from 0–13 mm to 575–600 mm. External micrometers with interchangeable anvils (Fig. 6.26) provide an extended range from two to six times greater than the fixed-anvil types. The smallest capacity is 0–50 mm and the largest 900–1000 mm. To ensure accurate setting of the interchangeable anvils, setting gauges are supplied with each instrument. Spindle and anvil measuring faces may be carbide tipped for greater wear resistance.

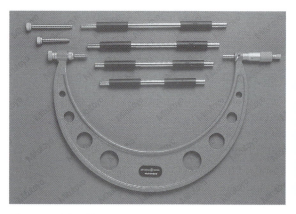

Figure 6.26 External micrometer with interchangeable anvils

Micrometers are available which give a direct mechanical readout on a counter (Fig. 6.27). The smallest model of 0–25 mm has a resolution of 0.001 mm and the largest of 125–150 mm a resolution of 0.01 mm.

Modern digital external micrometers (Fig. 6.28) use sensors and detectors. A tiny disc encoder is connected to and rotates with the spindle along the entire range of the spindle axis, while a

Figure 6.27 Mechanical counter micrometer

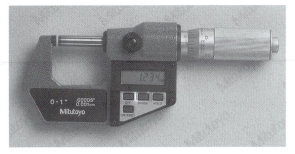

Figure 6.28 Digital external micrometer

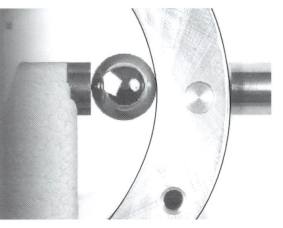

Figure 6.29 Measuring tube wall thickness

detector remains stationary to count the rotational signals and send data to a microprocessor and shows as a reading on an LCD.

An inch/metric conversion is built into the microprocessor so that a measurement in the desired units can be made. An additional benefit is that the zero point can be set anywhere within the micrometers range. With this ability the instrument can be used as a comparator to determine if the workpiece is above or below zero and by how much. The resolution of these instruments is 0.001 mm/0.00005″.

Standard micrometers are good for measuring flat and parallel features and outside diameters. However, if the feature to be measured is curved, as in the wall thickness of a tube, an accurate measurement cannot be achieved. In this case a ball or roller can be placed on the inside wall of the tube and a measurement taken (Fig. 6.29). By subtracting the diameter of the ball or roller, the wall thickness can be established. With a digital micrometer, the diameter of the ball or roller can be measured, the micrometer zeroed, and the measurement then made will be the wall thickness shown directly without the need for subtraction. For this type of measurement, an accessory is available called a 'ball attachment',

which is a ball (usually 5 mm diameter) held in a rubber boot to hold it safely in place on the micrometer anvil while a measurement is taken.

In order to facilitate a wide range of measuring requirements, micrometers are available with a variety of anvils for special purpose applications, a few of which are shown in Fig. 6.30:

a) Blade micrometer – for measuring diameter in a narrow groove (the spindle is non-rotational).

b) Tube micrometer – for measuring the wall thickness of a tube (anvil has a spherical surface).

c) Spline micrometer – for measuring spline-root shaft diameter.

d) Disc type micrometer – for measuring spur and helical gear root tangent.

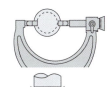

(a) Blade micrometer   (b) Tube micrometer

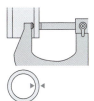

(c) Spline micrometer   (b) Disc type outside micrometer

Figure 6.30 Micrometer special anvils

Micrometers with an adjustable measuring force are available for applications requiring a constant low measuring force such as measuring thin wire, paper, plastics or rubber parts which avoids distortion of the workpiece. These are available in a range from 0–10 mm to 20–30 mm.

6.2.1 Notes on using a micrometer

1. Before use, clean the measuring surfaces by gripping a piece of clean paper (normal copier paper is ideal) between the anvil and the spindle and slowly pull it out. Ensure the instrument reads zero before taking a measurement. Always ensure digital models are zeroed.
2. Look straight at the index line when making a reading. If viewed from an angle, an error of reading can be made due to parallax effect (Fig. 6.12).
3. Ensure the surface being measured is clean and free of swarf and grit before taking a measurement.
4. Do not rotate the thimble using excessive force.
5. Use the ratchet device, if available, to ensure a consistent measuring force.
6. Be careful not to drop or bang the micrometer such as would cause damage to the instrument.
7. Always store the instrument in a clean place when not in use, preferably in a case or box.
8. Never leave the measuring faces clamped together when not in use. Always leave a gap between the measuring faces (say 1 mm).

6.2.2 Internal micrometer

The internal micrometer (Fig. 6.31) is designed for inside measurement and consists of a micrometer measuring head to which may be added external rods to cover a wide range of measurements and a spacing collar to make up for the limited range of the micrometer head. Micrometers of less than 300 mm are supplied with a handle to reach into deep holes. Each extension rod is marked with the respective capacity of the micrometer when that particular rod is used. The smallest size is 25–50 mm with a measuring range of 7 mm. The next size covers 50–200 mm with a measuring range of 13 mm, while the largest covers 200–1000 mm with a measuring range of 25 mm.

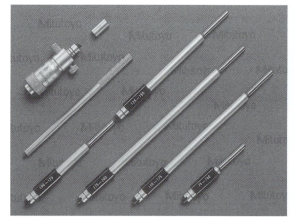

Figure 6.31 Internal micrometer

Readings are taken in the same way as described for the external micrometer, although, as already stated, the measuring range of the micrometer head is reduced.

Great care must be taken when using this instrument, as each of the measuring anvils has a spherical end, resulting in point contact. Experience in use is essential to develop a 'feel', and the instrument must be moved slightly back and forth and up and down to ensure that the measurement is taken across the widest point.

Digital versions are available, operating on the same principle except the micrometer measuring head has an LCD readout (Fig. 6.32). The resolution is 0.001 mm/0.0001".

Three-point bore micrometers are available with a standard micrometer head or as a digital head with

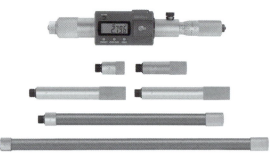

Figure 6.32 Digital inside micrometer

an LCD readout (Fig. 6.33). These gauges locate positively inside a bore by virtue of the three contact surfaces, making it easier to use and less likely to result in incorrect measurements. These

Figure 6.34   Depth micrometer

Figure 6.33   Digital three-point bore micrometer

are available in a range of instruments to measure bore sizes from 6 mm to 300 mm.

6.2.3 Depth micrometer

The depth micrometer (Fig. 6.34) is used for measuring the depths of holes, slots, recesses and similar applications. Two types are available: one with a fixed spindle and a capacity of 0–25 mm, the other with interchangeable rods giving a measuring capacity up to 300 mm. The interchangeable rods are fitted into the instrument by unscrewing the top part of the thimble and sliding the rod in place, ensuring that the top face of the thimble and the underside of the rod are perfectly clean. The top of the thimble is then replaced and holds the rod in place. Each rod is marked with its respective size.

The micrometer principle is the same as for the other instruments; however, the readings with this instrument increase as the thimble is screwed on, resulting in the numbering of sleeve and thimble

graduations in the opposite direction to those on the external and internal micrometers. To take a reading, you must note the 1 mm and 0.5 mm divisions covered by the thimble and add to this the hundredths of a millimetre indicated by the line on the thimble coinciding with the longitudinal line on the sleeve.

In the reading shown in Fig 6.35(a) the thimble has covered up 13 mm and not quite reached 13.5. The line on the thimble coinciding with the longitudinal line on the sleeve is 44, so $44 \times 0.01 = 0.44$ mm is added, giving a total reading of 13.44 mm. Similarly, in Fig. 6.35(b) the thimble has just covered 17 mm and line 3 on the thimble coincides, giving a total reading of 17.03 mm.

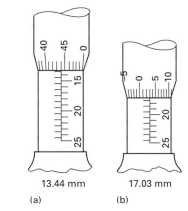

Figure 6.35   Depth micrometer readings

6

Figure 6.36 Mechanical counter depth micrometer

Depth micrometers are available giving a direct mechanical readout on a counter with a resolution of 0.01 mm (see Fig. 6.36).

Depth micrometers are also available as a digital model with an LCD readout (Fig. 6.37) giving direct readings in imperial or metric units with a resolution of 0.0001″ or 0.001 mm.

Figure 6.37 Digital depth micrometer

6.3 Dial indicators

Dial indicators magnify small movements of a plunger or lever and show this magnified movement by means of a pointer on a graduated dial. This direct reading from the pointer and graduated dial gives the operator a quick, complete and accurate picture of the condition of the item under test. Dial indicators are used to check the dimensional accuracy of workpieces in conjunction with other equipment such as gauge blocks, to check straightness and alignments of machines and equipment, to set workpieces in machines to ensure parallelism and concentricity and for a host of other uses too numerous to list completely.

The mechanism of a dial indicator is similar to that of a watch and, although made for workshop use, care should be taken to avoid dropping or knocking it in any way. Slight damage to the mechanism can lead to sticking which may result in incorrect or inconsistent readings.

6.3.1 Plunger-type instruments

The most common instrument of this type is shown in Fig. 6.38. The vertical plunger carries a rack which operates a system of gears for magnification to the pointer. The dial is attached to the outer rim, known as the bezel, and can be rotated so that zero can be set irrespective of the initial pointer position. A clamp is also supplied to prevent the bezel moving once it has been set to zero. The dial divisions are usually 0.01 or 0.002 mm, with an operating range between 8 and

Figure 6.38 Plunger-type dial indicator

Figure 6.39 Dial test indicator and stand

20 mm, although instruments with greater ranges are available.

In conjunction with a robust stand or surface gauge (Fig. 6.39) this instrument can be used to check straightness, concentricity, as well as workpiece heights and roundness.

It may not always be possible to have the dial of this type facing the operator, which may create problems in reading the instrument or safety problems if the operator has to bend over equipment or a machine. An instrument which can be used to overcome these difficulties is the back plunger-type shown in Fig. 6.40. The readings can be seen easily by viewing above the instrument. The direction of the plunger movement restricts the range to about 3 mm.

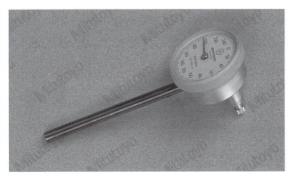

Figure 6.40 Back plunger-type dial indicator

Figure 6.41 Digital-type indicator

Modern-design digital plunger-type instruments (Fig. 6.41) are available with an LCD digital readout giving direct readings in imperial or metric units with a resolution of 0.0005″ or 0.01 mm. These instruments can be zeroed at any point within the range.

6.3.2 Lever-type instruments

The lever-type of instrument is shown in Fig. 6.42. Due to the leverage system, the range of this type is not as great as that of the plunger-type and is usually 0.5 or 0.8 mm. The dial divisions are 0.01 or 0.005 mm, and again the dial is adjustable to set zero. The greatest advantage of this type is the small space within which it can work. Another added advantage is an automatic reversal system which results in movement above or below the contact stylus registering on the dial pointer. This facility, together with the contact stylus being able to swing at an angle, means that checks can be made under a step as well as on top (Fig. 6.43).

Figure 6.42 Lever-type dial indicator

6

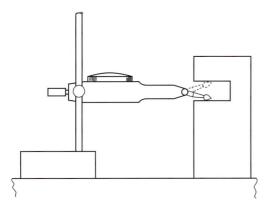

Figure 6.43 Lever-type dial indicator application

Table 6.1 Shows a summary of measuring instruments.

6.4 Modern measuring techniques

6.4.1 Laser scan micrometer

The laser scan micrometer is a high-precision laser measuring system that performs non-contact dimensional measurement using a high-speed scanning laser beam.

With the non-contact measurement capability, this system features high-precision measurement of workpieces that are difficult to measure using conventional measuring systems. These include hot workpieces, brittle or elastic workpieces, workpieces that must be kept free from contamination, and soft workpieces which would be affected by measuring forces.

The basic principle is that a laser beam is directed across the workpiece. The part of the beam not obstructed by the workpiece reaches a receiver which sends a signal for processing and the dimension is displayed digitally on the connected display unit.

Data software is available which allows import of measurement data from one or more display units to a PC for statistical and quality control purposes.

A typical measuring unit and display unit is shown at Fig. 6.44. The unit shown has a range from 0.1mm to 25mm with resolutions selectable

Table 6.1

Instrument	Advantages	Limitations
Vernier calipers	Large measuring range on one instrument. Capable of internal, external, step and depth measurements. Resolution of LCD model 0.01 mm	Resolution 0.02 mm. Point of measuring contact not in line with adjusting nut (Abbe's principle). Jaws can spring. Lack of 'feel'. Length of jaws limits measurement to short distance from end of workpiece. No adjustment for wear
Vernier height gauge	Large range on one instrument. Resolution of LCD model 0.01 mm	Resolution 0.02 mm. No adjustment for wear
Vernier depth gauge	Large range on one instrument. Resolution of LCD model 0.01 mm	Resolution 0.02 mm. Lack of 'feel'. No adjustment for wear
Bevel protractor	Accuracy 5 minutes over range of 360°. Will measure internal and external angles	Can be difficult to read the small scales except with the aid of a magnifying lens
External micrometer	Resolution 0.01 mm or, with vernier 0.002 mm or LCD model 0.001 mm. Adjustable for wear. Ratchet or friction thimble available to aid constant 'feel'	Micrometer head limited to 25 mm range. Separate instruments required in steps of 25 mm or by using interchangeable anvils
Internal micrometer	Resolution 0.01 mm. Adjustable for wear. Can be used at various points along length of bore	Micrometer head on small sizes limited to 7 and 13 mm range. Extension rods and spacing collar required to extend capacity. Difficult to obtain 'feel'
Depth micrometer	Resolution 0.01 mm or with LCD model 0.001 mm. Adjustable for wear. Ratchet or friction thimble available to aid constant 'feel'	Micrometer head limited to 25 mm range. Interchangeable rods required to extend capacity
Dial indicator	Resolution as high as 0.001 mm. Measuring range up to 80 mm with plunger types. Mechanism ensures constant 'feel'. Easy to read. Quick in use if only comparision is required	Has to be used with guage blocks to determine measurement. Easily damaged if mishandled. Must be rigidly supported in use.

Figure 6.44 Laser scan micrometer

from 0.00001mm to 0.01mm. Measuring units are available with a large measuring range of 1–160 mm.

6.4.2 Coordinate measuring machine (CMM)

With the advent of numerical controlled machine tools, the time required in removing metal was greatly reduced as the machine tools were controlled, initially by punched tape (NC), and subsequently, with the introduction of computers, by computer numerical control (CNC). As well as the increased speed of production, the consistent accuracy of the finished work also increased dramatically. Manufacturing tolerances had been reduced in order to satisfy improvements in product reliability and improved performance. Checking these parts was initially carried out by inspection departments using conventional measuring techniques which was slow and created bottlenecks in the production system. The resulting need for high-speed methods of accurate and reliable measurement brought about the introduction of the coordinate measuring machine (CMM). The introduction of computer aided design (CAD) systems together with CNC machine tools has allowed the manufacture of forms so complex that only a computer controlled CMM can economically assess the accuracy.

Basically the CMM is a mechanical system that moves a measuring probe to determine the co-ordinates of points on the surface of a workpiece. The CMM comprises the computer controlled machine itself, the measuring probe, the control system and the measuring software.

A popular design is the moving bridge type where the slides move freely on air bearings about a granite worktable. The machine shown in Fig. 6.45 is a CNC CMM and comprises three axes X, Y, Z, each of which is orthogonal, or at right angles to

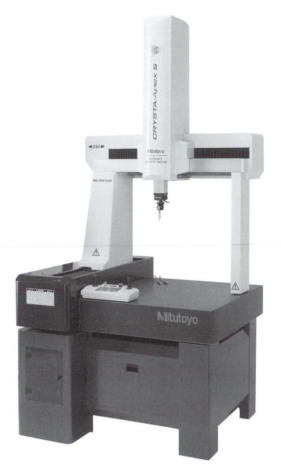

Figure 6.45 CNC co-ordinate measuring machine

each other as in a typical 3D co-ordinate system. Each axis has a scale system that locates the position of that axis. The probe is directed either by the operator or by a computer program. The machine then utilises the X, Y, Z coordinates of each of the discrete points provided by the probe to determine the size and position with great accuracy. The machine shown has a measuring range of 400 mm, 400 mm, 300 mm on the X, Y, Z axes respectively, with a resolution of 0.5 µm, but machines having much larger measuring ranges are available.

The measuring probe, mounted at the end of the quill, is usually a touch-trigger type used to measure discrete points. The probe has a stylus attached, at the end of which is a stylus tip, commonly a synthetic industrial ruby which makes contact with the surface being measured.

With volume production, the complete inspection routine is directly controlled from the measuring software. The measuring software can be used to generate the inspection routine directly from the computer aided design (CAD) data and verify, on-screen, the probe movement. It is essential that these programs be written off-line so that the CMM time is dedicated to its measuring activities. The overall measuring capability of the CMM and its ease of use depend almost entirely on the measurement software.

During the checking procedure, the probe is sent to its correct position, carries out the desired checks, and sends the data back to the inspection software before moving on to its next programmed position to repeat the process. The part geometry from the CAD model is compared with the part produced and any resultant deviation identified. This is shown graphically by the software. This information can be used in identifying problems in manufacture, used to modify tool offsets as well as being used in quality control functions. All these functions can be interlinked to create a computer integrated manufacturing (CIM) system.

As well as the larger floor-mounted CMM outlined above, portable co-ordinate measuring machines, also known as portable measurement arms or articulated measurement arms, are used to generate very precise geometric information. The portable measurement arm is an easily transportable, mobile instrument used for the measurement of high-precision 3D co-ordinates within a specific spherical measuring volume. The user attaches the base of the measurement arm in an absolutely stable position near the object to be measured. This rigid positioning is essential to obtain high-precision measurement results. The other end is fitted with a measuring probe in the same way as the floor-mounted CMM. The operator guides the measurement arm and triggers the actual measurement as soon as the probe tip is in the correct position. The model shown in Fig. 6.46 uses a series of rotary joints and arms, which are ideal for quick and accurate measurements of parts. Being portable, they can be moved around and used anywhere on the workshop floor, which makes them ideal for measuring large parts and are common in the automobile, shipbuilding and railway industries. Because portable CMM arms

Figure 6.46 Portable measurement arm

imitate the flexibility of the human arm, they are often able to reach inside complex parts that could not be probed by fixed bed machines. They are less expensive than the larger floor mounted machines and are suited to manual operation.

6.4.3 Vision measuring machine (VMM)

The vision measuring system provides edge sensing technology and in operation lies somewhere between the profile projector and a CMM, giving highly accurate measurement without making physical contact with the workpiece being measured. This non-contact facility means that quick accurate measurement can be obtained from thin, soft, brittle or elastic parts which is clean, safe and very versatile.

The larger CNC vision measuring machine with a moving bridge looks similar in construction and appearance to a CMM except that the mechanical touch trigger probe is replaced by a vision system comprising an objective lens to magnify the area under inspection, a camera (colour CCD or black and white) with interface device to transfer images to the computer vision system software and an illumination system. Small basic machines measuring on X, Y axes are two dimensional (2D) but the majority are (3D) also able to measure on the Z axis. Fig 6.47 shows a CNC vision measuring system with a measuring range of 200, 250, 100 mm in X, Y, Z respectively.

All vision systems use some sort of light source to illuminate the workpiece surface, usually LED

Figure 6.47 CNC vision measuring system

or halogen. Various objective lenses are available, e.g. 0.5X, 1X, 2.5X, 10X and 25X. This allows for 16X to 4800X magnification on screen.

All vision systems perform measurements by detecting changes in light intensity and hence the contrast of the feature, e.g. a black spot against a white background. First, the camera has to be calibrated to convert the pixel spacing into units of length. Once the image is captured it can be processed using the functions of the system software. The results of the measurement can be shown as a graphical display on a screen to show any deviation between the measured dimension and that of the imported CAD data. The data can also be output to other devices as in the case of in-process control or in a form required for quality control statistical analysis purposes similar to the CMM systems.

Although the description given relates to engineering measurement, the range of applications for vision systems is vast, due to its non-contact function, especially in continuous production processes where the parts or product is moving.

Vision systems are used to check:

▶ shape – whether a part is the correct shape or is properly positioned;
▶ flaw detection – detecting flaws during continuous production of sheet metal, paper, glass, plastics, and ceramic tiles;
▶ colour confirmation – used in the pharmaceutical industry to ensure the correct colour tablet has been placed in appropriate packing;
▶ single dimensional measuring – measuring width of continuous rolled or extruded products;
▶ character recognition – reading bar codes or incorrectly formed letters.
▶ part recognition – check whether a vehicle is an estate, saloon or hatchback and instruct paint robots accordingly;
▶ guidance – instruct robot where to pick up/put down an object. Controlling the path of robot spreading sealants or adhesive.

6

Review questions

1. State the main function of a dial indicator.
2. What is the measuring accuracy of a micrometer without a vernier scale?
3. Describe the basic principle of the vernier scale.
4. What is the difference between the graduations of an external micrometer and a depth micrometer?
5. State the types of measurement which can be carried out using a vernier caliper.
6. Name two types of dial indicator and state a typical use for each.
7. What measuring accuracy can be obtained when using a vernier bevel protractor?
8. State four limitations of a vernier caliper.
9. What do the initials CMM stand for and give a brief description of its main functions.
10. State two main advantages of the use of articulated measuring arms.

CHAPTER 7

Cutting tools and cutting fluids

The study of metal cutting is complex, due to the number of possible variables. Differences in workpiece materials and cutting-tool materials, whether or not a cutting fluid is used, the relative speed of the work and cutting tool, the depth of cut, and the condition of the machine all affect the cutting operation. However, certain basic rules apply, and when you know and can apply these you will be in a better position to carry out machining operations effectively.

The material from which the workpiece is made is not usually your choice. The operation to be carried out decides which machine you will use. This narrows the problem – knowing the operation and the machine, you can select the type of cutting tool; knowing the workpiece material you can decide the cutting-tool material, the cutting angles, the speeds at which to run the workpiece or cutting tool and whether to use a cutting fluid. Finally, you must be able to maintain the cutting tools in good condition as the need arises, and this requires a knowledge of regrinding the tool usually by hand, known as off-hand grinding.

7.1 Cutting-tool materials

7.1.1 Properties of cutting materials

To be effective, the material from which a cutting tool is made must possess certain properties, the most important of which are red hardness, abrasion resistance and toughness.

7.1.1.1 Red hardness

It is obvious that a cutting tool must be harder than the material being cut, otherwise it will not cut. It is equally important that the cutting tool remains hard even when cutting at high temperatures. The ability of a cutting tool to retain its hardness at high cutting temperatures is known as red hardness. Current practice often uses, the term 'hot hardness'.

7.1.1.2 Abrasion resistance

When cutting, the edge of a cutting tool operates under intense pressure and will wear due to abrasion by the material being cut. Basically, the harder the cutting-tool material the better its resistance to abrasion.

7.1.1.3 Toughness

A cutting-tool material which is extremely hard is unfortunately also brittle. This means that a cutting edge will chip on impact if, e.g. the component being machined has a series of slots and the cut is therefore intermittent. To prevent the cutting edge from chipping under such conditions, it is necessary that the material has a certain amount of toughness. This can be achieved only at the expense of hardness; i.e. as the toughness is increased so the hardness decreases.

It can be readily seen that no one cutting-tool material will satisfy all conditions at one time. A cutting tool required to be tough due to cutting conditions will not be at its maximum hardness and therefore not be capable of fully resisting abrasion. Alternatively, a cutting tool requiring maximum hardness will have maximum abrasion resistance but will not be tough to resist impact loads. The choice of cutting-tool material is governed by the type of material to be cut and the conditions under which cutting is to take place, as well as the cost of the tool itself. Remember that cutting tools are expensive, and great care should be taken to avoid damage and consequent wastage both in use and during any subsequent resharpening.

7.1.2 High-speed steels (HSS)

High-speed tool steels consist of iron and carbon with differing amounts of alloying elements such as tungsten, chromium, vanadium and cobalt. When hardened, these steels are brittle and the cutting edge will chip on impact or with rough handling. They have a high resistance to abrasion but are not tough enough to withstand high shock loads. These steels will cut at high speeds and will retain their hardness even when the cutting edge is operating at temperatures around 600 °C.

High-speed steels are alloys that gain their properties from either tungsten or molybdenum or with a combination of both. The tungsten-based grades are classified as T1, T15 etc. while the molybdenum-based grades are classified as M2, M42 etc. which are the most commonly used for cutting tools. Composition of the M2 grade is 6% tungsten, 5% molybdenum, 4% chromium and 2% vanadium, has high wear resistance, and is widely used in the manufacture of drills, reamers, taps, dies, milling cutters and similar cutting tools. The M42 grade is also molybdenum based but with the addition of 8% cobalt. Its composition is 1.5% tungsten, 9.5% molybdenum, 3.75% chromium, 1.15% vanadium and 8% cobalt and is referred to as 'super cobalt high speed steel' or 'super HSS' and often identified by the letters HSCo. A selection of hard surface ceramic

coatings is available, e.g. titanium nitride (TiN) giving a gold finish, titanium aluminium nitride (TiAlN) giving a black/grey finish and titanium carbon nitride (TiCN) giving a blue/grey finish. These are applied to enhance the performance of drills, taps and end milling cutters offering different levels of surface hardness, thermal and frictional properties.

Apart from being used to manufacture the cutting tools already mentioned, high-speed steel is available as 'tool bits' in round or square section already hardened and tempered. The operator has only to grind the required shape on the end before using.

To save on cost, cutting tools such as lathe tools are made in two parts, instead of from a solid piece of expensive high-speed steel. The cutting edge at the front is high-speed steel and this is butt-welded to a tough steel shank. These tools are known as butt-welded tools, Fig. 7.1. The cutting edge can be reground until the high-speed steel is completely used and the toughened shank is reached. At this stage the tool is thrown away.

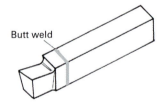

Butt weld

Figure 7.1   Butt-welded lathe tool

7.1.3 Cemented carbides

Cemented carbides are produced by a powder-metallurgy technique, i.e. by using metals in their powder form. The final mixture of powders consists of various amounts of hard particles and a binding metal. The hard particles give the material its hardness and abrasion resistance while the binding metal provides the toughness. They are produced as tips and inserts of various sizes, shapes and geometry.

The most common hard particle used is tungsten carbide, but titanium, tantalum and niobium carbides are often added in varying amounts. The binding metal used is cobalt, and various grades of cemented carbides are obtained for cutting different groups of materials by mixing in different proportions which determine the hardness and toughness of the material. In general, increasing the amount of cobalt binder together with increasing the tungsten carbide grain size contributes to increasing toughness, but also lowers the hardness, which reduces the wear resistance.

Because each workpiece material has its own unique characteristics, influenced by its alloying elements, heat treatment etc., this strongly influences the choice of cutting-tool material, grade and geometry. As a result, workpiece materials have been grouped into six major groups, in accordance with ISO standards, relative to their machineability. Each group is designated by a letter and colour.

▶ P (blue) – steel;
▶ M (yellow) – stainless steel;
▶ K (red) – cast iron;
▶ N (green) – non-ferrous, e.g. aluminium, copper, brass;
▶ S (orange) – heat-resisting super alloys;
▶ H (grey) – hard steels and chilled cast iron.

Subsequently the various grades of insert are available with corresponding colour coding to provide ease of choice of application.

Cemented carbides are used in cutting tools for turning, boring, milling, drilling etc. Solid carbide drills, taps and milling cutters are available for cutting a wide range of materials. Cemented carbides are also available in the form of tips brazed to a suitable tool shank, or as an insert clamped in an appropriate holder Fig. 7.2. The inserts are produced by mixing the metal powders in the correct proportions, pressing them to the required shape under high pressure and finally heating at a temperature in the region of 1400°C, a process known as sintering. The sintering stage

results in the cobalt binding metal melting and fusing with the hard particles, or cementing to form a solid mass – hence the term 'cemented carbides'. During the sintering process, the insert will shrink around 18% in all directions which represents around 50% of its original volume.

Brazed tip tools can be reground to give a keen cutting edge but due their extreme hardness silicon carbide wheels have to be used. To finish the cutting edge it may be necessary to use a diamond lap to maintain a fine cutting edge. Great care must be taken to avoid overheating, which can lead to surface cracking of the tip and subsequent breakdown of the cutting edge. A disadvantage of this type of tool is that it must be removed from the machine in order to regrind the edge and therefore must be reset in the machine once it has been resharpened.

Inserts which are clamped in a holder do not require regrinding. When the cutting edge wears and needs replacing, it is simply unclamped moved to the next fresh cutting edge (there may be up to eight cutting edges on the insert) and machining proceeds without the need for any resetting. When all the available cutting edges have been used, the insert is thrown away and replaced by a new insert.

Inserts are available in a range of cutting materials, sizes, shapes, cutting geometries and grades suited to a vast range of applications, a selection is shown in Fig. 7.3.

Figure 7.3 Cutting tool inserts

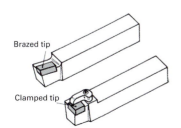

Figure 7.2. Tipped lathe tools

103

The range of cutting tool materials from the toughest to the hardest is:

▶ uncoated tungsten carbide;
▶ coated tungsten carbide;
▶ cermet;
▶ ceramic;
▶ cubic boron nitride;
▶ diamond.

7.1.3.1 Uncoated tungsten carbide

Uncoated cemented carbide inserts (without any additional coating) are used in moderate to difficult applications, cutting steel, heat-resistant super alloys, titanium, cast iron and aluminium in turning, milling and drilling at low speeds. The inserts produce sharp cutting edges and give a good combination of abrasive wear resistance and toughness.

7.1.3.2 Coated tungsten carbide

With the introduction of CNC drilling machines, lathes and machining centres capable of high speeds and high metal-removal rates has led to the development of coated cemented carbides which currently represents 80–90% of cutting-tool inserts. Coated cemented carbide combines cemented carbide with a surface coating which offers improved wear resistance and toughness, giving longer tool life with the use of higher cutting speeds and feed rates and are the first choice for a wide variety of tools and applications. Modern grades are coated with different carbide, nitride and oxide layers often in combination. The most common coatings are titanium nitride, aluminium oxide, titanium carbonitride and titanium aluminium nitride. These inserts are for general use with all types of workpiece material in turning, milling and drilling operations. They are available in a large variety of grades giving an extremely good combination of wear resistance and toughness resulting in very good wear characteristics together with long tool life.

7.1.4 Cermet

A cermet is a cemented carbide with titanium-based hard particles. The name cermet is a combination of ceramic and metal. In comparison to cemented carbide, cermet has improved wear resistance and reduced smearing tendencies (tendency of the workpiece material to smear or cling to the surface of the tool). These properties are offset through having a lower compressive strength as well as inferior thermal shock which can be avoided by machining without the use of coolant. Cermets can also be coated for improved wear resistance. Typical applications are in finishing operations using low feeds and depth of cut where close tolerances and good surface finish is required in workpiece material such as stainless steels, nodular cast iron and low-carbon steels.

7.1.5 Ceramic

All ceramic cutting tools have excellent wear resistance at high cutting speeds. There are a range of ceramic grades available for a range of applications. Oxide ceramics are aluminium oxide based with added zirconia for crack inhibition. This produces a material that is chemically stable but lacks resistance to thermal shock.

Mixed ceramics are available where the addition of cubic carbides or carbonitrides improves toughness and thermal conductivity.

Whisker-reinforced ceramics use silicon carbide whiskers to dramatically increase toughness and enable the use of a cutting fluid.

Ceramic grades can be applied to a broad range of applications and materials, most often in high-speed turning operations as well as in grooving and milling operations and are generally wear resistant with good hot hardness. They are mainly used in machining cast iron and steel, hardened materials and the heat-resisting super alloys, e.g. high alloyed iron, nickel, cobalt and titanium-based materials. The general limitation of ceramics is their low thermal shock resistance and fracture toughness.

7.1.6 Cubic boron nitride

Polycrystalline cubic boron nitride (CBN) is a material with excellent hot hardness and can be used at very high cutting speeds. It also has good toughness and resistance to thermal shock. CBN consists of boron nitride with ceramic or titanium nitride binder and is brazed onto a cemented carbide carrier to form an insert. CBN grades are largely used for finish turning hardened steel and high-speed rough machining cast iron by turning and milling operations. CBN is used in applications that require extreme wear resistance and toughness and is the only cutting material that can replace traditional grinding methods. CBN is referred to as a superhard cutting material.

7.1.7 Diamond

Polycrystalline diamond (PCD) is a composite of diamond particles sintered together with a metallic binder. Diamond is the hardest and therefore the most abrasion resistant of all materials. As a cutting-tool material, it has good wear resistance but lacks chemical stability at high temperatures. Usually in the form of a brazed-in corner tip on an insert or as a thin diamond-coated film on a carbide substrate, they are limited to non-ferrous materials, such as high silicon aluminium and non-metals such as carbon reinforced plastics in turning and milling operations. Used with flood coolant, PCD can be used in titanium super-finishing applications. Like CBN, PCD is also referred to as a superhard cutting material.

Natural and synthetic diamonds have been used for many years to true up and dress abrasive wheels. These are held in a holder of the required shape by vacuum brazing or held in a powder metal matrix and can be single point tools, multi-point grit tools, blade tools or cluster tools depending on the application (see Fig. 10.10).

7.2 Cutting tools

7.2.1 Clearance

All cutting tools, whether held by hand or in a machine, must possess certain angles in order to cut efficiently. The first essential is a clearance angle, which is the angle between the cutting edge and the surface of the material being cut. This prevents any part of the cutting tool other than the cutting edge from coming in contact with the work, and so eliminates rubbing.

If the end of a cutting tool is ground parallel to the workpiece as shown in Fig. 7.4(a), the tool will skid along the work surface. If the back of the tool, or heel, is ground below the level of the cutting edge, it will rub on the work surface as shown in Fig. 7.4(b). The correct angle is shown

in Fig. 7.4(c), where the heel of the tool is above the level of the cutting edge, thus leaving only the cutting edge in contact with the work.

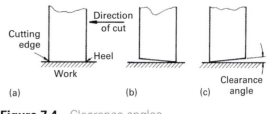

Figure 7.4 Clearance angles

The clearance angle should be kept at an absolute minimum, 8° being quite adequate for most purposes. Grinding an excessive clearance angle should be avoided – it is a waste of expensive cutting-tool material, a waste of time and money in grinding it in the first place, and, finally and most important, it weakens the cutting edge. In some cases, however, a greater clearance angle may be required, e.g. where holes are being machined using a boring tool, Fig. 7.5(a). If this additional clearance is provided up to the cutting edge, Fig. 7.5(b), serious weakening will result, so it is customary in these instances to provide the usual clearance angle for a short distance behind the cutting edge, known as primary clearance, followed by a second angle known as a secondary clearance angle, Fig. 7.5(c).

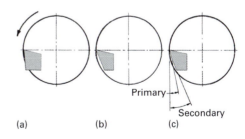

Figure 7.5 Primary and secondary clearance

7.2.2 Rake

For effective cutting, a second angle known as the rake angle is required. This is the angle between the tool face and a line at right angles to the surface of the material being cut.

The face upon which this angle is ground is the face along which the chip slides as it is being removed from the work. This angle therefore varies with the material being cut, since some materials slide more easily than others, while some break up into small pieces. Brass, for instance, breaks up into small pieces, and a rake angle of 0° is used. Aluminium, on the other hand, has a tendency to stick to the face of the tool and requires a steep rake angle, usually in the region of 30°.

For the majority of purposes, the rake angle used is positive, as shown in Fig. 7.6(a). When machining tough materials using the cemented-carbide cutting tools, it is necessary, due to the brittle nature of the carbide, to give maximum support to the tip. To achieve this, a negative rake is used so that the tip is supported under the cutting edge, Fig. 7.6(b).

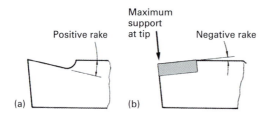

Figure 7.6 Positive and negative rake

Figure 7.7 identifies the rake and clearance angles on various cutting tools. The cutting angles of many cutting tools are established during their manufacture and cannot be changed by the user.

Such tools include reamers, milling cutters, taps and dies.

These cutting tools can of course be resharpened, but a specialised tool-and-cutter grinding machine is required. The basic cutting tools used on centre lathes and those used on the shaping machine are ground by hand, to give a variety of angles and shapes to suit different materials and applications.

Twist drills, although their helix angle is established during manufacture, are resharpened by grinding the point, the angle of which can be varied to suit different materials.

7.2.3 Turning tools

Cutting tools for use in turning may be required to cut in two directions. Such tools must therefore be provided with a rake and clearance angle for each direction of feed movement. This is illustrated in Fig. 7.8 which shows a turning tool for facing and turning.

Facing will require the back rake and front clearance, since cutting takes place when the tool is feeding in the direction shown in Fig. 7.9(a). Turning will require side rake and side clearance, since cutting takes place when the tool is feeding in the direction shown in Fig. 7.9(b). Having the back and side rake on the same surface results in

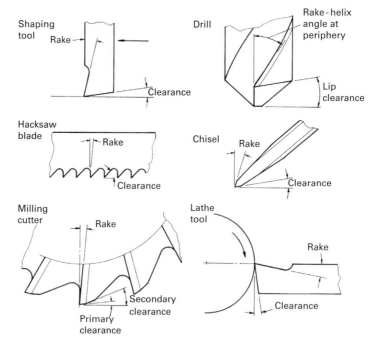

Figure 7.7 Identification of rake and clearance on various cutting tools

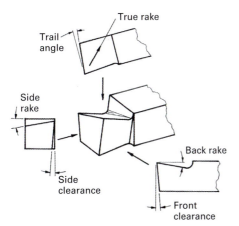
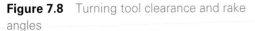

Figure 7.8 Turning tool clearance and rake angles

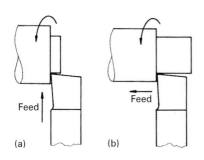

Figure 7.9 Direction of feed

a true rake angle somewhere between the two, which is the angle along which the chip will flow when cutting in either direction. The trail angle is required to prevent the rear or trailing edge of the tool from dragging on the workpiece surface.

Cutting tools which are used to cut in only one feed direction require only one rake angle, although a number of clearance angles may be required to prevent rubbing. The knife tool shown in Fig. 7.10(a) acts in the direction shown, and a rake and a clearance angle are required in the

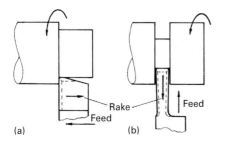

Figure 7.10 Direction of rake angle on knife and undercut tools

same direction. Front clearance is also required to clear the workpiece surface.

A tool used to part-off or form undercuts requires rake and clearance in the direction of feed but also requires side clearance to prevent rubbing in the groove produced, Fig. 7.10(b).

A lathe tool is considered to be right hand when it cuts from the right and left hand when it cuts from the left, Fig. 7.11.

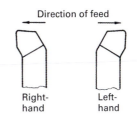

Figure 7.11 Right-hand and left-hand lathe tools

It is difficult to give precise values of rake angles, due to the number of variables encountered during machining. The values in Table 7.1 are offered as a guide for high-speed-steel cutting tools.

Table 7.1 Typical rake angles for high-speed-steel cutting tools

Material being cut	Rake angle
Brass	0°
Soft bronze	5°
Cast iron	8°
Mild steel	12°
Copper	20°
Aluminium	30°

7.2.4 Twist drills

The nomenclature of the twist drill is shown in Fig. 7.12. The helix angle of the twist drill is the equivalent of the rake angle on other cutting tools and is established during manufacture. The standard helix angle is 30°, which, together with a point angle of 118°, is suitable for drilling steel and cast iron, Fig. 7.13(a).

Drills with a helix angle of 20° – known as slow-helix drills – are available with a point angle of 118° for cutting brass and bronze, Fig. 7.13(b), and with a point angle of 90° for cutting plastics materials.

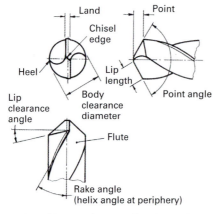

Figure 7.12 Nomenclature of twist drill

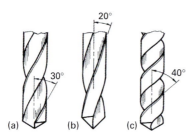

Figure 7.13 (a) Standard, (b) slow, and (c) quick helix drills

Quick-helix drills, with a helix angle of 40° and a point angle of 100°, are suitable for drilling the softer materials such as aluminium alloys and copper, Fig. 7.13(c).

Drills are available with parabolic flutes where the heel of the flutes is rolled over, which provides additional flute space for chip evacuation and produces a thinner land. This is advantageous particularly with deep hole drilling on CNC machines where constant withdrawl of the tool to clear chips can be minimised thus reducing cycle times. They have a 130° point angle and a quick helix around 38°. These are available with a nitride coating to increase wear resistance and improve tool life.

Drills are also available ground with a split point and a point angle of 135°. The chisel edge is removed and results in a single central point. Fig 7.14 shows the difference between a 118° conventional point and a 135° split point. Used with high performance CNC machines, the split point is self-centring to ease the start of a drilling operation and requires less force during drilling.

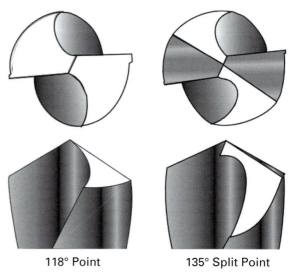

| 118° Point | 135° Split Point |

Figure 7.14 Conventional and split point

Drills are available as jobber drills for general purpose use and are of standard length, stub drills for short holes where rigidity is a requirement and long and extra long series drills for deep drilling. Depending on the diameter of the drill they may have a parallel shank or, on the larger diameters, a Morse taper shank. To aid wear resistance, drills are available with a range of oxide and nitride coatings.

Although usually made from HSS or HSCo, drills are also available in solid carbide from 1 mm to 12 mm with a point angle of 118° suitable for cutting a wide range of materials including carbon and alloy steels, stainless steels, cast iron and non-ferrous metals. High-performance solid carbide drills are available with titanium nitride (TiN) and titanium aluminium nitride (TiAlN) coatings which give increased wear resistance and improved tool life. They have a 130° or 140° point and are available from 3 to 20 mm diameter.

7.2.5 Drill grinding

To produce holes quickly, accurately and with a good surface finish, a correctly ground drill is required. In grinding the correct drill point, three important items must be controlled: the point angle, the lip length and the lip clearance angle.

When a great deal of drilling is done, it is economical to use a drill-point grinding machine, which ensures correct point angle and lip clearance and equal lip lengths. However, it is often necessary to regrind a drill by hand, and you should be able to do this so that the drill cuts correctly.

The lip clearance angle is required to prevent rubbing. Too much will weaken the cutting edge; too little or none at all will result in rubbing and will produce excessive heat. In general, an angle of 10° to 12° gives the best results, but this can be increased to 15° for aluminium alloys and copper.

When grinding, it is important to ensure that the angle and length of each lip are equal. A drill having unequal angles or unequal lengths of lip, or both, results in an oversize hole. The effect of unequal lip lengths is shown in Fig. 7.15. As shown, the point is ground at the correct angle, but the unequal lengths of lip have the effect of placing the centre of the web off the centre line of the drill. The result is a hole which is oversize by an amount equal to twice the offset x, i.e. an offset as small as 0.25 mm results in a hole 0.5 mm oversize.

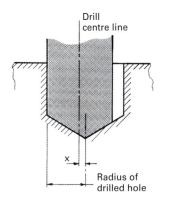

Figure 7.15 Effect of unequal lip lengths on a drill

To produce a point angle of 118°, hold the drill at 59° to the face of the grinding wheel. Hold the cutting edge to be sharpened horizontal, with the back end of the drill lower than the front to create the correct lip clearance. Push the drill forward against the grinding wheel, and at the same time rock it slightly away from the cutting edge to give the lip clearance. Turn the drill round and repeat for the second lip. Repeat this a little at a time on each lip to get the correct angle and lip length.

A simple gauge as shown in Fig. 7.16(a) can be used as a guide to the correct point angle. Correct

lip length can be checked using the simple gauge shown in Fig. 7.16(b). With the shank of the drill supported, a line is scratched with each lip. When the lines coincide, the lip lengths are equal.

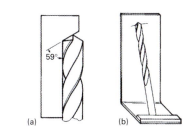

Figure 7.16 Angle and lip-length gauges

7.3 Cutting-tool maintenance

An off-hand grinding machine is used with a workpiece or cutting tool held by hand and applied to the grinding wheel. Because of this, its use requires stringent safety precautions.

The off-hand grinding machine is basically an electric motor having a spindle at each end, each carrying a grinding wheel, also referred to as an abrasive wheel. This arrangement allows a coarse wheel to be mounted at one end and a fine wheel at the other. All rough grinding is carried out using the coarse wheel, leaving the finishing operations to be done on the fine wheel.

These machines may be mounted on a bench, when they are often referred to as bench grinders (Fig. 7.17), or on a floor-mounted pedestal and referred to as pedestal grinders. Bench grinders should be securely anchored to a stout bench and pedestal grinders should be heavily built and securely bolted on good foundations.

Figure 7.17 Bench grinder

A guard of adequate strength must be securely attached to the machine frame. The aim is to enclose the wheel to the greatest possible extent and keep the opening as small as possible.

The guard has two main functions:

(i) to contain the wheel parts in the event of the wheel bursting;

(ii) to prevent as far as possible the operator coming in contact with the wheel.

The machine should never be run without these guards in position.

The wheel enclosure angles for a bench or pedestal grinder are shown in Fig. 7.18.

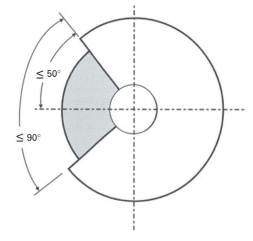

Figure 7.18 Wheel enclosure angle

A fixed shield with toughened or reinforced glass is provided which can be adjusted manually as the wheel wears down. Eye protection in the form of eye shields, goggles or safety spectacles must be worn by any person operating an off-hand grinder.

An adjustable work rest is fitted to the front and sometimes to the sides of the grinding wheel to support the cutting tool during grinding. The work rest is adjustable for angle so that cutting tools rested on it can be ground at a specific angle. The work rest must be adjusted so that at all times, the gap between it and the surfaces of the wheel is at a minimum, and it must be properly secured. This prevents the possibility of the workpiece, cutting tool or fingers becoming trapped between the wheel and the work rest. The work rest must be adjusted frequently to compensate for wheel wear and maintain the

gap at a minimum. Adjustments of the work rest should only be done when the grinding wheel is stationary and the machine isolation switch is in the 'off' position. All the above precautions are required by law under various regulations, and failure to comply with them not only can result in possible personal injury, but may also lead to prosecution.

When using an off-hand grinder, use the outside diameter of the wheel and keep the cutting tool moving across the wheel surface. Grinding in one place results in a groove being worn on the wheel surface making it impossible to grind cutting edges straight. Grinding on the side of a straight-sided wheel is dangerous, particularly when the wheel is appreciably worn or when sudden pressure is applied. Frequent dressing will be required to keep the wheel face flat and cutting correctly.

Do not grind for a long period without cooling the cutting tool. Grind off small amounts and cool frequently to avoid overheating and the possibility of cracking.

The wheels fitted to off-hand grinders are usually chosen for general use to cover a range of materials. These general grade wheels will not grind cemented carbides for which a green-grit silicon-carbide wheel is necessary.

The floor space around the machine should be kept free of obstructions and slippery substances. The machine should only be used by one person at the time and should not be kept running when not in use. Never stand directly in front of the wheel whenever the machine is started. Never stop the wheel by applying pressure to the periphery.

Changing a wheel should only be carried out by an appropriately trained person. It is recommended that a record of training in the safe mounting of abrasive wheels is kept, showing the trainee's name and date of training.

Abrasive wheels should always be mounted on the type of machine for which they were designed and it is essential that the speed of the spindle does not exceed the maximum speed marked on the wheel.

If there is a noise problem, it will be necessary to wear ear protection.

In order to avoid problems with vibration, do not grip the workpiece too tightly, do not exert excessive force, avoid continuous use and take action if you experience tingling, pins and needles or numbness.

7.3.1 Wheel dressing

Dressing abrasive wheels on an off-hand grinder can be carried out using a star-wheel dresser, comprising a handle, at the end of which is fitted a series of star-shaped wheels and two lugs on the underside as shown in Fig. 7.19.

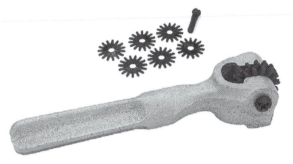

Figure 7.19   Star-wheel dresser

The tool rest is moved back to allow space for the dresser. The machine is then started and the lugs of the dresser are hooked in front of the tool rest, the handle tilted up and the dresser moved across the face of the abrasive wheel. When the star-wheels are pressed against the face of the rotating abrasive wheel, they rotate and fracture the bond and abrasive grains, clearing away loaded particles of material, exposing new sharp cutting grains and leaving a straight and true surface. These tend to be very aggressive and are used on the larger diameter wheels.

A dressing stick as shown in Fig. 7.20 can be used for light dressing. These are usually 25 mm square by 150 mm long and manufactured in the same way as an abrasive wheel. The stick is rested on the tool rest, and, using a corner, is pressed against the periphery of the rotating abrasive wheel, at the same time moving across the face. This gently removes worn grains and embedded material and gives a flat true surface. Dressing sticks made from boron carbide are also used for fine dressing. These are usually 12 mm × 5 mm × 76 mm, are extremely hard, and have a long service life, but are very expensive. Diamond dressers are also used for fine dressing.

Figure 7.20   Dressing stick

7.4 Cutting speed

The relative speed between the cutting tool and workpiece is known as the cutting speed and is expressed in metres per minute. In turning, the cutting speed is the surface speed of the work, i.e. the speed of a point on the circumference. In milling and drilling, it is the surface speed of the cutting tool.

As an example, we shall find the cutting speed S of a workpiece of diameter d mm being turned at N rev/min as shown in Fig. 7.21.

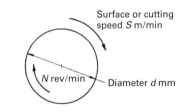

Figure 7.21   Cutting speed

The distance travelled by a point on the circumference in one revolution = the circumference = πd mm.

In one minute the workpiece turns through N revs, so the distance travelled by the same point in one minute = πdN mm/min.

But cutting speed S is expressed in metres per minute, so, dividing by 1000, we get

$$\text{cutting speed} \quad S = \frac{\pi dN}{1000} \text{ m/min}$$

7.4.1 Example 7.1

Find the cutting speed of a 50 mm diameter bar being turned with a spindle speed of 178 rev/min.

$$S = \frac{\pi dN}{1000} = \frac{\pi \times 50 \times 178}{1000} \quad 28 \text{ m/min}$$

7.4.2 Example 7.2

Find the cutting speed of a 15 mm diameter drill running at 955 rev/min.

$$S = \frac{\pi dN}{1000} = \frac{\pi \times 15 \times 955}{1000} \quad 45 \text{ m/min}$$

The cutting speed is dictated mainly by the type of cutting-tool material being used and the type of material being machined, but it is also influenced by the depth of cut and the feed rate. Using high-speed-steel tools, aluminium can be machined at higher cutting speeds than steel, and at even higher cutting speeds using cemented-tungsten-carbide tools.

The manufacturers of cutting-tool materials give recommendations for the cutting speeds at which their tool materials will cut various work materials such as brass, aluminium, steel and so on. A selection of these values is given for guidance in Table 7.2.

Table 7.2 Typical cutting speeds with HSS and tungsten-carbide tools

Material being cut	Cutting speed (m/min)	
	High-speed steel	Tungsten carbide
Cast iron	20	160
Mild steel	28	250
Bronze	35	180
Hard brass	45	230
Copper	60	330
Aluminium	100	500

Knowing the value of the cutting speed for a particular combination of cutting-tool material and work material, and knowing the diameter of the work to be produced (or the tool diameter), the unknown value is the spindle speed N at which the work or tool should be run. From the cutting-speed equation, this is given by

$$N = \frac{1000S}{\pi d}$$

7.4.3 Example 7.3

At what spindle speed would a 200 mm diameter high-speed-steel milling cutter be run to machine a steel workpiece, if the cutting speed is 28 m/min?

$$N = \frac{1000S}{\pi d} = \frac{1000 \times 28}{\pi \times 200} = 45 \text{ rev/min}$$

7.4.4 Example 7.4

What spindle speed would be required to turn a 150 mm diameter cast iron component using cemented-tungsten-carbide tooling at a cutting speed of 160 m/min?

$$N = \frac{1000S}{\pi d} = \frac{1000 \times 160}{\pi \times 150} = 340 \text{ rev/min}$$

You should adopt the habit of carrying out this calculation before starting any machining operation, even if only as a rough mental calculation. This way you at least have a basis for running the machine at a speed which is much nearer the correct speed than if you guess. When you gain experience, you will find that the spindle speed can be adjusted to suit different conditions by watching the way in which the chip is removed from the work.

7.5 Cutting fluids

During metal cutting, the metal immediately ahead of the cutting tool is severely compressed, which results in heat being generated. The metal then slides along the tool face, friction between the two surfaces generating additional heat. Any rubbing between the tool and the cut surface, which would occur with tool wear when the clearance angle is reduced, also produces heat. This heat is usually detrimental, especially to high-speed-steel cutting tools. Some metals, as they are cut, have a tendency to produce a chip which sticks or welds to the tool face, due chiefly to the high pressure between the metal and the tool. This has the effect of increasing the power required for cutting, increasing the friction and therefore heat, and finally, as the chip breaks away from the tool face and reforms, it creates wear on the tool face and a bad surface finish on the work. Excessive heat generated during the cutting may be sufficient to cause the work to expand. Work measured under these conditions may be undersize when it cools.

The basic role of a cutting fluid is to control heat, and it may do this by direct cooling of the work, chip and tool or by reducing friction by lubricating between the work, chip and tool. To cool effectively, a cutting fluid should have a high specific heat capacity and a high thermal conductivity.

The fluids most readily associated with cooling and lubricating are water and oil. Water has a higher specific heat capacity and thermal conductivity than oil, but unfortunately will promote rust and has no lubricating properties. Oil does not promote rust, has good lubricating properties, but does not cool as well as water. To benefit from the advantages of each, they can be mixed together with various additives to give a required measure of cooling and lubrication. With the high cost of oil, the cost savings of water-based fluids are so great that a great deal of development is being carried out to provide fluids which have good lubricating properties when mixed with water.

In general, the use of cutting fluids can result in

- ► less wear on cutting tools;
- ► the use of higher cutting speeds and feeds;
- ► improved surface finish;
- ► reduced power consumption;
- ► improved control of dimensional accuracy.

The ideal cutting fluid, in achieving the above, should

- ► not corrode the work or machine;
- ► have a low evaporation rate;
- ► be stable and not foam or fume;
- ► not injure or irritate the operator.

7.6 Types of cutting fluid

7.6.1 Neat cutting oils

These oils are neat in so much as they are not mixed with water for the cutting operation. They are usually a blend of a number of different types of mineral oil, together with additives for extreme-pressure applications. Neat cutting oils are used where severe cutting conditions exist, usually when slow speeds and feeds are used or with extremely tough and difficult-to-machine steels. These conditions require lubrication beyond that

which can be achieved with soluble oils. Such operations include heavy-duty turning, gear hobbing, broaching, tapping, threading and honing.

In some cases soluble oil cannot be used, due to the risk of water mixing with the hydraulic fluid or the lubricating oil of the machine. A neat oil compatible with those of the machine hydraulic or lubricating system can be used without risk of contamination. Neat cutting oils do not have good cooling properties and it is therefore more difficult to maintain good dimensional accuracy. They are also responsible for dirty and hazardous work areas by seeping from the machine and dripping from workpieces and absorbing dust and grit from the atmosphere.

Low-viscosity or thin oils tend to smoke or fume during the cutting operation, and under some conditions are a fire risk.

The main advantages of neat cutting oils are their excellent lubricating property and good rust control. Some types do, however, stain non-ferrous metals.

7.6.2 Soluble oils

Water is the cheapest cooling medium, but it is unsuitable by itself, mainly because it rusts ferrous metals. In soluble oils, or more correctly emulsifiable oils, the excellent cooling property of water is combined with the lubricating and protective qualities of mineral oil. Oil is, of course, not soluble in water, but with the aid of an agent known as an emulsifier it can be broken down and dispersed as fine particles throughout the water to form an emulsion. These are often referred to as 'suds' or coolant.

Other ingredients are mixed with the oil to give better protection against corrosion, resistance to foaming and attack by bacteria, and prevention of skin irritations. Under severe cutting conditions where cutting forces are high, extreme-pressure (EP) additives are incorporated which do not break down under these extreme conditions but prevent the chip welding to the tool face.

Emulsions must be correctly mixed, otherwise the result is a slimy mess. Having selected the correct ratio of oil to water, the required volume of water is measured into a clean tank or bucket and the appropriate measured volume of soluble oil is

113

added gradually at the same time as the water is slowly agitated. This will result in a stable oil/water emulsion ready for immediate use.

At dilutions between 1 in 15 and 1 in 20 (i.e. 1 part oil in 20 parts water) the emulsion is milky white and is used as a general-purpose cutting fluid for capstan and centre lathes, drilling, milling and sawing.

At dilutions from 1 in 60 to 1 in 80 the emulsion has a translucent appearance, rather than an opaque milky look, and is used for grinding operations.

As can readily be seen from the above, when the main requirement is direct cooling, as in the case of grinding, the dilution is greater, i.e. 1 in 80. When lubrication is the main requirement, as with gear cutting, the dilution is less.

The advantages of soluble oils over neat cutting oils are their greater cooling capacity, lower cost, reduced smoke and elimination of fire hazard. Disadvantages of soluble oils compared with neat cutting oils are their poorer rust control and that the emulsion can separate, be affected by bacteria and become rancid.

7.6.3 Synthetic fluids

Sometimes called chemical solutions, these fluids contain no oil but are a mixture of chemicals dissolved in water to give lubricating and anti-corrosion properties. They form a clear transparent solution with water, and are sometimes artificially coloured. They are very useful in grinding operations, where, being non-oily, they minimise clogging of the grinding wheel and are used at dilutions up to 1 in 80. As they are transparent, the operator can see the work, which is also important during grinding operations.

They are easily mixed with water and do not smoke during cutting. No slippery film is left on the work, machine or floor. They give excellent rust control and do not go rancid. At dilutions of between 1 in 20 and 1 in 30 they can be used for general machining.

7.6.4 Semi-synthetic fluids

Unlike synthetic fluids, these fluids, sometimes referred to as chemical emulsions, do have a small amount of oil emulsified in water. When mixed with water they form extremely stable transparent fluids, with the oil in very small droplets. Like the synthetic types, they are often artificially coloured for easy recognition.

They have the advantage over soluble oil of increased control of rust and rancidity and a greater application range. They are safer to use, will not smoke, and leave no slippery film on work, machine, or floor. Depending upon the application, the dilution varies between 1 in 10 and 1 in 30.

7.6.5 Vegetable oils

This range of oils is based on specially refined vegetable oils which exploit oil technology from the agricultural revolution in non-food crops. Advances in crop oil technology means that lubricants are derived from renewable sources and offer a wide range of viscosities. Vegetable oils have a number of performance advantages over mineral oil-based products. These include high natural lubricating properties and a higher flash point which reduces smoke formation and fire hazard. Vegetable oils have a high natural viscosity as the machining temperature increases, and as the temperature falls, they remain more fluid resulting in quicker drainage from metal chips and workpieces. Vegetable oils have a higher melting point, giving less loss from vaporisation and misting and improved safety for operators. They are used as water-based emulsions at concentrations from 1–10 to 1–100 depending upon application and level of performance and can be used for ferrous and non-ferrous metals. They are used in the automobile industry in high-production CNC machining centres as an advanced coolant for central systems, i.e. feeding various machines from a central source where the concentrations are controlled independently to maximise cutting performance and minimise running costs.

7.7 Application of cutting fluids

Having selected the correct type of cutting fluid, it is equally important to apply it correctly. This is best done by providing a generous flow at low pressure to flood the work area. Flooding has the added advantage of washing away the chips produced. Fluid fed at high pressure is not recommended, since it breaks into a fine spray or mist and fails to cool or lubricate the cutting zone. To cope with the large flow of fluid, the machines must have adequate splash guards, otherwise the operator tends to reduce the flow and the resulting dribble does little to improve cutting.

Many methods have been used to direct the fluid into the cutting zone and from every possible direction. The shape of the nozzle is important but depends largely on the operation being carried out and on the shape of the workpiece. The nozzle may be a simple large-bore pipe or be flattened as a fan shape to provide a longer stream. The main flow may be split into a number of streams directed in different directions – up, down or from the sides – or, by means of holes drilled in a length of pipe, create a cascade effect. In some cases, especially with grinding, where the wheel speed creates air currents which deflect the cutting fluid, deflector plates are fitted to the pipe outlet. Where the cutting tool is vertical, it can be surrounded by a pipe having a series of holes drilled into the bore and directed towards the cutting tool. Whatever the method used, the fundamental need is to deliver continuously an adequate amount of cutting fluid where it is required.

With the introduction of high metal removal CNC machining centres, the need to provide the cutting fluid where it is required, i.e. directly at the tool cutting edge, has led to the introduction of toolholders complete with nozzles which deliver high-pressure cutting fluids through the toolholder directly to the cutting region. Fig. 7.22 shows a holder with built-in nozzles which can deliver high-pressure cutting fluid to the correct portion of the cutting-tool insert. The nozzles are fixed and pre-directed so that the high-velocity coolant hits the right place on the insert and at the correct angle on the cutting edge so the operator is not required to make any settings. This has the advantage of improved chip control and ensures prolonged tool life. Similar arrangements are available for various other holders including grooving and boring bars. An in-process application is shown in Fig. 7.23. Drills are also available with internal holes throughout their length, which enable the cutting fluid to be fed through the holes to the cutting region thus lubricating and cooling and assisting

metal removal by washing the chips back up the flutes.

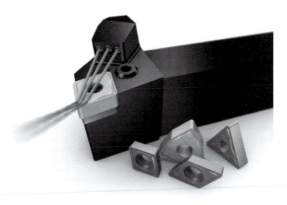

Figure 7.22 Toolholder with fixed coolant nozzles

Figure 7.23 Application of high pressure cutting fluid

7.8 Safety in the use of cutting fluids

Control of Substances Hazardous to Health (COSHH) Regulations require exposure to metalworking fluids (or cutting fluids), by inhalation, ingestion or skin contact to be prevented where reasonably practicable, or, failing that, adequately controlled.

To comply, the employer must:

▶ carry out a suitable and sufficient risk assessment;
▶ tell employees about the risks and precautions necessary to protect their health;

▶ train employees in the use of control measures and any personal protective equipment required;

▶ maintain fluid quality and control bacterial contamination of fluids;

▶ minimise skin exposure to fluids;

▶ prevent or control airborne mists;

▶ where there is exposure to fluid or mist, carry out health surveillance, e.g. regular skin inspections, usually about once a month.

Employees must:

▶ co-operate with their employer;

▶ make full use of any control measures, use personal protective equipment and report any defective equipment;

▶ attend and participate in health surveillance programmes at their workplace where appropriate.

The exposure to metalworking fluids can cause:

▶ irritation of the skin or dermatitis through the presence of bacteria;

▶ occupational asthma, bronchitis, irritation of the respiratory tract and breathing difficulties.

The exposure can also cause irritation to the eyes, nose and throat.

Metalworking fluids are mostly applied by continuous jet and can enter your body:

▶ if you inhale the mist generated during machining operations;

▶ through contact with unprotected skin, particularly hands, forearms and head;

▶ through cuts and abrasions or other broken skin;

▶ through the mouth if you eat or drink in work areas, or from poor personal hygiene, e.g. not washing hands before eating.

The following precautions, if observed, will reduce or eliminate the likely hazards:

▶ Follow the instructions and training given by the employer.

▶ Use splash guards to control splashing and misting.

▶ Use local exhaust ventilation to remove or control any mist or vapour produced.

▶ Allow a time delay before opening doors on machine enclosures to ensure all mist and vapour has been removed.

▶ Report any defective splash guards, ventilation or other control equipment.

▶ Open doors and windows where practicable, to improve natural ventilation.

▶ Do not use compressed air to remove excess metalworking fluids.

▶ Avoid direct skin contact with fluids.

▶ Wear adequate protective clothing.

▶ Take care not to contaminate the inside of your gloves with cutting fluid when putting them on or taking then off.

▶ Only use disposable wipes or clean rags.

▶ Contaminated rags and tools should never be put into overall pockets.

▶ Apply a pre-work cream to hands and exposed areas of arms before starting work and on resuming work after a break. These help to make removing contaminants easier. These do not provide a barrier and not a substitute for gloves.

▶ Thoroughly wash hands using suitable hand cleaners and warm water and dry using a clean towel before, as well as after, going to the toilet, before eating and at the end of each shift.

▶ Avoid eating or drinking in areas where metalworking fluids are used.

▶ Apply after-work or conditioning cream, after washing, to replace natural skin oils and help prevent dryness.

▶ Store personal protective equipment in the changing facilities provided.

▶ Contaminated clothing, especially undergarments, should be changed regularly and be thoroughly cleaned before reuse.

▶ Overalls should be cleaned frequently. Avoid taking overalls home, e.g. for washing.

▶ Do not use paraffin, petrol and similar solvents for skin cleaning purposes.

▶ Do not discard unwanted food, drink or other debris into the sump.

▶ Follow good working practices when mixing fluids, cleaning or topping up sump.

▶ All cuts and abrasions must receive prompt medical attention.

▶ Seek prompt medical advice if you notice any skin abnormality or chest complaints.

Review questions

1. State four precautions to be observed when using an off-hand grinder.
2. Name three types of cutting fluid and state a typical application of each.
3. What is the result of using an incorrectly sharpened drill with unequal lip lengths?
4. State the three characteristics of a cutting tool material.
5. State five safety precautions to be observed when using cutting fluids.
6. State the relationship between the rake angle and the softness of the material being cut.
7. Name four cutting tool materials.
8. At what spindle speed would a 15 mm diameter drill be run in order to drill a hole in a mild steel workpiece at a cutting speed of 28 m/min? (ans. 594 revs/min).
9. State the two angles necessary on a cutting tool for cutting to take place.
10. At what speed would you turn a 150 mm diameter bronze component using a tungsten carbide tool at a cutting speed of 180 m/min? (ans. 382 revs/min).

7

CHAPTER 8

Drilling

The majority of drilling work is carried out on pillar drilling machines, so called because the machine elements are arranged on a vertical pillar. The machines in the heavy-duty range have power feed, are driven from the motor through a gearbox, and have a drilling capacity in steel up to 50 mm diameter. Smaller sensitive machines, (see Fig. 8.1) have a hand feed, giving the sensitivity, are belt driven from the motor through pulleys, and have a maximum drilling capacity in steel ranging from 5 mm up to 25 mm diameter. These machines may be bench- or floor-mounted.

8.1 The sensitive drilling machine

The main elements of a typical sensitive drilling machine are shown in Fig. 8.1.

1. Base – provides a solid foundation for the machine, into which the pillar is securely clamped.
2. Pillar – provides a solid support for the drill head and worktable.
3. Worktable – provides a flat surface in correct alignment with the drill spindle upon which the workpiece can be positioned. Tee slots are provided for clamping purposes. The worktable can be raised, lowered and swung about the pillar and be securely clamped in the required position.

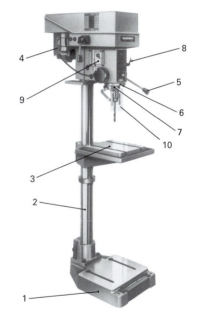

Figure 8.1 Sensitive drilling machine

4. Motor – provides the drive to the spindle through a five-step pulley system and a two-speed gearbox, Fig. 8.2. Thus five pulley speeds with A and B in mesh and five with C and D in mesh give a range of 10 spindle speeds from 80 to 4000 rev/min.
5. Handwheel – provides feed to the drill by means of a rack and pinion on the quill, Fig. 8.2.

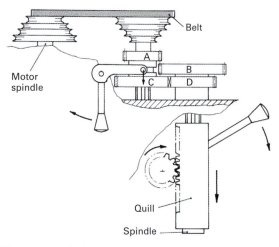

Figure 8.2 Drilling-machine drive system

Figure 8.3 Drilling-machine guard

6. Quill – this is the housing inside which the spindle rotates. Only the longitudinal movement is transmitted by the quill, which itself does not rotate.

7. Spindle – provides the means of locating, holding and driving the cutting tools and obtains its drive through the pulley.

8. Depth stop – provides a means of drilling a number of holes to a constant depth.

9. Stop/start – the machine shown is switched on by a shrouded push-button starter with a cover plate which can be padlocked to prevent unauthorised access. A mushroom-headed stop button is situated on the starter, and the machine can also be switched off using the emergency kick-stop switch at the front of the base. A safety switch is also incorporated under the belt guard and automatically stops the spindle should the guard be lifted while the machine is running.

10. Drill guard – provided to protect the operator from contact with the revolving chuck and drill while still retaining visibility of the operation. These guards range from simple acrylic shields to a fully telescopic metal construction with acrylic windows. A typical pedestal drill guard is shown in Fig. 8.3.

8.2 Tool holding

Drills and similar tools with parallel shanks are held in a drill chuck, Fig. 8.4. Many different types of chuck are available, each being adjustable over its complete range, and give good gripping power. By rotating the outer sleeve, the jaws can be opened and closed. To ensure maximum grip, the chuck should be tightened using the correct size of chuck key. This prevents the drill from spinning during use and chewing up the drill shank.

A hazard in the use of chucks is the possibility of leaving a chuck key in position. When the machine is then switched on, the chuck key can fly in any direction and cause serious injury. When you remove a drill from the chuck, always remember to remove the chuck key. Never leave it in the

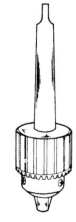

Figure 8.4 Drill chuck

chuck for even the shortest time. Better still, use a safety chuck key, Fig. 8.5, in which the central pin is spring-loaded and has to be pushed to engage. When the force is released, the pin retracts and the key falls from the chuck.

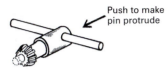

Figure 8.5 Safety chuck key

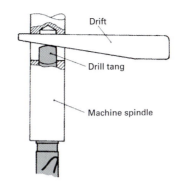

Figure 8.6 Drift in drill spindle

The chuck is fitted with a Morse-taper shank which fits into a corresponding Morse taper in the spindle. The size of Morse taper is identified from smallest to largest by numbers 1, 2, 3, 4, 5 and 6. The included angle of each taper is different but is very small, being in the region of 3°. If the two mating tapered surfaces are clean and in good condition, this shallow taper is sufficient to provide a drive between the two surfaces. At the end of the taper shank, two flats are machined, leaving a portion known as the tang. This tang fits in a slot on the inside of the spindle and its main purpose is for the removal of the shank.

To remove a shank from the spindle, a taper key known as a drift is used. The drift is inserted through a slot in the spindle as shown in Fig. 8.6.

Drills are available with Morse-taper shanks which fit directly into the spindle without the need for a chuck. The size of Morse taper depends on the drill diameter, and the range is shown in Table 8.1.

It is essential that tapers are kept clean and in good condition. As already stated, the drive is by friction through the tapered surfaces, and any damage to these surfaces puts some of the driving force on the tang. If this force is excessive, the tang can be twisted off. When this happens the drill has to be discarded, as there is no way of easily removing it from the spindle.

Where a cutting tool or chuck has a Morse taper smaller than that of the spindle, the difference is made up by using a sleeve. For example, a drill with a No. 1 Morse-taper shank to be fitted in a spindle with a No. 2 Morse taper would require a 1–2 sleeve, i.e. No. 1 Morse-taper bore and a No. 2 Morse taper outside. Sleeves are available from 1–2, 1–3, 2–3, 2–4 and so on over the complete range.

8.3 Clamping

Work is held on a drilling machine by clamping to the worktable, in a vice or, in the case of production work, in a jig. It is sufficient to say here that work held in a jig will be accurately drilled more quickly than by the other methods, but large quantities of the workpiece must be required to justify the additional cost of the equipment.

Standard equipment in any workshop is a vice and a collection of clamps, studs, bolts, nuts and packing. It should be stressed that work being drilled should never be held by hand. High forces are transmitted by a revolving drill, especially when the drill is breaking through the bottom surface, which can wrench the work from your hand. The resulting injuries can vary from a small cut to the loss of a finger.

Never take a chance – always clamp securely.

Small workpieces with parallel faces can be quite adequately held in a vice. The work is then

Table 8.1 Range of Morse tapers

Morse taper	No. 1	No. 2	No. 3	No. 4	No. 5	No. 6
Drill-diameter range (mm)	up to 14	14.25 to 23	23.25 to 31.75	32 to 50.5	51 to 76	77 to 100

positioned under the drill and the vice is clamped to the worktable.

Larger work and sheet metal are best clamped direct on to the worktable, care of course being taken to avoid drilling into the worktable surface. When required, the work can be raised off the worktable surface by means of suitable packing or on parallels. Tee slots are provided in the worktable surface into which are fitted tee bolts, or tee nuts in which studs are screwed, Fig. 8.7.

Various styles and shapes of clamp are available, one of which is shown in Fig. 8.8. The central slot enables it to be adjusted to suit the workpiece. To provide sound clamping, the clamp should be reasonably level, and this is achieved by packing under the rear of the clamp to as near as possible the same height as the workpiece, Fig. 8.9.

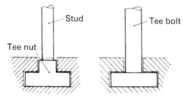

Figure 8.7 Tee nut and tee bolt

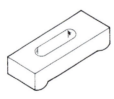

Figure 8.8 Clamp

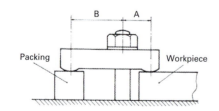

Figure 8.9 Clamping forces

The clamping bolt should be placed close to the work, since the forces on the work and packing are inversely proportional to their distances from the bolt. For greatest clamping force on the work, distance *A* in Fig. 8.9 must be less than distance *B*.

8.4 Cutting tools on drilling machines

Various cutting tools besides twist drills are used on a drilling machine, and some of them are described below.

8.4.1 Twist drill

Twist drills are available with parallel shanks up to 16 mm diameter and with taper shanks up to 100 mm diameter and are made from high-speed steel. Standard lengths are known as jobber-series twist drills, short drills are known as stub series, and long drills as long series and extra long series. Different helix angles are available for drilling a range of materials, as described in Chapter 7.

Combination drills known as Subland drills combine a number of operations in a single tool; e.g. drill and ream, drill two diameters, drill and chamfer, drill and spotface, drill and counterbore, Fig. 8.10. Each cutting edge has a separate land and flute, Fig. 8.11, which enables cutting to take place and resharpening to be easily carried out.

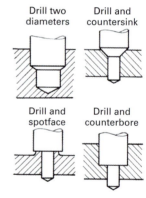

Figure 8.10 Examples of Subland-drill applications

Figure 8.11 Subland drill

8.4.2 Machine reamer

A reamer is used to produce a hole of greater accuracy than can be obtained using a drill. The hole is drilled undersize by an amount depending upon the diameter (see Table 8.2); the required finished size is then obtained with the reamer. Care should be taken with the position and alignment of the drilled hole, since the reamer will correct size, roundness and surface finish but will not correct errors in alignment.

Table 8.2 Reaming allowance

Size of reamed hole (mm)	Undersize amount (mm)
Below 4	0.1
Over 4–11	0.2
Over 11–39	0.3
Over 39–50	0.4

As a general rule, reaming is carried out at half the speed used for drilling.

Reamers are made from high-speed steel in sizes up to 50 mm diameter in a variety of types, the most common being the machine reamer, Fig. 8.12. Machine reamers have a Morse-taper shank, although reamers below 12 mm diameter are available with parallel shanks. The diameter is constant along its length, and cutting takes place on the chamfer or bevel at the front, which is usually 45°. This chamfer can be reground using a cutter-grinding machine.

Figure 8.12 Machine reamer

Flutes on reamers usually have a left-hand spiral or helix, opposite to that of a drill. This pushes the metal chips ahead of the reamer rather than back up the flutes and prevents scratching the bore. This feature also prevents the reamer from 'screwing' itself into the hole, which would tend to happen with a right-hand helix.

Always use a lubricant to enhance the life of the reamer. A 40:1 soluble solution is normally satisfactory. Do not allow the flutes to become blocked with swarf.

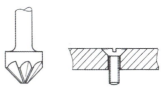

Figure 8.13 Countersink cutter

8.4.3 Countersink

Countersink cutters, Fig. 8.13, made from high-speed steel, are used to cut a large chamfer of the correct angle, usually 90°, as a seating for countersink-head screws. Countersinks should be run at a fairly slow speed to avoid chatter. They are available with parallel and taper shanks.

8.4.4 Counterbore

A counterbore cutter, Fig. 8.14, is used to enlarge an existing hole to provide a flat and square seating for a screw, bolt or nut under the workpiece surface. Teeth are provided on the end face and on the circumference, to permit cutting to a depth. A pilot is provided which locates in the existing hole and guides the tool during cutting. These pilots may be a solid part of the tool or detachable when the cutter is used on a series of different-size holes.

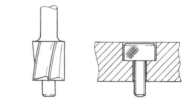

Figure 8.14 Counterbore cutter

Counterbore cutters are made from high-speed steel and may have parallel or taper shanks.

8.4.5 Spotface

A spotface cutter, Fig. 8.15, is used to provide a flat and square seating for a screw, bolt or nut on the surface of the workpiece, usually on the surface of a rough casting which would not otherwise provide a sufficiently accurate seating. The spotface is similar to a counterbore cutter, but has teeth on the end only. It will cut to only a very limited depth and cannot be used to counterbore.

Counterbore cutters, on the other hand, can be used to spotface.

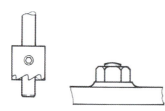

Figure 8.15 Spotface cutter

In some awkward places, back or underside spotfacing is required, Fig. 8.16. The pilot is fed through the hole and the cutter is fixed to the pilot, usually by some quick-locking mechanism or a simple grub screw. When the operation is finished, the cutter is removed and the pilot withdrawn.

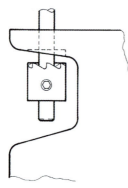

Figure 8.16 Back spotfacing

8.4.6 Trepanning tools

Where large-diameter holes are required in sheet metal, they can be conveniently cut using a trepanning tool, Fig. 8.17(a).

A small hole, to suit the pilot, is drilled at the centre of the required position. The adjustable arm is extended so that the edge of the high-speed-steel cutting edge produces the required size of hole. This may require a number of trials and adjustments before the correct size is reached. The pilot is located in the pilot hole and, as the tool rotates, it is fed through the work, Fig. 8.17(b).

The arm can be adjusted to any diameter within the range of the trepanning tool, which may be as large as 300 mm. The cutting tool can be ground to any angle to suit the material being cut.

8.4.7 Machine taps

Tapping can be carried out very efficiently on a drilling machine but requires the use of a special tapping attachment. The tapping attachment has a clutch which is preset according to the size of tap being used. When the tap hits the bottom of the hole or a hard piece of material, the clutch slips, causing the tap to remain stationary while the spindle keeps rotating. The machine is then switched to reverse and the tap is extracted. By using this attachment, broken taps are eliminated and the tapping operation is done more quickly and accurately than by hand.

A combined tap and drill is available in a range from M3 to M12. This comprises a spiral fluted tap with the front portion ground as a drill to a diameter equal to the appropriate tapping size (Fig. 8.18). These are used for 'through' holes where the drilling and tapping can be carried out in one operation.

Figure 8.18 Combined tap and drill

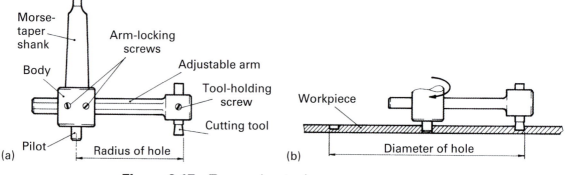

(a)

Morse-taper shank

Body

Pilot

Arm-locking screws

Adjustable arm

Tool-holding screw

Cutting tool

Radius of hole

(b)

Workpiece

Diameter of hole

Figure 8.17 Trepanning tool

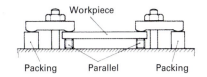

Figure 8.19 Component clamped on drilling machine

Although straight flute taps are the most commonly used type, other types of machine taps are available for use with high-production machines.

Spiral point taps, often referred to as gun-nosed taps, have an angle ground on the inside of the flute, to push the metal cuttings (or swarf) forwards down the hole ahead of the cutting edges and are recommended for through holes.

Taps with spiral flutes are intended for blind holes, where the spiral flute transports the swarf back out of the hole, away from the cutting edges. This avoids packing of the swarf at the bottom of the hole.

Taps having no flutes, known as cold-forming taps, produce the thread by plastic deformation of the material and therefore no swarf is produced by the action. This type of tap is used with ductile materials such as soft steels, aluminium and zinc alloys. Because the material is displaced, the tapping drill is larger than that for taps which cut, e.g. an M10 cut thread has a tapping drill of 8.5 mm, while an M10 cold-formed thread has a tapping drill of 9.3 mm.

Taps with through coolant holes are available, which results in reduced wear of the cutting edges and washes swarf out of the hole away from the cutting edges.

To enhance their cutting performance, some taps have surface coatings of titanium aluminium nitride (TiAlN) and titanium carbon nitride (TiCN).

8.5 Drilling operations

Unless the workpiece is held in a drill jig, the position of holes on a workpiece must be marked out. When the position of a hole is determined, its centre is indicated by means of a centre dot, using a centre punch. This centre dot is used to line up the drill and as a means of starting the drill in the correct position. The workpiece is set on the worktable, carefully positioned under the drill, using the centre dot, and clamped in position as shown in Fig. 8.19. Two clamps are usually required, one at each side of the component.

When held in a vice, the workpiece should be positioned and the vice be tightened securely. The workpiece is then positioned under the drill as before and the vice is clamped to the worktable.

In positioning the workpiece, take care to avoid drilling into the vice or worktable. If necessary, raise the workpiece on parallels, Fig. 8.19.

Before drilling, check if the drill is the correct size. Because you remove it from a space marked 5 mm does not mean it is a 5 mm drill – the person who used it before you may not have returned it to the correct space. Also, check the condition of the cutting edges and if necessary resharpen them.

Having carefully lined up and clamped the workpiece, begin drilling, taking care that the drill is still central with your required position. Small-diameter drills will start in the correct position with the aid of the centre dot; large-diameter drills with a long chisel edge require other means to assist in starting. The best method is to use a smaller diameter drill on the centre dot, but stop before it cuts to its full diameter. The larger drill will start in its correct position guided by the 118° dimple produced by the smaller drill.

Where the chisel edge is found to be too wide for a particular purpose, it can be reduced by point thinning, Fig. 8.20. This can be done using the edge of a well-dressed grinding wheel, but it is perhaps better left to a more experienced person.

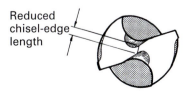

Reduced chisel-edge length

Figure 8.20 Drill-point thinning

A properly sharpened drill run at the correct speed will produce a spiral type of chip from each cutting edge. As the hole becomes deeper, the chips tend to pack in the flutes and the drill may have to be removed from the hole periodically to clear the chips.

Most trouble in drilling arises when the drill breaks through the far surface. The chisel edge of a drill centres and guides it through the workpiece, keeping the hole straight. When the chisel edge breaks through, it can no longer guide the drill and keep it central, and the drill will wobble and bounce in its own hole, an occurrence known as 'chatter'.

When the complete drill point has almost broken through, there is a tendency for the drill to 'snatch' or 'grab'. This happens when the metal still to be cut is so thin that it is pushed aside rather than cut and the drill pulls itself through due to the helix angle – in the same way as a screw thread advances. In the case of unclamped work, it would be the workpiece which would be pulled up the drill, wrenching it out of the hand holding it, with resulting injury or breakage, or both.

A repeat warning here – always clamp the workpiece.

To avoid these problems when breaking through, take great care and avoid too rapid a feed.

Holes can be drilled to a particular depth by setting the depth stop on the machine. The workpiece is positioned as already described, and drilling is started until the drill is just cutting its full diameter. The machine is switched off and the stationary drill is brought down into contact with the workpiece. The depth stop is set to the required dimension by adjusting it to leave the required space above the spindlehead casting and it is then locked in position, Fig. 8.21.

Where holes in two parts are required to line up with each other, a technique known as 'spotting' is carried out. The top part is marked out and drilled as already described. The two parts are then carefully positioned and clamped together. The holes in the bottom part are then transferred by 'spotting' through from the top part. Drilling of the bottom part can then proceed in the knowledge that both sets of holes are identical, which may not be the case if both parts are marked out and drilled individually.

When the two parts are to be screwed together, the bottom part requires to be tapped while the top part requires a clearance hole. The sequence is the same as for spotting except that, having positioned, clamped, and spotted with the clearance drill, the drill is changed to the tapping size. The hole is then drilled and tapped, Fig. 8.22.

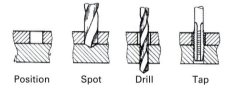

Position   Spot   Drill   Tap

Figure 8.22 Spotting through from existing workpiece

8.6 Drilling sheet metal

The same problems already discussed when the drill breaks through apply to drilling sheet metal. The problems are increased with thin sheet, since the chisel edge can break through before the drill is cutting its full diameter, due to the length of the drill point and the thinness of the material. In this case there is no guide at all – the drill will wander and produce a hole to some odd shape. Producing these odd-shaped holes is known as 'lobing'.

The same problem arises with 'snatching' or 'grabbing' – the thinner metal is pushed aside and the drill screws itself through. A further problem associated with this is damage to the metal sheet. A drill pushed with too much force tends to distort the thin sheet initially, rather than cut, and the resulting series of bulges around the holes is unacceptable.

These problems can be overcome by supporting the sheet on a piece of unwanted or waste metal plate. The support prevents distortion and the drill

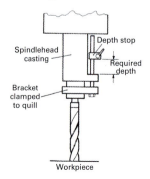

Spindlehead casting
Depth stop
Required depth
Bracket clamped to quill
Workpiece

Figure 8.21 Depth stop

point is guided until the hole is drilled through. There is no problem of breaking through, since the operation is the same as drilling a blind hole, Fig. 8.23. Large-diameter holes can be produced using a trepanning tool as shown in Fig. 8.17.

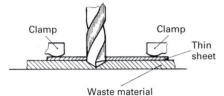

Figure 8.23   Thin sheet clamped to waste material

8.7 Drilling plastics

Plastics materials cover a wide range of types and applications, which are dealt with in detail in Chapter 15.

In general, plastics materials are easily machined using high-speed-steel cutting tools, although some plastics containing abrasive fillers wear out tools very quickly and the use of diamond tools is essential. Thermosetting plastics can be drilled using standard high-speed-steel twist drills. The chips from thermoplastic materials tend to stick and pack the flutes and cause overheating, which can affect the composition of the material. To prevent this, slow-helix drills with wide highly polished flutes are available. Point thinning can also be carried out to reduce friction and heat at the centre of the drill point. A better finish on breakthrough can be obtained by sharpening the point angle at 90°.

To avoid chipping on breakthrough when drilling the more brittle materials such as Perspex, the material should be held firmly against a solid backing such as a block of hardwood. Use of hardwood prevents damage to the drill point.

Large holes in sheet material can be produced using a trepanning tool.

8.8 Safety in use of drilling machine

Most accidents happen from:

▶ hair caught on rotating spindles, chucks and tools;
▶ entanglement of gloves, clothing, bandages, watches and rings;
▶ violent spinning of the workpiece due to poor clamping, causing broken bones, dislocations and even amputation;
▶ not wearing eye protection;
▶ swarf – causing cuts.

Remember that you have a duty under the various health and safety regulations already covered in Chapter 1. To avoid the risk of accident:

▶ Always follow the training provided by your employer.
▶ Always have guards properly fitted and in the correct position.
▶ Always wear eye protection and any other PPE required.
▶ Do not wear jewellery or loose clothing.
▶ Always have long hair tied back or in a hairnet.
▶ Never leave the chuck key in the chuck.

8

Review questions

1. By means of a sketch show how a clamp is positioned to give maximum clamping force to the workpiece.
2. What is the purpose of using a reamer?
3. What is the quill of a drilling machine?
4. State the purpose of a Morse taper.
5. Why is it necessary to clamp work during a drilling operation?
6. State the precautions necessary when drilling holes in sheet metal.
7. What piece of equipment is used to remove a shank from the drill spindle?
8. State two precautions to be taken when drilling plastics materials.
9. State the purpose of a counterbore.
10. When screwing two workpieces together why is it necessary to drill a clearance hole in the top part?

CHAPTER 9

Turning

Turning is carried out on a lathe of some description, the type depending on the complexity of the workpiece and the quantity required. All lathes are derived from the centre lathe, so called since the majority of work in the past was done between centres, to ensure concentricity of diameters. This is no longer the case, as accurate methods of workholding are now available.

Centre lathes are made in a variety of sizes and are identified by the maximum size of workpiece which can be machined. The most important capacity is the largest diameter which can be rotated over the bed of the machine, and this is known as the swing. A centre lathe with a swing of 330 mm will accept this diameter of workpiece without it hitting the machine bed. It should be noted that this maximum diameter cannot be accepted over the whole length of the bed, since the cross-slide is raised and will therefore reduce the swing. In the case of a 330 mm swing machine, the swing over the cross-slide is 210 mm.

The second important capacity is the maximum length of workpiece which can be held between the centres of the machine. A centre lathe with a swing of 330 mm may, e.g. accommodate 630 mm between centres.

9.1 Centre-lathe elements

A typical centre lathe showing the main machine elements is shown in Fig. 9.1.

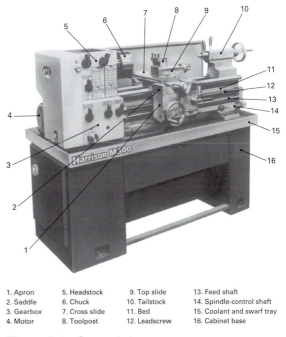

1. Apron	5. Headstock	9. Top slide	13. Feed shaft
2. Saddle	6. Chuck	10. Tailstock	14. Spindle-control shaft
3. Gearbox	7. Cross slide	11. Bed	15. Coolant and swarf tray
4. Motor	8. Toolpost	12. Leadscrew	16. Cabinet base

Figure 9.1 Centre lathe

9.1.1 Bed

The lathe bed is the foundation of the complete machine. It is made from cast iron, designed with thick sections to ensure complete rigidity and freedom from vibration. On the top surface, two sets of guideways are provided, each set consisting of an inverted vee and a flat, Fig. 9.2. The arrangement shown may vary on different machines. The outer guideways guide the saddle, and the inner guideways guide the tailstock and keep it in line with the machine spindle. The guideways are hardened and accurately ground.

Two styles of bed are available: a straight bed, where the guideways are continuous over the length of the bed, and a gap bed, where a section of the guideways under the spindle nose can be removed. Removal of this section increases the swing of the lathe, but only for a short distance, Fig. 9.3. For example, the 330 mm swing lathe with a gap bed increases its swing to 480 mm for a length of 115 mm.

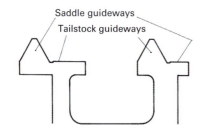

Figure 9.2 Lathe bed guideways

The bed is securely bolted to a heavy-gauge steel cabinet containing electrical connections and a tool cupboard, and provides a full-length cutting-fluid and swarf tray.

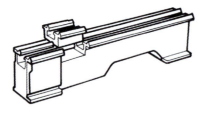

Figure 9.3 Gap bed

9.1.2 Headstock

The complete headstock consists of a box-shaped casting rigidly clamped to the guideways of the bed and contains the spindle, gears to provide a range of 12 spindle speeds, and levers for speed selection. The drive is obtained from the main motor through vee belts and pulleys and a series of gears to the spindle. The speed range is from 40 to 2500 rev/min.

The spindle is supported at each end by precision taper-roller bearings and is bored through to accept bar material. The inside of the spindle nose has a Morse taper to accept centres. The outside of the spindle nose is equipped with means of locating and securing the chuck, faceplate or other workholding device. The method shown in Fig. 9.4, known as a cam-lock, provides a quick, easy and safe means of securing work-holding equipment to the spindle nose. The spindle nose has a taper which accurately locates the workholding device, and on the outside diameter of the spindle nose are three cams which coincide with three holes in the face. The workholding device has three studs containing cut-outs into which the cams lock, Fig. 9.4(a) and (c).

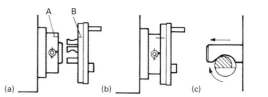

Figure 9.4 Cam-lock spindle nose

To mount a workholding device, ensure that the locating surfaces of both parts are clean. Check that the index line on each cam lines up with the corresponding line on the spindle nose, Fig. 9.4(a). Mount the workholding device on the spindle nose, ensuring that the scribed reference lines A and B on the spindle nose and the workholding device line up. These lines assist subsequent remounting. Lock each cam by turning clockwise, using the key provided. For correct locking conditions, each cam must tighten with its index line between the two vee marks on the spindle nose, Fig. 9.4(b); if this does not happen, do not continue but inform your supervisor or instructor who can then carry out the necessary adjustment. Since each workholding device is adjusted to suit a particular spindle, it is not advisable to interchange spindle-mounted equipment between lathes.

Removal of equipment is carried out by rotating each cam anticlockwise until the index lines coincide and then pulling the equipment away from the spindle nose.

The gearbox, fitted on the lower side of the headstock, provides the range of feeds to the saddle and cross-slide through the feed shaft, and the screw-cutting range through the leadscrew. By selecting the appropriate combination of lever positions in accordance with a table on the machine, a wide range of feed rates and thread pitches can be obtained.

9.1.3 Tailstock

The function of the tailstock is to hold a centre when turning between centres, or to act as a support at the end of long workpieces. Alternatively, the tailstock is used to hold drills and reamers when producing holes.

The tailstock can be moved on its guideways along the length of the bed and locked in any position. The quill contains a Morse-taper bore to accommodate centres, chucks, drills and reamers and is graduated on its outer top surface for use when drilling to depth. It can be fed in or out by means of the handwheel at the rear. Positive locking of the quill is carried out by means of a handle operating an eccentric pin.

9.1.4 Saddle

The saddle rests on top of the bed and is guided by two guideways which, for stability, are the two furthest apart. Accurate movement is thus maintained relative to the centre line of the spindle and tailstock for the complete length of the bed. The top surface contains the dovetail slideway into which the cross-slide is located and the cross-slide leadscrew, complete with handwheel and graduated dial, Fig. 9.5.

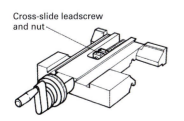

Cross-slide leadscrew and nut

Figure 9.5 Saddle

9.1.5 Cross-slide

Mounted in the dovetail slideway on the top surface of the saddle, the cross-slide moves at right angles to the centre line of the machine spindle. Adjustment for wear is provided by a tapered gib strip, which can be pushed further into the slide and slideway by the screw as wear takes place. Attached to the underside of the cross-slide is the leadscrew nut through which movement is transmitted from the leadscrew. Power feed is available to the cross-slide.

The top surface contains a radial tee slot into which two tee bolts are fitted. The central spigot locates the slideway for the top slide, which can be rotated and clamped at any angle by means of the tee bolts. Graduations are provided for this purpose, Fig. 9.6.

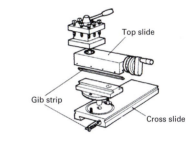

Top slide

Gib strip

Cross slide

Figure 9.6 Cross-slide and top slide

On the lathe shown, external dovetails are provided along each side of the cross-slide, for quick accurate attachment of rear-mounting accessories.

9.1.6 Top slide

The top slide shown in Fig. 9.6, often referred to as the compound slide, fits on its slideway and can be adjusted for wear by means of a gib strip and adjusting screws. Movement is transmitted by the leadscrew through a nut on the slideway. A toolpost, usually four-way hand-indexing, is located on the top surface and can be locked in the desired position by the locking handle. Movement of this slide is usually quite short, 92 mm on the machine illustrated, and only hand feed is available. Used in conjunction with the swivel base, it is used to turn short tapers.

9

9.1.7 Apron

The apron is attached to the underside of the saddle at the front of the machine and contains the gears for transmission of movement from the leadscrew and feed shaft. Sixteen feed rates from 0.03 to 1 mm per revolution are provided.

On the front are the handles to engage and disengage the leadscrew and feed shaft. Also mounted on the front is the handwheel for longitudinal traverse of the carriage along the bed, this movement being transmitted through gears to a rack fixed on the underside of the bed.

The complete assembly of apron, saddle and slides is known as the carriage. The spindle control on the apron is operated by lifting for spindle reverse, lowering for spindle forward and mid-position for stop.

9.2 Centre-lathe controls

The various controls of a typical centre lathe are shown in Fig. 9.7.

Before starting the machine, ensure that the feed-engage lever (20) and the thread-cutting lever (17) are in the disengaged position.

Select the feed axis required, i.e. longitudinal travel of carriage or cross-slide, by means of the apron push-pull knob (19).

Select the direction of feed by means of selector handle (7).

Select the feed rate required by referring to the charts on the headstock and selecting the appropriate position of selector dial (3) and handles (4), (5) and (6).

Select the spindle speed by means of selector handles (10) and (11).

Switch on the main electrical supply at the mains isolator (2).

Start the spindle by lifting the spindle-control lever (18) for reverse or lowering it for forward. The mid-position is 'stop'.

Start and stop the feed motion as required by means of the feed-engage lever (20).

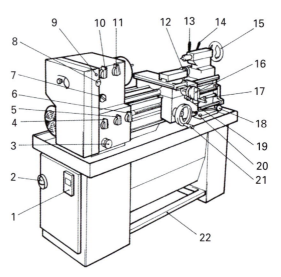

1. Coolant-pump starter	9. 'Supply on' lamp	17. Threadcutting
2. Mains isolator	10. Speed selector	engagement
3. Feed-selector dial	11. Speed selector	18. Spindle-control lever
4. Feed-selector handle	12. Top slide traverse	19. Feed-axis selector
5. Feed-selector handle	handle	20. Feed-engage lever
6. Feed-selector handle	13. Quill lock	21. Longitudinal-traverse
7. Feed-direction	14. Tailstock clamp	handwheel
selector	15. Quill-traverse handwheel	22. Emergency-stop and
8. Emergency stop	16. Cross-traverse handle	brake pedal

Figure 9.7 Machine controls

Do not attempt to change speeds and feeds when the spindle is running – always stop the machine first.

9.2.1 Stopping the machine

The machine can be stopped by returning the spindle-control lever (18) to its central stop position. Alternatively, press the emergency-stop push button (8) or depress the full-length foot brake pedal (22).

9.3 Guards

Guards are physical barriers which prevent access to the danger zone. The Provision and Use of Work Equipment Regulations 1998 (PUWER) require employers to take effective measures to prevent access to dangerous parts of machinery. These regulations also apply to contact with a rotating bar which projects beyond the headstock of a lathe.

For small manually operated lathes, it may be sufficient to provide simple protection for the operator from contact with the chuck and from swarf and coolant. A typical chuck guard is shown in Fig. 9.8. This comprises a metal frame incorporating a high-impact transparent material providing maximum operator protection and good visibility. This is mounted at the rear of the headstock on a pivot so that the guard may be lifted out of the way to give quick and easy access to the work. These can be fitted with an electrical safety interlock so that the machine cannot be started until the guard is in position or the machine will stop if the guard is lifted while the machine is running.

Larger lathes are best fitted with sliding shield guards shown in Fig. 9.9. These are made from high-quality steel with a polycarbonate window to provide maximum operator protection from chuck, swarf and coolant and provide excellent visibility. The guard slides over the headstock and out of the way when access to the chuck or tooling is required. These can also be fitted with an electrical safety interlock to prevent the machine being started when the guard is in the open position and to stop the machine should the guard be opened when the machine is running.

Figure 9.8   Lathe chuck guard

Figure 9.9   Sliding lathe shield

9.4 Workholding

Workpieces can be held in a centre lathe by a variety of methods depending on the shape and the operation being carried out.

The most common method of holding work is in a chuck mounted on the end of the spindle. Several types of chuck are available, the most common being the three-jaw self-centring chuck, the four-jaw independent chuck and the collet chuck.

9.4.1 Three-jaw self-centring scroll chuck

This chuck, Fig. 9.10, is used to hold circular or hexagonal workpieces and is available in sizes from 100 mm to 600 mm. It operates by means of

133

a pinion engaging in a gear on the front of which is a scroll, all encased in the chuck body. The chuck jaws, which are numbered and must be inserted in the correct order, have teeth which engage in the scroll and are guided in a slot in the face of the chuck body. As the pinion is rotated by a chuck key, the scroll rotates, causing all three jaws to move simultaneously and automatically centre the work.

Figure 9.12   Four-jaw independent chuck

Figure 9.10   Three-jaw chuck

Two sets of jaws are usually supplied: those which grip externally while turning, facing, and boring, Fig. 9.11(a), and those which grip internally while the outside diameter or face is machined, Fig. 9.11(b).

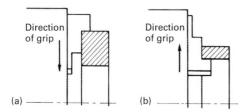

Figure 9.11   (a) Outside and (b) inside jaws

9.4.2 Four-jaw independent chuck

The four-jaw independent chuck, Fig. 9.12, is used to hold square, rectangular and irregular-shaped work which cannot be held in the three-jaw self-centring type. It is available in sizes from 150 mm to 1060 mm. As the name implies, each jaw is operated independently by means of a screw – the jaws do not move simultaneously.

Although the jaws are numbered and must be replaced in the appropriate slot, they are reversible, due to the single-screw operation.

Concentric rings are machined in the front face to aid setting up the work, and tee slots are sometimes provided on the front face for additional clamping or packing of awkward workpieces.

9.4.3 Collet chuck

This type of chuck, Fig. 9.13, fits on the spindle nose and is convenient for bar and the smaller diameter workpieces. Having fewer moving parts than the moving-jaw types makes it more accurate. It is more compact and does not have the same overhang from the spindle nose, and work can be machined up to the front of the collet. All-round gripping of the component makes it ideal for holding tube and thin-walled workpieces which tend to collapse in the three- or four-jaw chucks.

Figure 9.13   Collet chuck

In the model shown in Fig. 9.14, each collet is produced with a number of blades and will accommodate slight size variation up to 3 mm.

Figure 9.14 Multi-size collet

9.4.4 Chuck keys

Accidents occur when chuck keys are left in the chuck and the machine is inadvertently switched on.

No matter for how short a period, *never* leave the chuck key in the chuck. Safety chuck keys, Fig. 9.15, are now available which are spring-loaded and, if left in position, pop out and fall from the chuck.

Figure 9.15 Safety chuck key

9.4.5 Faceplate

The faceplate, Fig. 9.16, is used for workpieces which cannot be easily held by any of the other methods. When fixed to the machine, the face is square to the machine-spindle centre line. A number of slots are provided in the face for clamping purposes. Workpieces can be clamped to the faceplate surface but, where there is a risk of machining the faceplate, the workpiece must be raised from the surface on parallels before clamping. Positioning of the workpiece depends upon its shape and the accuracy required.

Flat plates which require a number of holes are easily positioned by marking out the hole positions and using a centre drill in a drilling machine to centre each position. A centre in the tailstock is then used to locate the centre position and

hold the workpiece against the faceplate while clamping is carried out, Fig. 9.16.

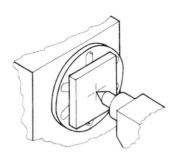

Figure 9.16 Locating workpiece on faceplate

Workpieces which already contain a hole which is to be enlarged, e.g. cored holes in a casting, can be marked out to produce a box in the correct position, the sides of which are the same length as the diameter of the required hole. Roughly positioned and lightly clamped, the workpiece can be set accurately using a scriber in a surface gauge resting on the cross-slide surface. The faceplate is rotated by hand and the workpiece is tapped until all of the scribed lines are the same height, indicating that the hole is on centre, Fig. 9.17. The workpiece is then securely clamped.

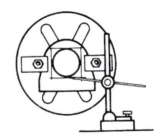

Figure 9.17 Setting workpiece on a faceplate

Accurate positioning of holes in a plate can be done with the aid of toolmaker's buttons. These consist of a hardened and ground steel bush of known diameter with the ends ground square and a flanged screw. The required hole positions are marked out and a hole is drilled and tapped to suit the screw. The accuracy of the drilled and tapped hole is not important, as there is plenty of clearance between the screw and the bore for the button to be moved about. The button is then held on the work by the screw and is accurately positioned by measuring across the outside of adjacent buttons using a micrometer. The buttons

9

are adjusted until the required distance is reached and are then securely tightened by means of the screw, Fig. 9.18(a). The centre distance of the hole to be produced is $x - d$.

The workpiece is then lightly clamped to the faceplate and is accurately positioned using a dial indicator on the button, Fig. 9.18(b). The workpiece is securely clamped and the button is removed. The hole is drilled and bored in the knowledge that it is in the correct position. This is repeated for the remaining holes.

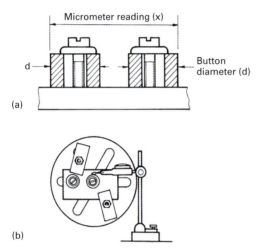

(a)

(b)

Figure 9.18 Using toolmaker's buttons

9.4.5.1 Precautions for faceplate work

When the workpiece has been clamped, check each nut and screw to ensure it is tight. Turn the faceplate by hand and check that all bolts and clamps are clear of the bed, cross-slide or toolpost. To ensure this, avoid using excessively long clamping bolts. Check for 'out of balance' of the faceplate – a counterbalance may be required.

9.4.6 Centres

Components having a number of diameters which are required to be concentric can be machined between centres. A centre is inserted in the spindle nose, using a reducing bush supplied for this purpose. This centre rotates with the spindle and workpiece and is referred to as a 'live' centre. A centre inserted in the tailstock is fixed, does not rotate and is referred to as a 'dead' centre. Great care must be taken to prevent overheating of

'dead' centres due to lack of lubrication or too high a pressure. Keep the centre well lubricated with grease, and do not overtighten the tailstock.

In order to drive the workpiece, a work-driver plate must be mounted on the spindle nose and the drive is completed by attaching a work carrier to the workpiece, Fig. 9.19.

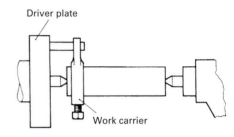

Driver plate

Work carrier

Figure 9.19 Workpiece between centres

Where the size of workpiece requires greater pressure and where considerable time is required for the operation, 'dead' centres will 'burn out', i.e. overheat and the point wear out. To overcome this, live or rotating tailstock centres are available, the centres of which run in bearings which will withstand high pressures without overheating.

Tailstock centres are often required for long work which is held in the chuck but requires support owing to its length.

9.4.7 Steadies

If unsupported, long slender work may tend to be pushed aside by the forces of cutting. To overcome this, a two-point travelling steady is used which provides support to the workpiece opposite the tool as cutting is carried out along the length of the work, Fig. 9.20.

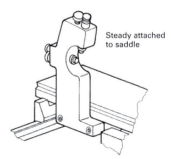

Steady attached to saddle

Figure 9.20 Two-point travelling steady

Work of a larger diameter than can be accepted through the machine spindle and yet requiring work to be carried out at one end can be supported using a three-point fixed steady. This steady is clamped to the machine bed and the points are adjusted so that the workpiece is running true to the spindle centre line before the machining operation is carried out, Fig. 9.21.

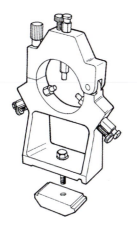

Figure 9.21   Three-point fixed steady

9.4.8 Mandrel

Work which has a finished bore and requires the outside to be turned concentric to it can be mounted on a mandrel. The mandrel is then put between centres and the work is machined as already described for 'between-centres' work.

A mandrel, Fig. 9.22, is a hardened and ground bar with centres in each end and a flat machined at one end to accept the work carrier. The diameter is tapered over its length, usually about 0.25 mm for every 150 mm length. When the work is pushed on, this slight taper is enough to hold and drive the work during the machining operation.

The flat for the carrier is machined on the end having the larger diameter, so that the carrier

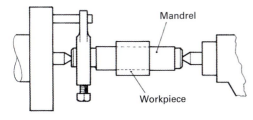

Figure 9.22   Mandrel

does not have to be removed to load and unload workpieces.

9.5 Centre-lathe operations

9.5.1 Turning

Accurate turning of plain diameters and faces can be simply carried out on a centre lathe. Wherever possible, diameters should be turned using the carriage movement, as the straightness of the bed guideways ensures parallelism of the workpiece and power feed can be used. Avoid using the top slide for parallel diameters, since it is adjustable for angle and difficult to replace exactly on zero without the use of a dial indicator. It has also to be hand fed.

When a number of diameters are to be turned on a workpiece, they should be produced at one setting without removing the workpiece from the chuck, in order to maintain concentricity between them. Accuracy is lost each time the workpiece is removed and put back in the chuck. Accurate sizes can be produced by measuring the workpiece when the final size is almost reached, then using the graduated dial on the handwheel to remove the required amount.

Where only diameters are being turned and a square shoulder is required, a knife tool is used, Fig. 9.23A, which cuts in the direction shown. Where facing and turning are being carried out in the same operation, a turning and facing tool is used, Fig. 9.23B. The slight radius on the nose produces a better surface finish, but the radius will be reproduced at the shoulder.

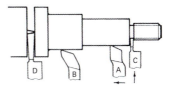

Figure 9.23   Turning-tool applications

Where a relief or undercut at the shoulder is required, e.g. where a thread cannot be cut right up to the shoulder, an undercut tool is used. This tool is ground to the correct width, the face parallel to the work axis, and is fed in the direction shown, Fig. 9.23C.

9

Work produced from bar can be cut to length in the lathe, an operation known as 'parting off'. The face of the parting-off tool is ground at a slight angle, so that the workpiece is severed cleanly from the bar, Fig. 9.23D.

It is essential that all cutting tools used on a lathe be set on the centre of the workpiece. A tool set too high reduces the clearance and will rub, while one set too low reduces the rake angle, Fig. 9.24. Cutting tools can be set relative to a centre inserted in the tailstock and then be raised or lowered using suitable thicknesses of packing, Fig. 9.25. A good stock of varying thickness of packing should be available which, when finished with, should always be returned for future use.

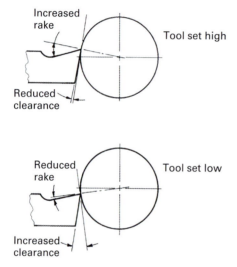

Figure 9.24 Effect of tool set above and below centre

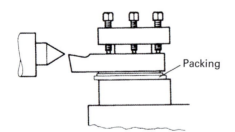

Figure 9.25 Setting tool on centre

9.5.2 Drilling

Drilling is carried out on a lathe by holding the drill in a chuck or mounting it directly in the quill of the tailstock, which contains a Morse taper for this purpose. As with all drilling operations, some guide is required to enable the drill to start central, and a centre drill, Fig. 9.26, is commonly used.

Figure 9.26 Centre drill

Centre drills are available in various sizes, and their purpose is to produce centres in a workpiece for turning between centres. Due to its rigid design, a centre drill is convenient in providing a suitable guide to start the drill in the centre of the bar.

When using a centre drill, great care must be taken to prevent breakage of the small point which, because of its size, does not have deep flutes to accommodate swarf. Feed in gently a short distance at a time, using the tailstock handwheel, winding the drill out frequently to remove swarf before it packs the flute and snaps off the point. Use high spindle speeds for the small point diameter. The centre drill should be fed in just deep enough to give the drill a start.

Drilling is then carried out to the required depth, which can be measured by means of the graduations on the quill. Relieve the drill frequently, to prevent swarf packing the flutes.

9.5.3 Tapping

If the hole drilled on the lathe requires threading, this can be carried out by hand, using hand taps. Having drilled a hole of the correct tapping size (see Chapter 2), it is essential that the tap is started parallel to the workpiece axis. The first step is to isolate the machine. This is a hand operation and power rotation of the machine is not required. Hold a taper or first tap in a tap wrench and offer it into the start of the hole. Slide up a centre located in the tailstock until it locates in the centre at the rear of the tap. This set-up is shown in Fig. 9.27. The tap wrench is then rotated while at the same time light force is applied to keep the tailstock centre in contact with the rear centre of the tap by rotating the tailstock handwheel. Once the tap has started to cut the first few threads, the tailstock centre can be withdrawn out of the way and tapping continued until the required depth is reached. Do remember to withdraw the tap at regular intervals in order to remove swarf and prevent the tap from clogging, which could lead to breakage of the tap. A proprietary tapping compound should be used to ease cutting and produce good quality threads. Depending on whether you are tapping a through hole or blind hole, it may be necessary to change the tap in stages to a second or intermediate, or to a bottoming or plug tap (refer to Chapter 2). For production use, the machine power is used together with tapping heads where a clutch mechanism can set depending on the size of the tap and will slip when the bottom of the hole is reached or a restriction is encountered. The machine is then reversed and the tap withdrawn.

Figure 9.27 Hand tapping on lathe

9.5.4 Reaming

Holes requiring a more accurate size and better surface finish than can be achieved with a drill can be finished by reaming. The hole is drilled smaller than required (see Table 8.2) followed by the reamer, using a spindle speed approximately half that used for drilling. A reamer will follow the hole already drilled, and consequently any error in concentricity or alignment of the hole axis will not be corrected by the reamer. Where accurate concentricity and alignment are required on larger sizes, the hole should be drilled a few millimetres undersize, bored to the required size for reaming, thus correcting any errors and finally reamed to achieve the finished size and finish. A suitable lubricant should be used to enhance the life of the

reamer; soluble oil is normally satisfactory at 40:1 dilution. The flutes of the reamer should not be allowed to become blocked with swarf.

9.5.5 Boring

As already stated, boring can be used to correct errors in concentricity and alignment of a previously drilled hole. The hole can be finished to size by boring without the use of a reamer, as would be the case when producing non-standard diameters for which a reamer was not available. Boring is also used to produce a recess which may not be practical by drilling and reaming, Fig. 9.28.

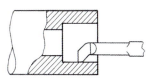

Figure 9.28 Boring tool

A boring tool must be smaller than the bore it is producing, and this invariably results in a thin flexible tool. For this reason it is not usually possible to take deep cuts, and care must be taken to avoid vibration. In selecting a boring tool, choose the thickest one which will enter the hole, to ensure maximum rigidity. Ensure also that adequate secondary clearance is provided in relation to the size of bore being produced, as shown in Fig. 7.5.

9.5.6 Knurling

Knurling is a process whereby indentations are formed on a smooth, usually round, surface to allow hands or fingers to get a better grip. Fig. 9.29 shows diamond knurling on a tap wrench. Knurling can also be used purely for decorative purposes. The indentations are formed by pressing hardened knurling wheels (known as knurls) held in a knurling tool, Fig. 9.30, onto the workpiece and deforming the surface. The knurls may produce a straight, diagonal or criss-cross (known as a diamond knurl) pattern on the workpiece, the form of which can be coarse, medium or fine depending on the knurls used.

In operation, the workpiece is held in the lathe chuck while the knurling tool is held in the toolpost on the cross-slide.

Figure 9.29 Diamond knurl

9.6 Taper turning

The method used to turn a taper depends upon the angle of taper, its length and the number of workpieces to be machined. Three methods are commonly used: with a form tool, with the top or compound slide and with a taper-turning attachment.

9.6.1 Form tool

Short tapers of any angle can be produced by grinding the required angle on the cutting tool, Fig. 9.31. The cutting tool is then fed into the work until the desired length of taper is produced.

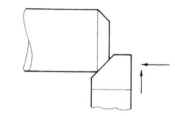

Figure 9.31 Angle form tool

This method is normally used for short tapers such as chamfers, both internal and external. The long cutting edge required by long tapers has a tendency to chatter, producing a bad surface finish.

9.6.2 Top or compound slide

Taper turning can be carried out from the top slide by swivelling it to half the included angle required on the work, Fig. 9.32. Graduations are provided on the base plate, but any accurate angle must be determined by trial and error. To do this, set the top slide by means of the graduations, take a trial cut, and measure the angle. Adjust if necessary,

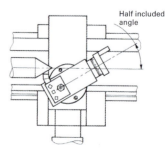

Figure 9.30 Basic knurling tool

The cross-slide is positioned so that the knurls are above the centre of the workpiece where they are adjusted to contact the workpiece surface using the adjusting screw. Slender work may have to be supported using a centre in the tailstock.

With the lathe running at a low speed and cutting fluid applied, the knurls are tightened until the required form appears on the workpiece surface.

The knurls can then be fed along the workpiece to give the desired length.

If a deeper form is required, the knurls are released and moved back to the start before picking up in the original indents, increasing the force on the knurls and the operation repeated.

Figure 9.32 Top slide set at half included angle

take a second cut, and remeasure. When the correct angle is obtained, ensure that the clamping nuts are securely tightened.

Turning the angle is done by winding the top slide handle by hand. The tool will feed at the angle to which the top slide is set. After the first cut, the tool is returned to its starting position by rewinding the top slide. The feed for the second cut is achieved by moving the cross-slide.

This method can be used for any angle, internal or external, but the length is restricted by the amount of travel available on the top slide.

9.6.3 Taper-turning attachment

Taper-turning attachments can be fitted at the rear of the cross-slide and can be used to turn included angles up to 20° over a length of around 250 mm, both internally and externally. A plan view of a typical taper-turning attachment is shown in Fig. 9.33. The guide bar, which swivels about its centre, is mounted on a base plate which carries the graduations. The base plate is attached to the connecting rod, which passes through a hole in the clamp bracket where it is held tightly by a clamping screw. The clamp bracket is clamped to the bed of the machine. Thus the guide bar, base plate, connecting rod and clamp bracket are securely fixed to each other and to the machine bed.

The guide block slides on the guide bar and is located in the sliding block by a spigot. This gives a solid location and at the same time allows the guide block to take up the angle of the guide bar.

The sliding block is attached to the end of the cross-slide leadscrew and is guided in a bracket which is bolted to the rear face of the saddle.

It can therefore be seen that, if the carriage is traversed along the bed and the guide bar remains stationary (i.e. clamped to the bed), the sliding block can only push or pull the cross-slide leadscrew. For this movement to be transmitted to the cross-slide and so to the cutting tool, a special leadscrew is required, Fig. 9.34. The front end of the leadscrew has a spline which slides up the inside of the handwheel spindle. When the sliding block pushes the leadscrew, the leadscrew moves back and, since it passes through the leadscrew nut which in turn is screwed to the cross-slide, the cross-slide and the cutting tool mounted on it will also move back, pushing the spline up the inside of the handwheel spindle.

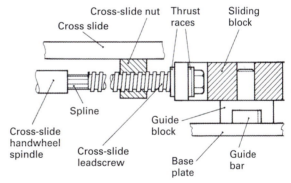

Figure 9.34 Cross-slide leadscrew for taper-turning attachment

By this method, a cut can be put on merely by rotating the handwheel, driving through the spline to the leadscrew and nut without interfering with the taper-turning attachment. To revert to a normal operating condition, the connecting rod is unclamped and the clamp bracket removed and, since the complete attachment moves with the carriage, the cross-slide can then be used in the normal way.

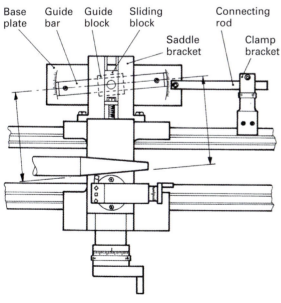

Figure 9.33 Taper-turning attachment

9.7 Screw-cutting

The thread now standardised in British industry is the ISO metric thread, ISO being the International Organization for Standardization. Terminology of this thread is shown in Appendix 1.

The ISO metric thread has a 60° truncated form, i.e. the thread does not come to a sharp point but has a flat crest. The root of the thread also has a small flat.

A single-point tool sharpened as shown in Fig. 9.35 produces the thread angle and the flat at the root, the major diameter being produced at the turning stage. To cut an accurate thread requires a definite relationship between the rotation of the work in the spindle and the longitudinal movement of the carriage by means of the leadscrew. All modern centre lathes have a gearbox through which a wide range of pitches can be obtained by referring to a chart on the machine and turning a few knobs.

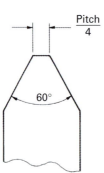

Figure 9.35   Screw-cutting tool for metric thread

The longitudinal travel of the carriage is obtained from the leadscrew through a split nut housed in the apron and operated by a lever on the apron front, Fig. 9.36. By closing the split nut, the drive can be started at any position.

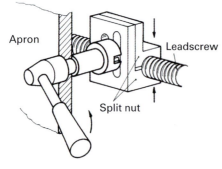

Figure 9.36   Split nut for screw cutting

The position of engagement of the split nut on the leadscrew for each cut is important in order that the tool will travel along the same path as the previous cut. To achieve this accuracy of engagement, a thread indicator dial is fitted at the end of the apron, Fig. 9.37. The dial is mounted on a spindle at the opposite end of which is a gear in mesh with the leadscrew. These gears are interchangeable, are stored on the spindle, and are selected by referring to a chart on the unit. They are arranged to give a multiple of the pitch required, relative to the 6 mm pitch of the leadscrew.

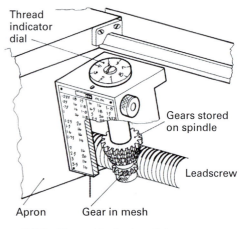

Figure 9.37   Thread indicator dial

The chart shows the gear used for a particular pitch of thread and the numbers on the thread indicator dial at which the split nut may be engaged. To cope with the different diameters of gears, the unit pivots and is locked in position when the gear is in mesh. To avoid unnecessary wear, the unit is pivoted back out of mesh when not in use for screw-cutting.

9.7.1 Method

Having turned the workpiece to the correct diameter, the following procedure should be followed. This procedure is for screw-cutting a right-hand external metric thread on a machine having a metric leadscrew.

1. Carefully grind the tool to 60° with the aid of a screw-cutting gauge, Fig. 9.38, leaving a flat on the tool nose.
2. Mount the tool in the toolpost on the centre of the workpiece.

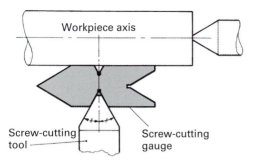

Figure 9.38 Screw-cutting gauge and positioning of tool

3. Set the tool relative to the axis of the workpiece, using the screw-cutting gauge, Fig. 9.38.
4. Calculate the required thread depth.
5. Select the required pitch.
6. Select the correct gear on the thread indicator dial and mesh with the leadscrew.
7. Engage a slow spindle speed.
8. Start the machine.
9. Wind in the cross-slide until the tool just touches the outside of the workpiece and move the carriage so that the tool is clear of the end of the workpiece.
10. Stop the machine.
11. Set the dial on the cross-slide to zero.
12. Restart the machine.
13. Wind the cross-slide to give a small cut of 0.05 mm.
14. Wait until the appropriate number on the thread indicator dial comes round to the mark and engage the split nut.
15. Take a trial cut.
16. When the end of the workpiece is reached, unwind the cross-slide to remove the tool from the work and disengage the split nut. This is done in one movement. At all times during screw-cutting, one hand should be resting on the cross-slide handwheel, the other on the split-nut lever.
17. Stop the machine.
18. Check the thread to make sure the correct pitch has been cut.
19. Rewind the carriage to the starting point.
20. Restart the machine.
21. Rewind the cross-slide back to the original graduation and put on a further cut.

22. Wait for the correct number on the thread indicator dial, engage the split nut, and repeat until the final depth is reached.

Depending upon the accuracy of thread required, final checking should be carried out by means of a gauge or by checking against a nut or the mating workpiece.

Internal threads are cut in exactly the same manner, except that the tool is similar to a boring tool ground to give a 60° thread form.

Left-hand threads are produced in the same manner by reversing the rotation of the leadscrew and starting from the opposite end of the workpiece.

9.7.2 Imperial threads

Imperial threads are designated not by their pitch but by the number of threads per inch (t.p.i.). The leadscrew of a metric centre lathe has a pitch of 6 mm and, since the number of t.p.i. cannot be arranged as a multiple of the leadscrew pitch, the split nut, once it is engaged, must never be disengaged during the thread-cutting operation. This also means that the thread indicator dial is of no use when cutting imperial threads on a metric lathe.

The procedure when cutting imperial threads on a metric lathe is the same as before up to the point when the split nut is disengaged and a trial cut taken.

16. When the end of the workpiece is reached, withdraw the tool and stop the machine but do not disengage the split nut.
17. Reverse the spindle direction so that the carriage moves back to the starting point.
18. Stop the machine, put on a further cut, and restart the machine spindle in a forward direction.
19. Repeat until the thread has been cut to size before disengaging the split nut.

9.8 Safety in use of lathe

Most accidents happen from:

▶ entanglement on workpieces, chucks, carriers and unguarded bar protruding from rear of spindle;

9

- direct contact with moving parts (especially when adjusting coolant supply or removing swarf);
- eye injuries and cuts from machine cleaning and swarf removal;
- chuck keys ejected from rotating chuck;
- health issues with metalworking fluids and with noise (e.g. bars rattling inside the machine spindle).

Remember that you have a duty under the various health and safety regulations already covered in Chapter 1. To avoid the risk of accident:

- Always follow the training provided by your employer.

- Ensure chuck guard is in position.
- Ensure splash guard is in position to protect from swarf and metalworking fluid.
- Always wear eye protection and any other PPE required.
- Take great care before engaging automatic feeds and when screw-cutting.
- Do not attempt to adjust the coolant supply or remove swarf when the machine is running.
- Do not wear jewellery or loose clothing.
- Always have long hair tied back or in a hairnet.
- Never use emery cloth by hand on a rotating workpiece.
- Never leave the chuck key in the chuck.

Review questions

1. What is the purpose of the thread-cutting dial during screw-cutting?
2. Name two pieces of equipment used to prevent long slender workpieces flexing during machining on a centre lathe.
3. What is a mandrel used for?
4. State the two important capacities of a centre lathe.
5. What is the purpose of using a centre drill?
6. Name four types of workholding equipment used on a centre lathe.
7. Why is it essential to have the cutting tool set at the centre height of the workpiece?
8. Give two reasons why a boring operation would be carried out on a centre lathe.
9. Why is it necessary to have a direct relationship between rotation of the work and the longitudinal movement of the tool during screw-cutting?
10. Describe three methods used to turn a taper.

CHAPTER 10

Surface grinding

Surface grinding is used to produce flat accurate surfaces and can be carried out on all materials, hard or soft. There may be no other way of removing metal from a hardened workpiece. It is normally considered a finishing operation, but large machines are used in place of milling and shaping machines to remove large amounts of material.

A typical surface grinder is shown in Fig. 10.1 and uses a 300 mm diameter by 25 mm wide grinding wheel. The reciprocating table and cross-slide movements are hydraulically operated, although alternative hand operation is provided.

The capacity of such a machine is the maximum length and width of surface which can be ground, in this case 500 mm × 200 mm, and the maximum height which can go under a grinding wheel of maximum diameter. Using a 300 mm diameter wheel, the maximum height of workpiece on the machine shown is 400 mm.

10.1 Elements of a surface-grinding machine

The main elements of a typical surface-grinding machine are shown in Fig. 10.2.

10.1.1 Base

The base is a heavily ribbed box-section casting to ensure rigidity and complete freedom from

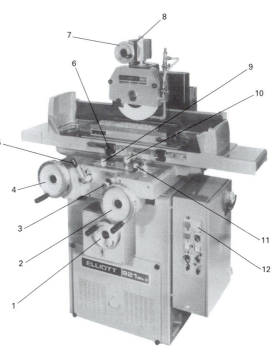

Figure 10.1 Surface grinding machine

vibration. The bottom of the base houses the hydraulic pump and fluid reservoir. At the rear of the base is a vertical dovetail slideway which guides the column. Two vee slideways on top of the base guide the saddle and are widely spaced to maintain accuracy and rigidity.

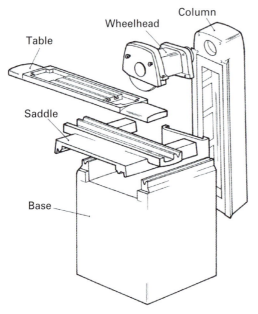

Figure 10.2 Main elements of surface grinder

10.1.2 Column

The column, guided on a dovetail slide, carries the wheelhead at its top end and contains the motor and belt drive to the wheel spindle. The column and wheelhead are raised and lowered through a screw and nut from a handwheel on the front of the machine. A telescopic guard is fitted to prevent grinding dust coming between the slide surfaces.

10.1.3 Wheelhead

The wheelhead carries the wheel spindle, which is mounted in precision bearings. The complete grinding-wheel collet assembly is fitted on a taper on the end of the spindle. Drive to the spindle is by vee belt and pulley from the motor mounted in the bottom of the column.

10.1.4 Saddle

The saddle is fitted on top of the base in the two vee slideways and provides the cross-traverse movement. The cross traverse can be applied automatically in continuous or incremental feed by hydraulic power or, alternatively, with a manually operated handwheel. The automatic cross movement is infinitely variable up to a maximum of 10 mm. The increment of cross movement is

applied at each end of the table stroke, resulting in complete grinding of the workpiece surface in the manner shown schematically in Fig. 10.3. The top surface carries a vee-and-flat slideway to guide the table at right angles to the saddle movement.

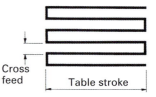

Figure 10.3 Schematic diagram of cross-feed movement to table stroke

10.1.5 Table

The table is guided by the vee-and-flat slideway on the saddle and can be manually operated with a handwheel. Automatic reciprocation of the table is transmitted through a hydraulic cylinder at infinitely variable speeds from 0.6 to 30 m/min. Reversal of the table movement is achieved automatically by trip dogs operating a direction-reversing valve. The trip dogs can be set to give the required length of table stroke and position of reversal.

A simplified diagram of the hydraulic circuit is shown in Fig. 10.4. When the direction-reversing valve A is in the position shown, the sliding valve

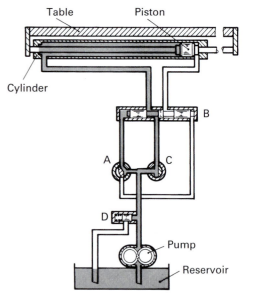

Figure 10.4 Simplified diagram of table hydraulics

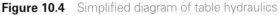

B moves to the right. This allows hydraulic fluid into the left of the table cylinder which is attached to the saddle. The fluid moves the piston, and the table attached to it, to the right.

When the end of the table stroke is reached, the trip dog moves the direction-reversing valve, causing fluid to move the sliding valve to the left. This allows fluid into the right of the table cylinder, moving the piston and table to the left. Thus automatic continuous reciprocating movement of the table is achieved. Other connections into this circuit are made to give automatic cross movement at the end of each table stroke.

Control valve C meters the amount of fluid reaching the cylinder and so controls the speed of the table movement. Relief valve D allows any pressure build-up to be released. This prevents mechanical damage in the event of accidental overload or jamming of the table, as the fluid is merely returned to the reservoir.

A series of tee slots is provided on the top table surface to enable clamping of workpieces or workholding equipment.

10.2 Controls

Controls of a typical surface grinder are shown in Fig. 10.1.

Handwheel (1) raises and lowers the column. By lowering the column, a cut is put on by the wheel. Since the accuracy of the workpiece depends on how much metal is removed, the graduations on this handwheel represent very small increments of movement, in this case 0.0025 mm.

Handwheel (2) provides cross movement of the saddle, graduations on this handwheel representing increments of 0.01 mm.

Handwheel (4) is used to reciprocate the table by hand.

Length of stroke and position of table reversal are controlled by trip dogs (6) striking the direction-reversing-valve lever (10).

The table-speed control knob (11) can be adjusted to give infinitely variable speeds from 0.6 to 30 m/min.

Lever (9) is used to select continuous cross feed or incremental feed at the end of each table stroke.

Where continuous cross feed is selected, lever (5) controls the speed, which is infinitely variable from 0 to 5 m/min.

The rate of incremental feed is controlled by lever (3) and is infinitely variable from 0.28 to 10 mm.

The switch panel at the right side of the machine controls the motors for the hydraulic pump, wheel spindle, cutting fluid, etc. and carries the main isolator and a large mushroom-headed stop button (12).

Although the basic principles are the same, current surface grinder models have differing control systems. A typical model is shown in Fig. 10.5. These control systems were developed to offer the operator the advantages of computer control without the complexity of full CNC systems. Full CNC grinding machines are of course available.

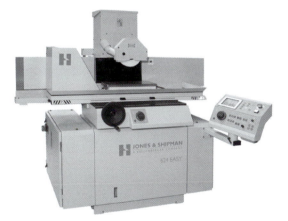

Figure 10.5 Surface grinder with computer control

The table movement (x axis) is manual via the handwheel, or hydraulic for automatic reciprocation. The saddle cross-slide movement (y axis) and wheel downfeed (z axis) are controlled by digital AC servo motors. Table reversal is set by adjustable table dogs and proximity switches in the saddle.

At start-up, the operator is given three options: manual, dress and grind cycle.

In manual mode the machine operates as a hand-controlled machine where the slideway is operated manually by the handwheel or by automatic reciprocation. Work speeds and feeds are selected via the control touch screen and electronic handwheel. Feed per graduation of the electronic handwheel is 0.001, 0.01 or 0.1 mm and is operator selectable between metric and imperial units. A typical control panel is shown on Fig. 10.6.

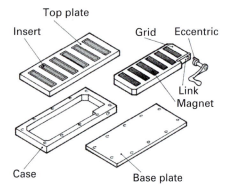

Figure 10.7   Permanent-magnet chuck

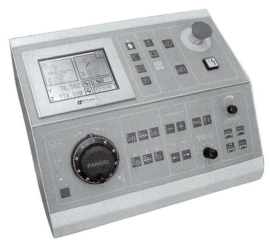

Figure 10.6   Control panel

Dress mode allows dressing of the abrasive wheel to a number of different forms, the basic mode being straight across the periphery of the wheel.

Grind cycle mode allows a preset automatic grinding cycle to be selected via the touch screen on the control panel.

10.3 Workholding

The basic method of workholding in surface grinding is the permanent-magnet chuck, used to hold workpieces having flat surfaces. These chucks will not hold non-magnetic materials such as the non-ferrous range. The complete chuck consists of a top plate containing inserts separated from the top plate by a non-magnetic epoxy-resin filler, a non-magnetic case, a moving grid containing the permanent magnets insulated from the grid and magnetised vertically and a base plate, Fig. 10.7.

The principle upon which permanent-magnet devices operate is to establish the magnetic lines of force or flux from the permanent magnets through the workpiece when switched on, and

to divert or 'short circuit' the flux when switched off. This is achieved by moving the magnets in line with the top plate and so completing the circuit to the insert, grid and base plate through the workpiece, Fig. 10.8(a). For a workpiece to be gripped, it is therefore necessary for it to bridge the top plate and an insert. To switch off, the magnets are moved out of line with the top plate, diverting the flux so that the circuit is completed not through the workpiece but through the top plate, insert and base plate, Fig. 10.8(b). The workpiece is thus deprived of flux and is released.

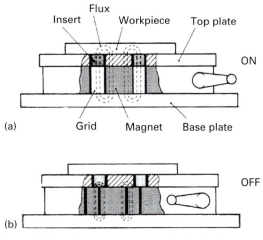

Figure 10.8   Permanent-magnet chuck showing flux lines in (a) ON and (b) OFF positions

Other methods of workholding are used when the shape or the material from which the workpiece is made does not allow direct holding on the permanent-magnet chuck. However, the devices used to hold the workpieces are invariably themselves held on the permanent-magnet chuck.

Vices are used to hold workpieces, but it should be remembered that grinding is usually a finishing operation and so any vice used should be accurate and if possible kept only for use in grinding. Care must also be taken to avoid distortion of the component, as this will be reflected in the finished workpiece.

Surfaces required to be ground at right angles can be clamped to the upright surface of an angle plate.

Vee blocks are used to hold circular workpieces.

10.4 Grinding wheels

All machining operations are potentially dangerous – lack of understanding or undue care have resulted in many accidents. The use of grinding wheels, also known as abrasive wheels, which are described fully at the end of this chapter, is potentially one of the most dangerous for two reasons.

▶ A grinding wheel is made of small abrasive particles held together by a bonding material. Compared with metal it is extremely fragile.
▶ Grinding wheels are run at high speeds. A 300 mm diameter wheel is run at about 2000 rev/min, giving a speed at the diameter of almost 1900 m/min. Compare this with a piece of steel of the same diameter being cut with a high-speed-steel cutting tool on a lathe at 30 m/min.

The Provision and Use of Work Equipment Regulations 1998 (PUWER) require, amongst other things, that all machinery is suitable for its intended use and is properly maintained, and that employees, including those using, mounting and managing the operation of abrasive wheels, are fully informed and properly trained in their use. Any training programme should cover, amongst others, aspects such as hazards and risks, uses, marking, storage, mounting and dressing of abrasive wheels. No attempt is to be made here to cover the detailed instruction requirements of the Regulations, merely to draw attention to them.

10.4.1 Dressing

A grinding wheel is made up of a large number of tiny teeth. The teeth are formed by the tiny

grains of hard abrasive, held together by a bonding material. As with any other metal-cutting operation, the 'teeth' or grains must be kept sharp. To some extent a grinding wheel is self-sharpening. The ideal situation during grinding is that, as the grains which are cutting become blunt, greater force is exerted which tears the blunt grains from the bonding material, exposing fresh sharp ones.

It follows that, when grinding a hard material, the grains become blunt quickly and will require to be torn from the bonding material quickly. To allow this to take place, less bonding material is used to hold the grains, the grains tear away easily when blunt, and the wheel is referred to as 'soft'. The opposite is true when grinding a soft material: the grains do not blunt so readily and can be held in position longer, and therefore more bonding material is used. These wheels are referred to as 'hard'.

Sharpness and trueness of the grinding wheel face can be achieved by 'dressing' using an industrial diamond dresser. Dressing with diamonds should always be carried out using a copious supply of coolant which should be turned on before the diamond touches the wheel. The wheel is lowered until it touches the diamond whereupon the diamond is moved across the surface of the wheel, Fig. 10.9. The wheel is then lowered a little and the operation repeated until all the worn grains have been removed, exposing fresh grains and leaving the face flat and true.

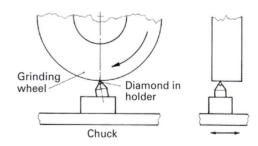

Figure 10.9 Dressing the grinding wheel using a diamond

The diamond can be single point or multi-point and by varying the depth of cut and the traverse rate of dressing, different wheel surfaces and hence different cutting actions can be achieved. The size of diamond used is important and is dependent upon the size of wheel, e.g. a 250 mm diameter by 25 mm wide wheel would typically require a 0.5 carat single diamond while a multi-point would typically require to be 1.3 carat. To ensure maximum life, the diamond dresser should be regularly rotated in its holder to ensure a keen edge is always presented to the wheel. A selection of diamond dressers is shown in Fig. 10.10.

Figure 10.10   Diamond dressers

The machine shown in Fig. 10.1 has a built-in dressing attachment mounted above the wheel. By turning the graduated dial (8) in Fig. 10.1, the diamond is lowered in contact with the top of the wheel and is traversed across the wheel surface using the handwheel (7). An attachment of this kind saves time through not having to position and remove the diamond for each dressing nor having to lower the wheel for dressing and to raise it again to continue grinding.

10.4.2 Balancing

It is impossible to produce good-quality work on any grinding machine if the wheel is out of balance, thus setting up vibrations through the spindle.

The wheel is mounted on a collet, the complete assembly being removable from the spindle, Fig. 10.11. The wheel spigot upon which the wheel is located has a taper bore to accurately locate on the spindle nose. The wheel flange locates on the wheel spigot and is held by three screws which, when tightened, securely hold the wheel between the two surfaces. On the outer

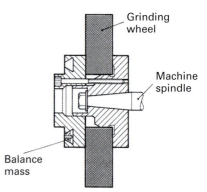

Figure 10.11   Wheel-collet assembly

face of the wheel flange is an annular dovetail groove which holds balance masses, usually two. These masses can be locked in any position round the groove.

When the wheel is mounted correctly and the periphery has been dressed, the collet assembly is removed from the spindle. The balance masses are then removed. A balancing arbor is inserted in the bore. The balancing arbor has a taper identical to that on the spindle and has equal size parallel diameters at each end, Fig. 10.12(a). This assembly is placed on a balancing stand, Fig. 10.12(b), which has previously been set level. The wheel is allowed to roll on the knife edges and is left until it comes to rest, which it will do with the heaviest portion at the bottom.

A chalk mark is made at the top of the wheel, opposite the heaviest portion. The masses are then replaced in a position opposite to each other and at right angles to the chalk mark. The masses can now be moved equally a little way towards the

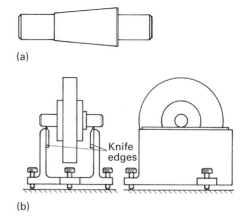

Figure 10.12   (a) Balancing arbor and (b) stand

light portion and locked. The wheel is then rolled and allowed to come to rest. If the same heavy portion again comes at the bottom, both masses are moved a little closer to the light portion and again the wheel is allowed to roll and come to rest. This process is repeated until the wheel will stop in any position and show no tendency to roll along the balancing stand. The wheel is then ready for use and should be replaced on the spindle in the correct manner.

Electromechanical grinding wheel balancing systems are available which measure the vibrations caused by an unbalanced grinding wheel and suitably move weights flanged to the outside of the grinding wheel, or inserted in the machine spindle. This eliminates the need to manually pre-balance.

10.4.3 Guarding

Due to the fragile nature of a grinding wheel and the high speeds at which it runs, it is possible for a wheel to burst. Under the Regulations, all grinding wheels must be adequately guarded and the guard be fitted at all times before the wheel is run.

A guard has two main functions: first to contain the wheel parts in the event of a burst, and second to prevent, as far as possible, the operator coming in contact with the wheel. Abrasive wheels should be enclosed to the greatest extent practicable and this will depend on the nature of the work. On surface grinders, the guard consists of a plate which encloses the front of the wheel to retain fragments, thus protecting the operator should the wheel burst. Details of the wheel enclosure angles for peripheral surface grinders supplied after the publication of BS EN 13218 in 2008 are shown in Fig. 10.13.

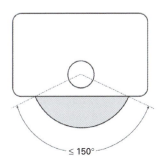

Figure 10.13   Wheel enclosure angle

Full width guards are fitted to the front of the machine table to protect the operator in the event of a workpiece flying off a magnetic chuck and from coolant spray.

10.4.4 Bonded-abrasive grinding wheels

A bonded-abrasive wheel consists of two main essentials: the abrasive, which does the actual cutting, and the bonding material or bond, which holds the abrasive together and forms the wheel shape, Fig. 10.14.

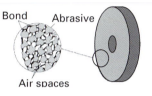

Figure 10.14   Abrasive-wheel features

The ideal cutting condition of an abrasive wheel is that, when they have done their work and become dull, the abrasive grains fracture or are released from the bonding material to expose new sharp cutting grains in their place. This is repeated continuously during cutting and gives the wheel a self-sharpening effect.

If the grains are released before they have done their work and become dull, the wheel wears rapidly and is said to act 'soft'.

When the strength of the bonding material is too great to allow the grains to be released when they have become dull, the wheel is said to act 'hard' and is recognisable by a glazed and shiny appearance on the cutting face.

In order to grind a range of materials efficiently under a variety of cutting conditions, different abrasives of varying grain size are arranged within different bonding materials to give a number of features or characteristics.

10.4.5 Abrasives

Two abrasive materials are in general use: aluminium oxide and silicon carbide.

Aluminium oxide is produced by fusing bauxite in special electric furnaces. When cool, it is

151

crushed to produce shaped particles which are then graded in a series of grain sizes. Owing to its tough nature, aluminium oxide is suitable for grinding metals of high tensile strength, such as steel. This abrasive is designated by the letter A. Ceramic aluminium oxide, a patented ceramic form of aluminium oxide, is available which is harder and sharper than conventional aluminium oxide abrasives and gives high stock removal together with a cooler cutting action. Differing qualities of aluminium oxides are identified by different manufacturers using colours, e.g. white, pink, ruby and blue.

Silicon carbide is made by the chemical reaction, in an electric furnace, of high carbon coke and pure silica sand, with small amounts of salt and sawdust added to assist the reactions. The resultant mass is crushed to give grains of the correct shape which are then graded in a series of grain sizes. This is a hard and brittle abrasive and is most efficient for grinding materials of low tensile strength, such as cast iron, brass, copper, aluminium and glass as well as grinding extremely hard materials such as cemented carbides. These are usually green in colour and are thus often referred to as 'green grit' wheels. This abrasive is designated by the letter C.

10.4.6 Grain size

The size of the abrasive particles is indicated by a number representing the number of openings per linear inch (25 mm) in the screen used for sizing. The standard grain sizes, from the coarsest to the finest, are 4, 5, 6, 7, 8, 10, 12, 14, 16, 20, 22, 24, 30, 36, 40, 46, 54, 60, 70, 80, 90, 100, 120, 150, 180, 220, 230, 240, 280, 320, 360, 400, 500, 600, 800, 1000 and 1200. The sizes most widely used range from 10 to 120.

The grain size used affects the amount of material which will be removed and the final surface roughness. For rough grinding requiring a large amount of material removal, a large grain size, e.g. 10, is used, resulting in a rough surface. If a fine surface is required, a small grain size, e.g. 120, is used. The small grains will not, of course, be able to remove the same large amount of material as the large grains, but they produce a fine surface finish.

10.4.7 Grade

This is the strength of the bond holding the abrasive grains in place. It is a measure of the amount of bond present.

More bond material will have a greater hold on the abrasive grains, which will be less readily released. This is referred to as a 'hard' grade.

Less bond material will not have such a great hold on the abrasive grains, which will then be released more readily. This is referred to as a 'soft' grade.

When grinding a hard material, the abrasive grains dull more quickly than when grinding a soft material. This means that the grains must be released more quickly, to expose new sharp grains, so a wheel having a soft grade is used, which allows easy release of the grains as they become dull.

When grinding soft materials, the grains are not required to work as hard, do not dull quickly, and therefore do not need to be released as readily. A hard-grade wheel can be used which will retain the grains for a longer period.

Hence the common saying in industry: 'A hard wheel for soft materials; a soft wheel for hard materials.'

Grades are designated from the softest to the hardest by letters A to Z.

10.4.8 Structure

The structure of an abrasive wheel is determined by the proportions and arrangement of the abrasive and bond. The grains are spaced to leave smaller or larger air spaces or pores between them.

Wheels where the spacing is wide, and large air spaces exist, are classified as having an 'open' structure. The large air spaces provide clearance for metal chips as they are removed from the work. Since larger chips are removed from the softer materials, open-structure wheels are used on such materials.

In grinding conditions where a large area of contact exists between wheel and work, a greater amount of heat is generated. This is often the case in surface and internal grinding. As the open-structure wheel has large air spaces, fewer grains are in contact and therefore less heat is generated. This type of wheel used in these conditions is said to have a free cool-cutting action.

Wheels where the spacing is closer and smaller air spaces exist are classified as having a 'dense' structure. When hard metals are being ground, small chips are produced which do not need a large clearance in the wheel. Dense-structure wheels are therefore used for these types of material.

Due to the close spacing, dense-structure wheels generate more heat, which can result in burning the surface of the metal being ground. This is recognisable by brown patches on the metal surface. If soft materials are ground with a dense-structure wheel, the clearance for metal chips is insufficient, the air spaces clog with metal, and the wheel is said to be 'loaded'.

Structure is designated by a series of numbers from 0 to 18, indicating the densest to the most open spacing.

10.4.9 Bond

There are two main types of bonding agent: inorganic and organic. Inorganic bonds are mainly vitrified, i.e. the wheel is generally fired in a furnace. Organic bonds are not fired, but are cured at a low temperature, commonly resinoid and rubber.

10.4.9.1 Vitrified

The majority of wheels have vitrified bonds. The abrasive grains are mixed in the correct proportions with clay and fusible materials. This mixture is then pressed in moulds to produce the correct wheel shape. The wheels are passed through drying rooms, to remove any moisture, before being fired in a kiln. During the firing process, the clay and fusible materials melt to form bonds between adjacent abrasive grains. On cooling and solidification, a 'glass-like' material is formed.

This type of bond is designated by the letter V. Vitrified wheels are porous, strong and unaffected by water, oils and ordinary temperature conditions. In general, surface speeds do not exceed 1950 m/min.

10.4.9.2 Resinoid

To produce resinoid-bond wheels, the abrasive grains are mixed with a thermosetting synthetic resin, moulded to shape, and cured. This bond is very hard and strong, and wheels can be run at surface speeds from 2850 to 4800 m/min. These high surface speeds give rapid metal removal, due to the greater number of abrasive particles cutting in any given time, and wheels with this bond are therefore suitable where high rates of metal removal are required and are ideal for work in foundries and steel mills.

Resinoid-bonded wheels can be made extremely thin, can be used with safety at high speeds, and are ideal as cut-off wheels for cutting metal bars, tubes, etc. This type of bond is designated by the letter B.

Where a higher strength bond is required, with a degree of flexibility, an open-weave fabric reinforcement is incorporated in the resin, designated by the letters BF.

10.4.9.3 Rubber

To produce rubber-bond wheels, the abrasive grains are mixed with rubber and vulcanising agents running between heated rolls. After rolling to thickness, the wheels are cut to the correct diameter and then vulcanised or 'cured' – undergoing a chemical reaction in which the molecules of rubber are interlinked using heat and pressure, usually with the aid of sulphur, to give a high degree of resilience.

Very thin wheels can be made by this process, because of the elasticity of the material, and are ideal for cut-off wheels, where they are run at high surface speeds between 3000 and 4800 m/min. Rubber-bond wheels are also used for control wheels in centreless-grinding machines.

This type of bond is designated by the letter R.

10.4.10 Characteristics

The characteristics already described are designated by up to eight symbols (three of which are optional) arranged in the following order:

0 – type of abrasive (optional), manufacturer's own symbol

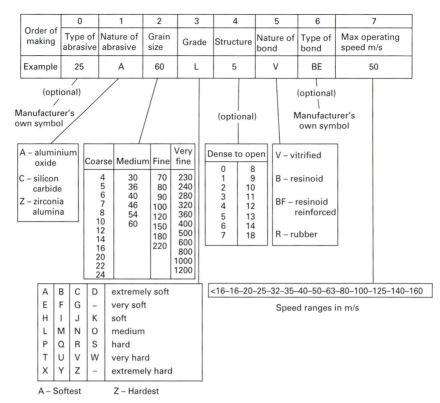

Order of making	0	1	2	3	4	5	6	7
	Type of abrasive	Nature of abrasive	Grain size	Grade	Structure	Nature of bond	Type of bond	Max operating speed m/s
Example	25	A	60	L	5	V	BE	50

(optional)

Manufacturer's own symbol

(optional)

(optional)

Manufacturer's own symbol

A – aluminium oxide
C – silicon carbide
Z – zirconia alumina

Coarse	Medium	Fine	Very fine
4	30	70	230
5	36	80	240
6	40	90	280
7	46	100	320
8	54	120	360
10	60	150	400
12		180	500
14		220	600
16			800
20			1000
22			1200
24			

Dense to open	
0	8
1	9
2	10
3	11
4	12
5	13
6	14
7	18

V – vitrified
B – resinoid
BF – resinoid reinforced
R – rubber

A	B	C	D	extremely soft
E	F	G	–	very soft
H	I	J	K	soft
L	M	N	O	medium
P	Q	R	S	hard
T	U	V	W	very hard
X	Y	Z	–	extremely hard

A – Softest   Z – Hardest

<16–16–20–25–32–35–40–50–63–80–100–125–140–160

Speed ranges in m/s

Figure 10.15 Standard symbols for marking of a grinding wheel

1 – nature of abrasive
2 – grain size
3 – grade
4 – structure (optional)
5 – nature of bond
6 – type of bond (optional), manufacturer's own symbol
7 – maximum permissable speed of rotation in metres per second (m/s).

The symbols are selected from the standard symbols set out in BS ISO 525 and shown in Fig. 10.15. These standard symbols are normally marked on the wheel by the manufacturer or, in the case of small wheels, on the box or package in which they are contained.

The example shows a specification 25A60L5VBE50 and without the optional symbols would be written as A60LV50.

Abrasive products shall be manufactured for maximum operating speeds (MOS) in the ranges shown in Fig. 10.15 where the speed is expressed in m/s.

10.4.11 Abrasive wheel marking system

The marking of abrasive wheels should be in accordance with BS EN 12413.

An independent international organisation known as the Organisation for the Safety of Abrasives (oSa) was set up by leading global producers of high-grade abrasive products to ensure that the safety and quality of their products remains consistently high worldwide. As well as conforming to BS EN 12413, the oSa label on abrasive products signifies the additional requirements of the oSa.

All markings must be:

▶ legible to the naked eye;
▶ indelible;
▶ marked on every abrasive wheel larger than 80 mm in diameter, or on a fixed blotter or on a fixed label.

The British Standard requires the following marking on bonded abrasives and is illustrated in Fig. 10.16:

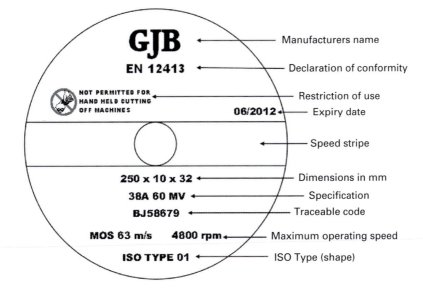

Figure 10.16 Marking on bonded abrasives

- Trade name: manufacturer's name.
- Declaration of conformity: in accordance with BS 12413 (and oSa logo for members).
- Dimensions: nominal dimensions in accordance with ISO 525.
- MOS: maximum permissible speed in revolutions per minute (rpm) and m/s. This speed must never be exceeded.
- Specification mark: in accordance with ISO 525. Minimum information required is abrasive type, grain size, grade and type of bond, e.g. 38A60MV (see Fig. 10.15).
- Restriction of use: restriction given as a full description or by pictogram. These are:
  - RE1 – not permitted for hand-held machine and manually guided grinding;
  - RE2 – not permitted for hand-held cutting-off machines;
  - RE3 – not suitable for wet grinding;
  - RE4 – only permitted for totally enclosed working area;
  - RE6 – not permitted for face grinding.
- Colour stripe: maximum permissible speeds of 50 m/s and above shown by a colour stripe as follows:
  - maximum 50 m/s – one blue stripe;
  - maximum 63 m/s – one yellow stripe;
  - maximum 80 m/s – one red stripe;
  - maximum 100 m/s – one green stripe;

- maximum 125 m/s – one blue and one yellow stripe.
- Traceability code: a code or batch number to enable the source and manufacturing details to be identified.
- ISO type: indicates the wheel shape in accordance with ISO 525.
- Date of expiry: organic bonds such as B and BF are subject to deterioration. The date of expiry shall be, at the longest, within 3 years of the date of manufacture. It is expressed as month and year, e.g. 06/2012. This does not apply to vitrified bonds which are not subject to deterioration.

10.4.12 Summary of cutting characteristics

A summary is shown in Table 10.1.

10.5 Surface-grinding operations

Surface grinding is used to produce flat accurate surfaces. This can be illustrated by considering the grinding of all surfaces of the component shown in Fig. 10.17.

It is important to first establish the datum faces A and B from which all faces are then ground.

Clamp face B against an angle plate supported on a parallel and grind face A to clean up, Fig. 10.18

155

Table 10.1 Summary of cutting characteristics

Grain Size	
Use large grain size	**Use small grain size**
For soft ductile materials, e.g. soft steel, aluminium	For hard brittle materials, e.g. hardened tool steels
For rapid stock removal, e.g. rough grinding	For small amounts of stock removal
Where surface finish is not important	Where a fine surface finish is required
For large areas of contact	Where small corner radii are required
	For small areas of contact
Grade	
Use soft grade	**Use hard grade**
For hard materials	For soft materials
For rapid stock removal	For increased wheel life
For large areas of contact	For small or narrow areas of contact
Structure	
Use open structure	**Use dense structure**
For soft materials	For hard materials
Large areas of contact	Small areas of contact

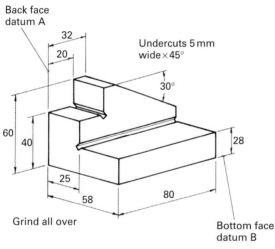

Figure 10.17 Workpiece

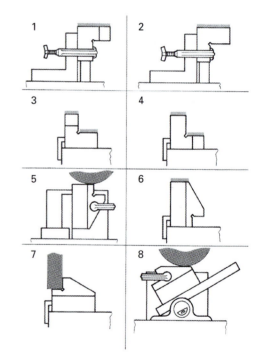

Figure 10.18 Sequence of operations

stage 1. Reclamp with face A against the angle plate, again supported on a parallel, and grind face B to clean up, Fig. 10.18 stage 2. This ensures that datums A and B are square to each other.

It is usual practice to fit a 'fence' to the rear of a magnetic chuck and to grind its face using the side of the grinding wheel. This face is then parallel to the table movement. The faces of a workpiece pushed against the fence can also be ground parallel using the side of the grinding wheel.

Place the workpiece on the magnetic chuck on face B, with face A against the 'fence'.

Grind the step face to 28 mm thickness, taking care to avoid hitting the adjacent face with the side of the grinding wheel.

Raise the wheelhead and grind the top face to the 60 mm dimension, Fig. 10.18 stage 3.

Reset the workpiece on the magnetic chuck on face A, with face B against the 'fence'. Grind the top face to the 58 mm dimension. Lower the wheelhead and grind the step face to the 25 mm dimension, taking care to avoid hitting the adjacent face with the side of the grinding wheel, Fig. 10.18 stage 4.

Clamp face A against an angle plate and set face B vertical, using a square.

Grind the end face to clean up, Fig. 10.18 stage 5.

Reset the workpiece on the magnetic chuck on the end face just ground.

Grind the opposite end face to the 80 mm dimension, Fig. 10.18 stage 6.

Reset the workpiece on the magnetic chuck on face B with the end face against the 'fence'.

Grind the step face to the 40 mm dimension and the adjacent face to the 20 mm dimension, using the side of the grinding wheel, Fig. 10.18 stage 7.

Clamp face A against the angle plate and tilted at 30°, using a protractor under face B.

Grind the 30° angle face, Fig. 10.18 stage 8, to achieve the 32 mm dimension.

10.6 Safety in the use of abrasive wheels

Most accidents happen from:

▶ contact with rotating abrasive wheel;
▶ impact injuries from bursting wheels (sometimes fatal);
▶ fires and explosions arising from poor control of grinding dust;
▶ eye injuries;
▶ health issues from metalworking fluids, noise and inhalation of harmful dust and fumes.

Remember that you have a duty under the various health and safety regulations already covered in Chapter 1. To avoid the risk of accident:

▶ Only use equipment you are authorised to use.
▶ Always follow the training and instruction provided by your employer.
▶ Always wear eye protection.
▶ Ensure abrasive wheel is correct for its intended use.
▶ Never exceed the maximum operating speed marked on the abrasive wheel.
▶ Always examine the abrasive wheel for damage or defects, and if found do not use.
▶ Never use a machine that is not in good working order.
▶ Always ensure guards are in position and in good working order.
▶ Always ensure wheel has stopped before leaving the machine unattended.
▶ Ensure that workpiece is secure and properly supported.
▶ Never stand directly in front of the wheel whenever the machine is started.
▶ Never start the machine with the workpiece in contact with the abrasive wheel.
▶ Avoid clogging and uneven wear. Dress frequently where appropriate.
▶ Always allow the abrasive wheel to stop naturally, not by applying pressure to its surface.
▶ Turn off coolant and allow excess to 'spin out' before stopping the machine.

Review questions

1. Name the two main essentials of a grinding wheel.
2. Why is it necessary to balance a grinding wheel?
3. For what types of operation would surface grinding be most suitable?
4. Explain the principle when a grinding wheel is said to act soft or hard.
5. What is the most basic method of workholding on a surface grinder?
6. Why is it essential that the correct guard be fitted on a surface grinder?
7. What effect does the structure of an abrasive wheel have on metal removal?
8. By means of a sketch show the principal of operation of a permanent magnet chuck.
9. What is the effect of grain size on metal removal?
10. Name two types of bonding material used in abrasive wheels.

CHAPTER 11

Milling

Milling is the machining of a surface using a cutter which has a number of teeth. The surface produced may be plain or, by using additional equipment or special cutters, formed surfaces may be produced.

There are many types and sizes of milling machines, but the most versatile in common use in the majority of workshops is the knee-and-column type, so called because the spindle is fixed in the column or main body and the table arrangement, mounted on a knee, is capable of movement in the longitudinal, transverse and vertical directions.

Knee-and-column machines are subdivided into the following models:

▶ plain horizontal, with the spindle located horizontally;
▶ universal, which is similar to the plain horizontal but equipped with a swivelling table for use when cutting helical grooves;
▶ vertical, with the spindle located vertically.

Typical plain horizontal and vertical knee-and-column milling machines are shown in Figs. 11.1 and 11.2.

The capacity of these machines is identified by the size of the working surfaces of the table, the length of travel of the longitudinal, transverse and vertical movements, and the maximum distance

Figure 11.1 Horizontal milling machine

from spindle to table surface on the horizontal model or from spindle to column on the vertical model.

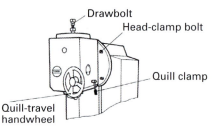

Drawbolt
Head-clamp bolt
Quill clamp
Quill-travel handwheel

Figure 11.4 Top of column of vertical milling machine

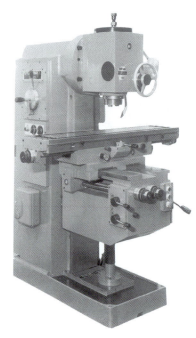

Figure 11.2 Vertical milling machine

11.1 Milling-machine elements

The main elements of a typical knee-and-column horizontal milling machine are shown in Fig. 11.3. The elements of a vertical machine are the same

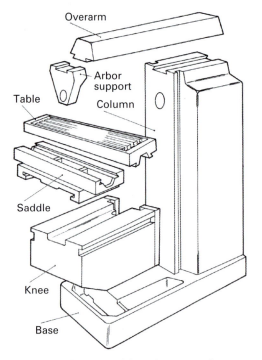

Overarm

Arbor support

Table

Column

Saddle

Knee

Base

Figure 11.3 Main machine elements of horizontal milling machine

except that the spindle head is mounted at the top of the column, as shown in Fig. 11.4.

11.1.1 Column and base

The column and base form the foundation of the complete machine. Both are made from cast iron, designed with thick sections to ensure complete rigidity and freedom from vibration. The base, upon which the column is mounted, is also the cutting-fluid reservoir and contains the pump to circulate the fluid to the cutting area.

The column contains the spindle, accurately located in precision bearings. The spindle is driven through a gearbox from a vee-belt drive from the electric motor housed at the base of the column. The gearbox enables a range of spindle speeds to be selected. In the model shown, 12 spindle speeds from 32 to 1400 rev/min are available. The front of the column carries the guideways upon which the knee is located and guided in a vertical direction.

11.1.2 Knee

The knee, mounted on the column guideways, provides the vertical movement of the table.

Power feed is available, through a gearbox mounted on the side, from a separate built-in motor, providing a range of 12 feed rates from 6 to 250 mm/min. Drive is through a leadscrew, whose bottom end is fixed to the machine base. Provision is made to raise and lower the knee by hand through a leadscrew and nut operated by a handwheel at the front. The knee has guideways on its top surface giving full-width support to the saddle and guiding it in a transverse direction.

A lock is provided to clamp the knee in any vertical position on the column.

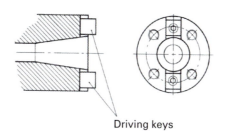

Figure 11.5 Standard milling-machine spindle nose

11.1.3 Saddle

The saddle, mounted on the knee guideways, provides the transverse movement of the table.

Power feed is provided through the gearbox on the knee. A range of 12 feeds is available, from 12 to 500 mm/min. Alternative hand movement is provided through a leadscrew and nut by a handwheel at the front of the knee.

Clamping of the saddle to the knee is achieved by two clamps on the side of the saddle.

The saddle has dovetail guideways on its upper surface, at right angles to the knee guideways, to provide a guide to the table in a longitudinal direction.

11.1.4 Table

The table provides the surface upon which all workpieces and workholding equipment are located and clamped. A series of tee slots is provided for this purpose. The dovetail guides on the undersurface locate in the guideways on the saddle, giving straight-line movement to the table in a longitudinal direction at right angles to the saddle movement.

Power feed is provided from the knee gearbox, through the saddle, to the table leadscrew. Alternative hand feed is provided by a handwheel at each end of the table. Stops at the front of the table can be set to disengage the longitudinal feed automatically in each direction.

11.1.5 Spindle

The spindle, accurately mounted in precision bearings, provides the drive for the milling cutters. Cutters can be mounted straight on the spindle nose or in cutter-holding devices which in turn are mounted in the spindle, held in position by a drawbolt passing through the hollow spindle. Spindles of milling machines have a standard spindle nose, shown in Fig. 11.5, to allow for easy interchange of cutters and cutter-holding devices. The bore of the nose is tapered to provide accurate location, the angle of taper being 16° 36'. The diameter of the taper depends on the size of the machine and may be 30, 40 or 50 IST (International Standard Taper). Due to their steepness of angle, these tapers – known as non-stick or self-releasing – cannot be relied upon to transmit the drive to the cutter or cutter-holding device. Two driving keys are provided to transmit the drive.

Cutters which are mounted directly on the spindle nose are located on a centring arbor, and four tapped holes are provided to hold the cutter in position. The two keys again provide the means of transmitting the drive.

The spindle of a horizontal machine is fixed and cannot be adjusted in an axial direction, i.e. along its axis. On vertical machines, provision is made for axial movement, which is controlled by a handwheel on the spindle head. The spindle runs in a quill which is moved through a rack and pinion in the same way as a drilling-machine spindle (see Fig. 8.2). A locking bolt is provided to lock the quill in any position along its operating length.

11.1.6 Overarm and arbor support

The majority of cutters used on horizontal machines are held on an arbor which is located and held in the spindle. Due to the length of the arbors used, support is required at the outer end to prevent deflection when cutting takes place. Support is provided by an arbor-support bracket, clamped to an overarm which is mounted on top of the column in a dovetail slide. The overarm is adjustable in or out for different lengths of arbor, or can be fully pushed in when arbor support is not required. Two clamping bolts are provided to lock the overarm in any position. The arbor support is located in the overarm dovetail and is locked by means of its clamping bolt. A solid bearing is provided in which the arbor runs during spindle rotation.

11

11.1.7 Guards

As well as providing individual guards for cutters (as shown in Fig. 1.4), machine guards are fitted which enclose the working area, giving protection to the operator from swarf and coolant as shown in Fig. 11.6. These are capable of sliding and swinging aside to provide access in order to load/unload the workpiece, change cutters or remove swarf. The high-impact polycarbonate panels provide all-round visibility. Electrical safety interlocks are available which prevent the machine being started if the guard is in the open position and stop the machine should the guard be opened when the machine is running.

Figure 11.6   Slide and swing-aside guard

11.2 Controls

The various controls of a typical horizontal milling machine are shown in Fig. 11.7. These are identical to those of a vertical machine.

Spindle speeds are selected through the levers (4), and the speed is indicated on the change dial (5). The speeds must not be changed while the machine is running. An 'inching' button (3) is situated below the gear-change panel and, if depressed, 'inches' the spindle and enables the gears to slide into place when a speed change is being carried out. Alongside the 'inching' button is the switch for controlling the cutting-fluid pump (1) and one for controlling the direction of spindle rotation (2). The feed rates are selected by the lever (9) and are indicated on the feed-rate dial.

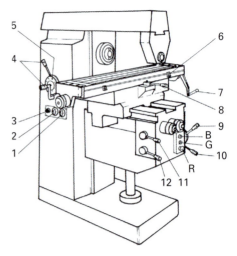

Figure 11.7   Milling-machine controls

To engage the longitudinal table feed, lever (8) is moved in the required direction – right for right feed, left for left feed. Adjustable trip dogs (6) are provided to disengage the feed movement at any point within the traverse range. Limit stops are incorporated to disengage all feed movements in the extreme position, to prevent damage to the machine in the event of a trip dog being missed.

To engage cross or vertical traverse, lever (12) is moved up or down. The feed can then be engaged by moving lever (11) in the required direction. With cross traverse selected, movement of lever (11) upwards produces in-feed of the saddle, moving it downwards produces out-feed of the saddle. With vertical traverse selected, movement of lever (11) upwards produces up-feed to the knee, moving it downwards produces down-feed to the knee.

Rapid traverse in any of the above feed directions is engaged by an upward pull of lever (10). Rapid traverse continues as long as upward pressure is applied. When released, the lever will drop into the disengaged position. Alternative hand feed is provided by means of a single crank handle (7), which is engaged by slight pressure towards the machine. Spring ejectors disengage the handle on completion of the operation, for safety purposes – i.e. the handle will not fly round when feed or rapid traverse is engaged. The single crank handle is interchangeable on table, saddle and knee movements.

11.2.1 Starting and stopping the machine

The switch panel, situated on the front of the knee, contains a black button (B) to start the feed motor, Fig. 11.7. This is provided to facilitate setting up when feed movements are required without spindle rotation.

The green button (G) starts the spindle and feed motors, while the mushroom-headed red button (R) provides the means of stopping the machine.

11.3 Milling cutters

There are many different types of milling cutters available, and for convenience they can be classified according to the method of mounting: those with a central hole for mounting on an arbor, those with a screwed shank for holding in a special chuck and the large facing cutters which mount directly on to the spindle nose.

11.3.1 Arbor-mounted types

11.3.1.1 Cylindrical cutter

This cutter has teeth on the periphery only, and is used to produce flat surfaces parallel to the axis of the cutter, Fig. 11.8(a). The teeth are helical, enabling each tooth to take a cut gradually, reducing shock and minimising chatter. Cylindrical cutters are made in a variety of diameters and lengths up to 160 mm diameter × 160 mm long.

11.3.1.2 Side-and-face cutter

This cutter has teeth on the periphery or face and on both sides. It is used to produce steps, cutting on the face and side simultaneously, Fig. 11.8(b), or for producing slots. The use of these cutters in pairs with their sides cutting is known as straddle milling, Fig. 11.8(c). The teeth are straight on cutters up to 20 mm wide but are helical above this thickness. Side-and-face cutters are available in a variety of sizes up to 200 mm diameter and 32 mm wide.

A staggered-tooth side-and-face cutter, also having teeth on the periphery and on both sides, is designed for deep-slotting operations. In order to reduce chatter and provide maximum chip clearance, the teeth are alternately right-hand and left-hand helix, and each alternate side tooth is removed, Fig. 11.8(d). Staggered-tooth cutters are available in the same sizes as the plain side-and-face type.

11.3.1.3 Angle-milling cutter

Angle cutters are available, made with single angle or double angle with teeth on the angled surfaces. A single-angle cutter also has teeth on the flat side. They are used on angle faces or for producing a chamfer on the edge of the workpiece, Fig. 11.8(e) and (f).

Single-angle cutters are available with angles of 60° to 85° in 5° steps, and the double-angle cutters with 45°, 60° and 90° included angle.

11.3.1.4 Single corner-rounding cutter

This cutter has a concave quarter circle on one side and is used to produce a corner radius on the edge of the workpiece, Fig. 11.8(g). Cutters are available with a variety of corner radii from 1.5 mm to 20 mm.

11.3.1.5 Shell end-milling cutter

More often referred to as a 'shell end mill', Fig. 11.8(h), this cutter has teeth cut on the circumference and on one end. The tooth end is recessed to receive a screwhead for holding the cutter on an arbor. A key slot on the back face provides the drive from two keys in the arbor. The teeth are helical and the cutter is used for work of a larger size than can be efficiently handled by the ordinary end mill. A range of sizes is available from 40 mm to 160 mm diameter.

11.3.2 Screwed-shank types

11.3.2.1 End-milling cutter

This cutter has helical teeth on the circumference and teeth on one end and is used for light operations such as milling slots, profiling and facing narrow surfaces, Fig. 11.8(i). The end teeth on non-centre-cutting cutters are not cut to the centre, so this cutter cannot be fed in a direction along its own axis. Centre-cutting types have teeth cut to the centre which allows drilling and plunging operations.

11

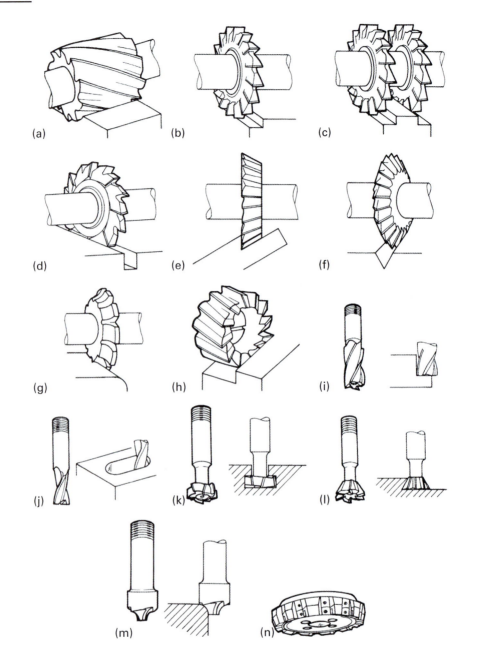

Figure 11.8 Milling cutters (a) cylindrical cutter (b) side-and-face cutter (c) straddle milling (d) staggered-tooth side-and-face cutter (e) and (f) angle-milling cutters (g) single corner-rounding cutter (h) shell end-milling cutter (i) end-milling cutter (j) slot drill (k) tee-slot cutter (l) dovetail cutter (m) corner-rounding cutter (n) face-milling cutter

Although the majority of end-milling cutters are manufactured from HSS and HSCo, some with hard surface coatings, solid carbide cutters are available, also with surface coatings which include titanium aluminium nitride (TiAlN) to further enhance their cutting performance.

11.3.2.2 Slot drill

Usually having two or three helical teeth cut in the circumference and teeth on the end, cut to the centre, this cutter can be fed along its own axis in the same way as a drill. It is used to produce keyways and blind slots with the cutter sunk into the material like a drill and fed longitudinally the length of the keyway or slot, cutting on its circumference, Fig. 11.8(j). It is available in a variety of sizes up to 50 mm diameter.

11.3.2.3 Tee-slot cutter

Designed for milling tee slots in machine tables, this cutter has teeth on its circumference and on both sides. To reduce chatter and provide maximum chip clearance, the teeth are alternately right-hand and left-hand helix, and each alternate side tooth is removed. To produce a tee slot, the groove is first cut using a side-and-face cutter, end mill or slot drill and finally the wide slot at the bottom is cut using a tee-slot cutter, Fig. 11.8(k). The shank is reduced to clear the initial groove. This cutter is available for standard tee slots to suit bolt sizes up to 24 mm.

11.3.2.4 Dovetail cutter

Designed for milling dovetail slides of machines, this cutter has teeth on its angle face and on the end face. To produce a dovetail slide, a step is machined to the correct depth and width; the angle is then finally machined using the dovetail cutter, Fig. 11.8(l). It is available in a variety of sizes up to 38 mm diameter, with 45° and 60° angles.

11.3.2.5 Corner-rounding cutter

Designed to produce a radius along the edge of the workpiece, this cutter has a quarter circle cut in the outer edge, Fig. 11.8(m). It is available in a variety of sizes with corresponding radii up to a maximum radius of 12 mm.

11.3.3 Direct-mounted types

11.3.3.1 Face-milling cutter

More usually referred to as a 'face mill', this cutter is used to face large surfaces. The cutter consists of a tough steel body with high-speed-steel or tungsten-carbide cutting edges in the form of inserts clamped in their correct position, Fig. 11.8(n). This construction results in a cheaper cutter than would be the case if the complete cutter were made from an expensive cutting-tool material – it also facilitates the replacement

of one insert in the event of damage to a single cutting edge. It is available in a variety of sizes from 100 mm to 450 mm diameter, the larger sizes being used only on the biggest machines, since the power requirement for such a cutter may be as high as 75 kW.

11.4 Cutter mounting

11.4.1 Arbor-mounted cutters

11.4.1.1 Standard arbor

Milling cutters having a hole through the centre are mounted on an arbor. The standard arbor used in horizontal milling machines is shown in Fig. 11.9. One end has an international taper to suit the machine spindle, for location. A threaded hole in the end provides the means of holding the arbor in position, by means of a drawbolt through the machine spindle. The flange contains two key slots to provide the drive from two keys on the spindle nose. The long diameter is a standard size, to suit the hole size of the cutter, and the thread carries the arbor nut to clamp the cutter. A keyway is cut along the length of this diameter into which a key is fitted, to provide a drive and prevent the cutter slipping when taking heavy cuts. To position the cutter along the length of the arbor, spacing collars are used. These are available in a variety of lengths, with the ends ground flat and parallel. Towards the end, a larger bush is positioned. This has an outside diameter to suit the bearing of the arbor support and is known as the 'running bush'.

To mount the arbor, the taper is inserted in the machine spindle, ensuring that the surfaces are free of all dirt and metal cuttings. The flange key slots are located in the spindle keys, and the arbor

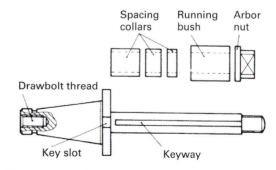

Figure 11.9 Standard milling-machine arbor

is securely held by the drawbolt. Spacing collars are slipped on the arbor, again ensuring that all faces are clean and free from dirt and metal cuttings. The cutter is positioned and spacing collars are added, together with the running bush, to make up the length of the arbor. The arbor nut is then screwed in position – hand-tight only.

The arbor support is now positioned on the overarm so that it is central on the running bush and is then clamped in position. The arbor nut can now be tightened with the appropriate spanner. Never tighten the arbor nut without the arbor support in position, as the arbor can be bent.

To prevent deflection of the arbor during heavy cutting operations, it is sometimes necessary to mount a second arbor support nearer the spindle nose. The cutter is then positioned between the two supports.

11.4.1.2 *Stub arbor*

Cutters which are used close to the spindle, such as shell end mills, are mounted on a stub arbor, Fig. 11.10. This arbor is located, held, and driven in the spindle in the same way as a standard arbor. The cutter is located on a spigot or stub and is held in position by a large flanged screw. Two keys on the arbor provide the drive through key slots in the back face of the cutter.

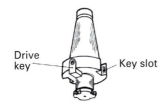

Figure 11.10   Milling-machine stub arbor

11.4.2 Screwed-shank cutters

Cutters having screwed shanks are mounted in a special chuck, shown in Fig. 11.11. The collet, which is split along the length of its front end and has a short taper at the front, is internally threaded at its rear end. Collets of different sizes are available to suit the shank diameter of the cutter used. The collet is inserted into the locking sleeve and the assembly is screwed into the chuck body until the flange almost meets the end face of the body.

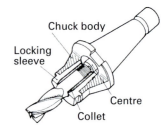

Figure 11.11   Milling chuck for screwed-shank cutters

The cutter is inserted and screwed into the collet until it locates on the centre inside the chuck body and becomes tight. The centre anchors the end of the cutter and ensures rigidity and true running. A spanner is used to give the locking sleeve a final tighten.

The cutter cannot push in or pull out during the cutting operation. Any tendency of the cutter to turn during cutting tightens the collet still further and increases its grip on the cutter shank. This type of collet chuck is located, held and driven in the machine spindle in the same way as the previously mentioned types.

11.4.3 Direct-mounted cutters

Large face mills are mounted directly on the spindle nose. To ensure correct location and concentricity, a centring arbor with the appropriate international taper is held in the spindle by the drawbar. The diameter on the end of the centring arbor locates the cutter, which is driven by the spindle keys through a key slot in the back face of the cutter. The cutter is held in position by four screws direct into the spindle nose, Fig. 11.12.

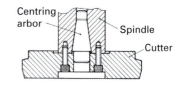

Figure 11.12   Direct-mounted cutter

11.5 Workholding

The simplest method of holding a workpiece for milling is to clamp it directly to the worktable. Adequate tee slots are provided for this purpose. Care should be taken to avoid machining the table – if necessary, the workpiece should be raised on a pair of parallels. Clamping should be carried out in the manner already described in the chapter on drilling (Section 8.3).

11.5.1 Vice

A vice is the most versatile piece of equipment for holding workpieces. It must be positioned to ensure accurate alignment with machine movements.

11.5.2 Rotary table

A rotary table, Fig. 11.13, is used where part of the surface being machined is of a circular nature. In this case, the table is moved to bring the workpiece in the correct position under the cutter. The workpiece is then rotated past the cutter to produce a circular profile.

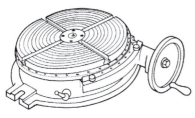

Figure 11.13 Rotary table

The rotary table consists of a base with lugs to clamp to the machine table, inside which a circular table is rotated by means of a handwheel at the front. A dial round the periphery of the table is graduated, usually in degrees. Some models have a vernier scale fitted to give more accurate readings, in some cases as small as 1 minute of arc.

The circular table is provided with tee slots, to enable clamping of the workpiece. A central hole enables setting of the workpiece about the centre of rotation and enables the rotary table to be set central with the machine spindle. Concentric circles on the table surface are provided to aid the initial setting of the workpiece roughly central.

The rotary table can also be used to hold a workpiece requiring a series of holes spaced on a pitch circle diameter (pcd) as shown in Fig. 11.14.

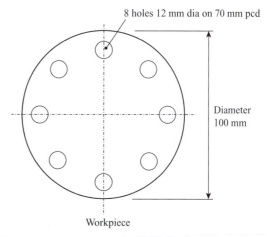

8 holes 12 mm dia on 70 mm pcd

Diameter 100 mm

Workpiece

Figure 11.14 Workpiece

11.6 The dividing head

Where slots, grooves or teeth are required to be spaced round the circumference of a cylinder or disk, as in the case of teeth on gears or flutes on a reamer, the dividing head is employed. Applied in a vertical position it can be used to drill holes on a pcd in a similar fashion to a rotary table. A high degree of accurate spacing is achieved using the dividing head. The dividing head, which is used to hold the workpiece, is clamped to the table of a milling machine using the central tee slot for alignment. If necessary, a tailstock to support the other end of the workpiece, also located in the central tee slot, is clamped at the other end of the machine table.

The head consists of a spindle which has a 40-tooth worm wheel attached to it. Meshing with this is a single start worm on a spindle which has a crank and handle attached and which projects from the front of the head (Fig. 11.15). An index plate containing a number of circles of different-spaced holes is attached to the front of the head. The spring-loaded pin attached to the crank locates in the required hole circle. The head spindle has a central hole to accept a morse taper centre while the outside is threaded to take a three- or four-jaw chuck (Fig. 11.16).

Since the gear ratio is 40:1, 40 turns of the crank will turn the spindle (and the workpiece attached to it) through one complete revolution. Alternatively, one turn of the crank will cause the spindle to rotate 1/40 of a revolution or 9°.

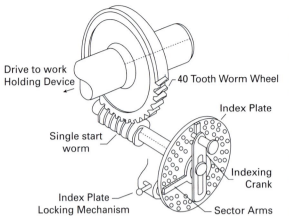

Figure 11.15 Worm and worm wheel

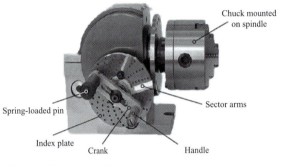

Figure 11.16 Dividing head

The purpose of the index plate is to further subdivide one revolution of the crank. The index plate contains a number of circles of holes, each containing a different number. The crank can be adjusted for radius so that the spring-loaded pin will fit in any circle of holes. Two sector arms on the index plate can be adjusted to reveal the number of holes required (Fig. 11.16). The number of subdivisions which can be achieved depends on the number of holes in each index plate supplied by the manufacturer.

For example, a manufacturer may supply three index plates containing circles with the following number of holes:

Plate 1 – 15, 16, 17, 18, 19 and 20 holes.
Plate 2 – 21, 23, 27, 29, 31 and 33 holes.
Plate 3 – 37, 39, 41, 43, 47 and 49 holes.

Since 40 turns of the crank cause 1 turn of the workpiece and we require x equal divisions, each division will require $1/x$ of its circumference and the crank will have to turn $40/x$.

Example: A workpiece is required to have 12 divisions.

Indexing $= \frac{40}{12} = 3\frac{4}{12} = 3\frac{1}{3} = 3\frac{5}{15}$, i.e. 3 complete turns of the crank and 5 holes in a 15-hole circle (or 11 holes in a 33-hole circle).

Similarly for 36 divisions, indexing $= \frac{40}{36} = 1\frac{4}{36} = 1\frac{1}{9} = 1\frac{3}{27}$, i.e. 1 complete turn and 3 holes in a 27-hole circle (or 2 holes in an 18-hole circle).

It may be necessary to cut a number of grooves or slots at a given angle to each other. The principle is the same since one turn of the crank is 1/40; then 360/40 will give us 9°. The crank turns required will therefore be the angle required divided by 9.

Example: A number of slots are required 38° apart.

Indexing $= \frac{38}{9} = 4\frac{2}{9} = 4\frac{6}{27}$, i.e. 4 complete turns of the crank and 6 holes in a 27-hole circle.

Having set the crank radius to the required hole circle, the sector arms need to be set to uncover the required number of holes. In the last example we need to uncover six holes in the 27-hole circle. The sector arms are adjusted to reveal six holes. Once the four complete turns of the crank have been completed, you then continue with the additional six holes revealed between the sector arms and allow the spring-loaded pin to enter the hole. The sector arms are then rotated against the spring-loaded pin, thus revealing the next six holes. This is repeated until the operation is complete.

11.7 Milling operations

Consider machining the workpiece shown in Fig. 11.17, using in the first instance a horizontal milling machine and second producing the same workpiece using a vertical milling machine. The two essential requirements are: opposite faces parallel to each other and square with their adjacent faces, and as many operations as possible done at a single setting. The most convenient method of holding such a workpiece is to grip it in a machine vice.

The machine vice is set relative to the machine movements using a parallel gripped in the jaws and is checked by means of a dial indicator attached to a fixed part of the machine. The machine vice, having been satisfactorily set up, is securely clamped to the machine table.

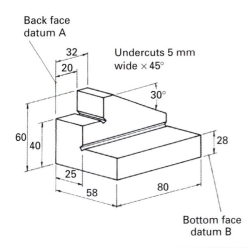

Back face
datum A

32

20

Undercuts 5 mm
wide × 45°

30°

60

40

28

25

58

80

Bottom face
datum B

Figure 11.17 Workpiece to be machined

11.7.1 Operations on a horizontal machine

Set the block in the vice on parallels and ensure that at least 32 mm is protruding above the vice jaws. Tighten the vice, ensuring that the workpiece is seated on the parallels. May require hitting with a soft faced mallet.

Machine face C to clean up, using a cylindrical cutter, Fig. 11.18 stage 1.

Release the workpiece and reset with face C against the fixed jaw. Tighten the vice, ensuring that the workpiece is seated on the parallels.

Machine face B to clean up, Fig. 11.18 stage 2. This ensures squareness of faces B and C.

Release the workpiece and reset with face B against the fixed jaw and face C seated on the parallels.

Machine face A to achieve the 58 mm dimension, Fig. 11.18 stage 3. This ensures parallelism of faces A and C and squareness with face B.

Release the workpiece and reset with face A against the fixed jaw and face B seated on the parallels, i.e. the workpiece is located with its datum faces against the fixed jaw and on the parallels.

Machine face D to achieve the 60 mm dimension, Fig. 11.18 stage 4.

Release the workpiece and tilt in the vice at 30° with the aid of a protractor. Machine the angle, leaving the 32 mm dimension approximately 1 mm

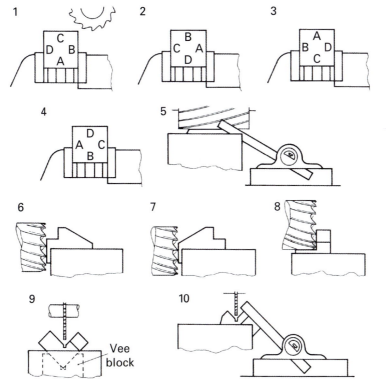

Figure 11.18 Sequence of operations on horizontal milling machine

169

too long, to allow for machining the end face, Fig. 11.18 stage 5.

Release the workpiece and reset with the end face protruding beyond the edge of the vice jaws, again seated on parallels. Mount a shell end mill on a stub arbor and load into the machine spindle.

Machine the end face to produce the 32 mm dimension to the start of the angle. At the same setting, machine the step to 20 mm and 40 mm dimensions, Fig. 11.18 stage 6.

Release the workpiece and reverse it, with the second end protruding beyond the edge of the vice jaws.

Machine to 80 mm length, Fig. 11.18 stage 7.

Release the workpiece and reset gripping on the 80 mm length.

Machine the step along the length to achieve the 25 mm and 28 mm dimensions, Fig. 11.18 stage 8.

Since the cutter on a horizontal machine cannot be inclined, milling the undercuts can be done only by tilting the workpiece.

Depending on the size and shape of the workpiece, setting can be simplified using a vee block. The vee block is set in the vice with the workpiece resting in the vee. The workpiece is then gripped in the vice across its ends, Fig. 11.18 stage 9.

Alternatively, a protractor is used to set the workpiece at 45° Fig. 11.18 stage 10. These two final stages are carried out using a 5 mm side-and-face cutter mounted on a standard arbor.

11.7.2 Operations on a vertical machine

The stages of machining the same workpiece using a vertical machine are shown in Fig. 11.19. Although different cutters are used, the same basic principles apply.

Stages 1, 2 and 3 employ the same set-up as for a horizontal machine, except that a shell end mill is used.

At stage 4, the step can be conveniently produced at the same setting and using the same cutter as for face D.

Stage 5 shows the angle being machined. The head of the vertical machine has been swivelled,

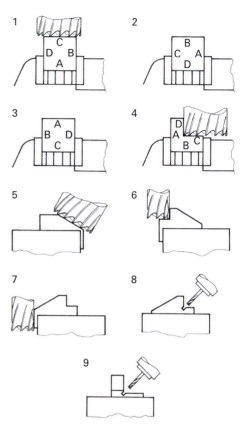

Figure 11.19 Sequence of operations on vertical milling machine

so allowing the workpiece to be held in the vice seated on the parallels. This is a more convenient method if more than one workpiece is being machined. Each workpiece can be held in the vice in the normal manner, making it easier and quicker to set, with an assurance that the angle is identical on each item.

Stages 6 and 7 are set as before, but using an end mill to give a cutting edge long enough to machine the height of the workpiece and of large enough diameter to machine the step.

Machining the undercut at stages 8 and 9 again makes use of the swivelling head, with a 5 mm diameter end mill. Alternatively, with the head vertical the workpiece can be held as for horizontal milling stages 9 and 10.

11.7.3 Use of a rotary table

With the rotary table clamped to the worktable, the first essential is to bring the centre of the rotary table in line with the centre of the machine spindle. This is done by inserting a plug of the correct diameter into the hole in the centre of the rotary table.

Attach a dial indicator to the spindle nose and disengage the main gearbox to allow free rotation of the spindle. You may require a spanner on the drawbolt in order to rotate the spindle. Move the table and saddle traverses by hand to bring the plug roughly central with the machine spindle. Rotate the spindle, and dial indicator attached to it, round the plug and adjust the table and saddle movements until a constant reading is obtained, Fig. 11.20. When this happens, the plug is central about the spindle centre line. Set the micrometer dial on each traverse movement to zero.

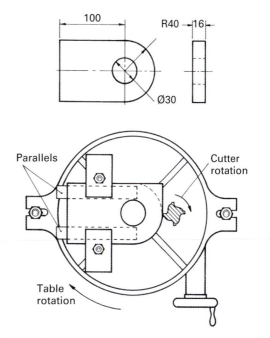

Figure 11.21 Workpiece set on rotary table

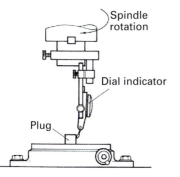

Figure 11.20 Centring a rotary table using a dial indicator

Having centred the rotary table, consider the workpiece shown in Fig. 11.21, which requires the 40 mm radius to be machined using an end-milling cutter in a vertical milling machine.

A bung is produced with one end to suit the hole in the rotary table and the other to suit the diameter of hole in the workpiece. Locate the bung in the rotary table and the workpiece on the bung. The workpiece is set on a pair of thin parallels, to raise the workpiece and avoid machining the table surface, Fig. 11.21.

Clamp the workpiece in position and mount the required size of end-mill cutter in the machine.

Lock the saddle movement in the already established central position. Move the machine table a distance equal to the radius to be machined plus half the cutter diameter. This distance is moved accurately by referring to the micrometer dial on the machine-table traverse.

Start the machine spindle, having selected the speed appropriate to the cutter size, and raise the knee to take a cut. Rotate the workpiece past the cutter until the complete radius is machined.

Raise the knee to take a further cut and repeat until all the surface is produced at the correct radius. Since a vertical milling machine is used, an alternative method of taking a cut can be used by lowering the spindle.

In order to drill the eight holes shown in Fig. 11.14, setting up the rotary table is exactly the same, and then the table is moved a distance equal to half the pcd, i.e. 35 mm. The graduations on the rotary table are set to zero and the first hole drilled. The rotary table is then rotated 45° (360/8) and the second hole drilled. This is repeated until all the holes are drilled, resulting in eight equi-spaced holes on a 70 mm pcd.

11.8 Safety in the use of milling machines

Most accidents happen from:
▶ Entanglement and contact with rotating cutters when:
  ▶ loading/unloading workpieces;
  ▶ removing swarf;

- ▶ measuring;
- ▶ adjusting coolant flow.

These cause entanglement injuries such as: broken bones, dislocations, lacerations and amputations.

- ▶ Eye injuries and cuts from machine cleaning and swarf removal.
- ▶ Health issues from metalworking fluids, noise and handling large workpieces.

Remember that you have a duty under the various health and safety regulations already covered in Chapter 1. To avoid the risk of accident:

- ▶ Always follow the training provided by your employer.
- ▶ Ensure guard is in position to avoid contact with cutter, swarf and coolant.
- ▶ Always wear eye protection and any other PPE required.
- ▶ Take great care before engaging automatic feeds and ensure table traversing handle is disengaged.

- ▶ Ensure workpiece and workholding equipment (e.g. vice) is securely clamped.
- ▶ Make sure safety limit stops are present and in the correct position.
- ▶ Always isolate the machine when changing the chuck, arbor or cutter.
- ▶ Do handle milling cutters carefully, as the cutting edges are very sharp.
- ▶ Always ensure cutter is rotating in the correct direction.
- ▶ Never use damaged cutters (e.g. chipped or broken teeth).
- ▶ Always ensure machine is fully stopped before taking a measurement.
- ▶ Do not attempt to adjust the coolant supply or remove swarf when the machine is running.
- ▶ Do not wear jewellery or loose clothing.
- ▶ Always have long hair tied back or in a hairnet.

Review questions

1. Name two types of work holding equipment used with milling operations.
2. What is the purpose of the 'running bush' used on the arbor of a horizontal milling machine?
3. State the characteristics used to identify the capacities of a milling machine.
4. Name two types of milling machine.
5. Describe how a milling cutter is directly mounted on the spindle of a vertical milling machine.
6. Name three types of cutter used on a horizontal milling machine and give a use for each.
7. What is the name of the taper used in a milling machine spindle?
8. Name three types of cutter used on a vertical milling machine and give a use for each.
9. What is the purpose of spacing collars on a horizontal milling machine arbor?
10. With the aid of a sketch show how a screwed shank cutter is held in a milling chuck.
11. A workpiece is required to have (a) 14 divisions and (b) a number of slots 23° apart. Calculate the dividing head settings if the supplied index plate has 21, 23, 27, 29, 31 and 33 holes. (Answer: (a) 2 turns and 18 holes in a 21-hole circle and (b) 2 turns and 15 holes in a 27-hole circle.)

CHAPTER 12

Introduction to computer numerical control

Computer numerical control or CNC refers to any machine tool, e.g. drilling machine, milling machine or lathe, which uses a computer to electronically control the motion of one or more axes on the machine through the use of coded instructions.

Simply put, instead of turning handwheels as is required by conventional machines the CNC machine control unit (MCU) sends a motion signal via a controller board to a servo motor attached to each machine axis. This causes the servo motor to rotate a ballscrew attached to the table, cross-slide or column, causing it to move. The actual position of the axis is continuously monitored and compared to the commanded position with feedback from a transmitter attached to the ballscrew. The ballscrews have almost no backlash, so when the servo reverses direction there is almost no lag between a commanded reversing motion and corresponding change in slide direction.

This chapter is intended to give an insight into CNC machining, i.e. drilling, milling and turning, although CNC is used in many industries in a wide range of applications including: grinding, plasma, laser, foam and waterjet cutting, tube benders, routers, turret press and punching machines, and electro discharge machining (EDM), as well as co-ordinate measuring machines and industrial robots.

The use of CNC machine tools has radically changed the manufacturing industry. Curves are as easy to machine as straight lines, as well as parts with complicated contours and complex 3D shapes. The increased automation has resulted in improvements in consistency and quality together with a reduced frequency of errors. CNC also allows for more flexibility in the way workpieces are held, often avoiding the use of expensive jigs or fixtures required on conventional machines. Dimensional modifications can be done quickly and the time required to change over to a different part is reduced. Small batches or short production runs as well as parts needed in a hurry can be easily accommodated.

All CNC machine tools have two or more programmable directions of motion called axes. An axis of motion can be linear, i.e. in a straight line, or rotary, i.e. along a circular path. The more axes a machine system is capable of simultaneously controlling, the more complex the machine tool.

A body in space has six degrees of freedom. Motion can therefore be resolved into six axes, namely three linear referred to as X, Y and Z and three corresponding rotational axes referred to as A, B and C as shown in Fig. 12.1.

The relationship of the axis designation of a milling type machine is shown in Fig. 12.2. A basic milling machine has X, Y, and Z axes. A basic lathe has

Workshop Processes, Practices and Materials, Fifth Edition. 9781138784727.

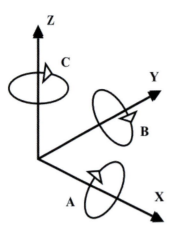

Figure 12.1 Degrees of freedom

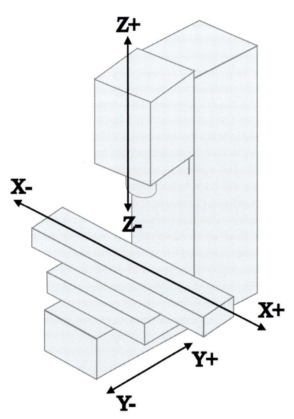

Figure 12.2 Milling machine axes

a point-to-point control system. With this system the tool is moved to a position over the workpiece in X and Y and then performs the operation at that point only, along the Z axis. The tool is not in contact with the workpiece while it is moving, e.g. moving the tool to a position over the workpiece and then drilling a hole.

Other systems can control the X and Y axes simultaneously and this is known as a continuous path control system. This system causes the tool to move from one position to another while the tool is in continuous contact with the workpiece, e.g. milling a slot or pocket.

A machine tool with a system capable of controlling X, Y and Z axes simultaneously is referred to as a three-axis machine. Similarly four- and five-axis machine tools are available which incorporate the fourth and fifth axes of rotation about one or two linear axes, e.g. a four-axis machine can move simultaneously in X, Y and Z directions and rotate the workpiece around the X or Y axis (i.e. the A or B axis).

Although basic machine tools are available, the most common machine tools used in metal cutting industries are vertical or horizontal machining centres and turning centres. These machines are highly sophisticated and can have a number of cutting tools in a tool magazine, each of which can be automatically selected as and when required. An example of a CNC vertical machining centre is shown in Fig. 12.3.

For any CNC machine to operate, it must be provided with information regarding the operations it will be required to perform. This information is provided by a part program, which is a series of step-by-step instructions, written in sequence, in a format which the machine control system will read, interpret and execute. The control system will first read, interpret and execute the very first command in the program. Only then will it go to the next command then read, interpret and execute. The control will then continue to execute the program commands in sequential order until it reaches the end of program.

Everything that an operator would be required to do on a conventional machine can be programmed on a CNC machine. For example, if you had to drill a hole on a manual drilling machine you would

two axes X and Z, the third axis obtained through rotation of the workpiece held in the chuck and is shown in Fig. 12.8.

Not all control systems are capable of operating a number of axes simultaneously. The most basic machine control system will only move the axes on a linear path on X and Y and then operate the Z axis separately and is referred to as

Figure 12.3 CNC vertical machining centre

have to load the workpiece, select and load the correct size drill, select the correct spindle speed, position the workpiece, switch on the spindle, manually operate the lever at an appropriate downward feed rate to drill the hole, withdraw, and finally switch off the machine. The CNC part program is written to replicate this same step-by-step approach in a more automated fashion.

In order to write a part program, the programmer must have a good knowledge of the machining operations required and the order in which they are to be carried out as well as the tooling required together with the necessary speeds and feeds and whether or not coolant is used.

Machine control systems are equipped with a verification system which enables a 'dry run' to be made to confirm the correctness of the part program as well as an editing facility so that any mistakes can be rectified.

For simple applications a part program can be developed manually. As applications have become more complicated, a lot of effort has been devoted to automate part programming. CAD/CAM

systems (computer aided design/computer aided manufacture) have been developed incorporating interactive graphics which can be integrated with the CNC part programming. Graphic-based software using menu-driven techniques improves user friendliness. The part programmer can create the geometry in the CAM package or directly extract geometry from a CAD/CAM database. Built-in tool motion commands can assist the programmer by automatically calculating the tool paths. The part programmer can then verify the tool paths through the graphic display using the animation function of the CAM system. This greatly enhances the speed and accuracy of tool path generation, avoiding any potential tool collisions and enables any required editing to be easily carried out.

12.1 Manual part programming for milling/drilling

As already stated, the part program is a sequence of instructions which describes all the machining data necessary to be carried out on the workpiece, in a format required by the machine control system. The machining data will include:

- ▶ machining sequence;
- ▶ cutting tools required;
- ▶ machine home, part zero, tool change positions;
- ▶ co-ordinate positional values along X, Y and Z axes;
- ▶ cutting depths, tool paths, tool offsets;
- ▶ cutting parameters – spindle speed and direction of rotation, feed rates, coolant.

12.1.1 Co-ordinate systems

Drawing co-ordinates can be specified in one of two ways:

- ▶ cartesian or rectangular co-ordinates – where the dimensions are taken relative to datums at right angles to each other (see Fig. 3.2);
- ▶ polar co-ordinates – where the dimension is measured along a radial line from a datum (see Fig. 3.1).

The most popular of these, used with CNC machines, is the rectangular co-ordinate system which relates directly to the axis motions at

12

right angles to each other, although systems are capable of dealing with both.

The relationship of the machine motions can be shown as a graph in Fig. 12.4.

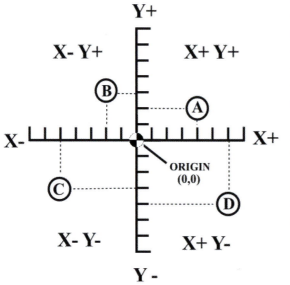

Figure 12.4 Graph of X, Y axes

Imagine you are looking from above onto the table and cross-slide of a vertical milling machine. The machine table is represented by the horizontal X axis, which can move in a positive or negative direction from zero, its origin. Similarly the cross-slide is represented by the vertical Y axis, which can also move in a positive or negative direction from zero. For CNC purposes, the origin point on the graph is commonly called the part origin point or part zero. This part zero establishes the point of reference for motion commands in a CNC part program and thus all movements are programmed from a common location. If the part zero is chosen wisely, the co-ordinates required for the part program can be taken directly from the part drawing and can also result in X and Y movements in a positive direction.

For example, if our graph origin is considered to be the part zero, then the programmed moves to positions A, B, C and D are as follows, dependent upon the quadrant in which each is located:

▶ From part zero to position A will be X+4 Y+2
▶ From part zero to position B will be X−2 Y+3
▶ From part zero to position C will be X−5 Y−3
▶ From part zero to position D will be X+6 Y−4.

It can be seen that the choice of the part zero will have a direct effect on the direction of the programmed moves, whether they will all be positive, all negative or a mixture of both. Direction of moves is also affected by incremental or absolute dimensioning (explained later in this chapter).

In practice, machine control systems assume plus and so there is no need to write the plus sign in a part program. This means, however, that it is absolutely essential the minus sign is included, otherwise plus is assumed and tool collisions can occur. It is worth noting that minus Z axis (−Z) moves in milling and both minus X and minus Z (−X, −Z) in turning move the cutting tool towards the workpiece, so that if the minus sign is omitted, the tool movement is away from the work into a safe position.

It is essential to always think of the programmed movements as the direction the cutting tool is moving, i.e. the spindle in milling or the cutting tool in turning. The positive and negative direction is the direction of the cutting tool, not the direction of the machine slides.

12.1.2 Dimensioning system

Drawings may be dimensioned in two different ways: incremental or absolute.

With incremental systems the dimensions are taken from the previous position (Fig. 12.5). The disadvantage of this system is that if an error occurs, it will accumulate throughout the length of the part. In the case of incremental programming, the programmer, as well as considering direction, must always consider 'how far should the tool move'. Referring to Fig. 12.5 programming can start from the bottom left corner, set as the part zero. To program movement to A, B, C, and D in sequence, the XY co-ordinates would be:

▶ From part zero to A will be X25 Y20
▶ From A to B will be X75 Y−10 (i.e. move from previous point)
▶ From B to C will be X−25 Y58
▶ From C to D will be X−20 Y−30.

With absolute systems, all the dimensional references are made from a single datum (Fig. 12.6). In this case the programmer has to consider 'to what position should the tool be moved'.

12

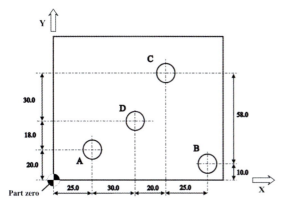

Figure 12.5 Incremental dimensioning

Referring to Fig. 12.6, programming again starts from the bottom left corner, set as part zero. To program movements to A, B, C and D in sequence, the XY co-ordinates would be:

- ▶ To A will be X25 Y20
- ▶ To B will be X100 Y10
- ▶ To C will be X75 Y 68
- ▶ To D will be X55 Y38.

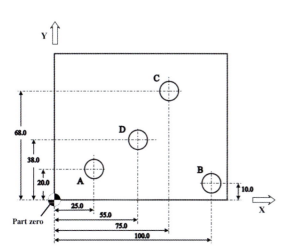

Figure 12.6 Absolute dimensioning

All control systems are capable of accepting incremental or absolute programming provided the appropriate command is programmed.

To sum up, programming in absolute is concerned with the position from a fixed zero reference point, while programming in incremental is concerned with a move from the previous position.

12.1.3 Part programming

Current CNC controls use a word address format for part programming.

CNC programs list instructions to be performed in the order they are written. They read like a book, left to right and from the top down. Each sentence is written on a separate line called a block. Each block should contain enough information to perform one operation. Each block must end with an end of block (EOB) character which tells the control where to separate the data into blocks. Usually the EOB character is a semi-colon generated by the EOB key on the machine control or return key on a keyboard.

A block comprises a series of words. Each word has a letter address followed by a numeric value. The letter address (G, M etc.) tells the control the kind of word and the numeric value tells the control the value of the word, e.g. a word can have an address G followed by a digital value 01 written G01. For example a block can be written:

N020 G01 X45.7 F150 S1200 M03

Common addresses include:

Address	Function
N	Sequence number
G	Preparatory function
X, Y, Z	Linear axis co-ordinate
I, J, K	Arc centre for circular interpolation
F	Feed rate
S	Spindle speed
T	Tool number
M	Miscellaneous function

Each block contains addresses and although there is no positional order, they can be shown as follows:

N, G, X, Y, Z, F, S, T, M;

The order should be maintained throughout each block in the program, although individual blocks may not necessarily contain all these addresses. The programmer needs to have an organised program format that is consistent and efficient, so that everyone involved can understand it.

12.1.4 Explanation of words

12.1.4.1 Sequence number (N address)

A sequence number is used to identify the block and is always placed at the beginning of the block and can be regarded as the name of the block. The numbers need not be consecutive, in fact

12

it aids programming if additional blocks need to be inserted, e.g. N0010 followed by N0020 then N0030 and so on. The program is executed according to the block sequence and not the sequence of the number.

12.1.4.2 Preparatory function (G address)

Preparatory functions known as G codes determine how the tool is moved to the programmed position. G codes are classified as modal or non-modal. Modal G codes remain active until another G code in the same group is programmed, e.g. G00 in a block will remain active throughout subsequent blocks until, say a G01, is programmed. A non-modal G code is active only in the block in which it is programmed and then is immediately forgotten by the control system. The most common G codes include:

Code	Function
G00	point to point positioning at rapid feed (set by machine tool manufacturer)
G01	linear interpolation (straight line feed move at programmed feed rate)
G02	circular interpolation, clockwise (CW)
G03	circular interpolation, counterclockwise (CCW)
G20	imperial data input (inches)
G21	metric unit input (millimetres)
G40	cutter compensation cancel
G41	cutter compensation, left
G42	cutter compensation, right
G90	absolute dimensioning
G91	incremental dimensioning
G92	programmed offset of tool change point

12.1.4.3 Co-ordinate word (X, Y, Z address)

A co-ordinate word specifies the target point of the tool movement (absolute dimension system – G90), or the distance to be moved (incremental dimension system – G91), e.g. X95.5 is a move to, or distance moved of, 95.5 mm.

12.1.4.4 Parameters for circular interpolation (I, J, K address)

These parameters specify the distance from the start point of an arc to its centre. The numerical value following I, J, K are in the X, Y and Z planes respectively.

12.1.4.5 Feed rate (F address)

A cutting tool feed rate is programmed as an F address except for rapid traverse. Rapid traverse is set by the manufacturer and is programmed with a G00 code. The feed rate is usually in mm/minute, e.g. F150 is 150 mm/min.

12.1.4.6 Spindle speed (S address)

Spindle speed is commanded by an S address and is in revolutions per minute (rev/min), e.g. S2000 is 2000 rev/min.

12.1.4.7 Tool number (T address)

Selection of the required tool is commanded by the T address, e.g. T03.

12.1.4.8 Miscellaneous function (M address)

A miscellaneous function or M code is used to control a machine operation other than the co-ordinate movements. Like G codes, M codes are classified as modal or non-modal. Modal M codes remain active until another M code in the same group is programmed, e.g. M03 in a block will remain active throughout subsequent blocks until, say, an M05 is programmed. A non-modal M code is active only in the block in which it is programmed and then is immediately forgotten by the control system, e.g. M00, M06, M30. The most common M functions include:

Code	Function
M00	program stop (stops program – will start at same position if cycle start is pressed)
M03	spindle ON clockwise (when viewed down on spindle)
M04	spindle ON counterclockwise
M05	spindle STOP
M06	tool change
M08	coolant ON
M09	coolant OFF
M30	program end, reset for another start

Before running a program, the machine control system needs to know where the machine axes are positioned and establish their relationship to the workpiece and the tool change position, i.e. the machine co-ordinate system and the machine control system must be synchronised.

The origin of the machine co-ordinate system is called 'machine home'. This is the position of the centre of the machine spindle with the Z axis fully retracted and the table moved to its limits on X and Y axes, e.g. near the back left corner. This position may vary with different machine tool manufacturers.

When a CNC machine tool is first turned on, it does not know where the axes are positioned in the workspace. Home position is found by the 'power up/restart' sequence initiated by the operator. This 'power up/restart' sequence drives all three axes slowly towards their extreme limits of travel, which can be −X, +Y, +Z. As each axis reaches its mechanical limit of travel, a limit switch is activated. This signals to the control system that the home position of each axis has been reached. Once all three axes have stopped moving, the machine is said to be 'homed'. The machine co-ordinates are synchronised with the machine control system.

The position of the part zero can now be determined relative the home position and is entered into the machine control system as an offset.

The final position required is the tool change position, i.e. the position the spindle needs to be for a tool to be changed whether or not this is a manual or automatic operation.

Having established the correct relative positions, the program can be run and the machining operations carried out.

With more advanced machining operations requiring a range of different tools, the different tool lengths and tool diameters have to be considered and are referred to as tool offsets.

Modern CNC control systems have specially designed functions to simplify the programming process mainly in avoiding repetitive or more complex commands and reducing the length of the program. Most of these functions are specific to a particular system, are beyond this introductory stage, and are therefore not covered in this chapter. These functions include mirror imaging, program repetition and looping, pocketing cycles and drilling, boring, reaming and tapping cycles.

The following example of a simple program is intended to show a number of machining operations using a single cutting tool and is intended to show the general arrangement only. It does not relate to any particular machine tool or any particular control system.

12.1.5 Example program of part shown in Fig 21.7

The program is written with the following points in mind:

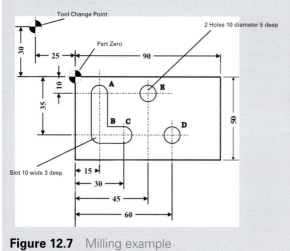

Figure 12.7 Milling example

▶ Programming is from part zero (XY top left corner, Z top surface of workpiece).
▶ Start at tool change point.
▶ 10mm slot drill is already loaded at tool change position.
▶ Programming is in absolute mode and metric units.
▶ The material is aluminium using a spindle speed of 2000 rev/min and feed rate of 250 mm/min.
▶ Mill the slot from A to B to C.
▶ Drill hole D then E.
▶ Return to tool change position.

12.1.5.1 Program

N005 G92 X-25 Y30 Z20 (offset of tool change point, Z axis 20mm above workpiece)
N010 G21 G90 (metric units, absolute dimensioning)

12

N015 S2000 M03 (spindle on clockwise at 2000 rev/min)
N020 M08 (coolant on)
N025 G00 X15 Y-10 Z1 (rapid traverse to 'A', cutter 1mm above workpiece)
N030 G01 Z-3 F250 (feed to 3mm deep at 250 mm/min)
N035 Y-35 (feed move to 'B')
N040 X30 (feed move to 'C')
N045 G00 Z1 (rapid traverse to 1mm above workpiece)
N050 X60 (rapid traverse to 'D')

N055 G01 Z-5 (feed to 5mm deep)
N060 G00 Z1 (rapid traverse to 1mm above workpiece)
N065 X45 Y-10 (rapid traverse to 'E')
N070 G01 Z-5 (feed to 5mm deep)
N075 G00 Z20 (rapid traverse to 20mm above workpiece)
N080 M09 (coolant off)
N085 X-25 Y30 (rapid traverse to tool change position)
N090 M05 (spindle off)
N095 M30 (end of program).

12.2 Manual part programming for turning

In the same way as described earlier for milling, a lathe also moves cutting tools to specific locations according to its co-ordinate system. On a lathe, the workpiece, held in a chuck, is rotating about the spindle axis and the cutting tool is moved about two axes X and Z. The X axis movement determines the diameter and the Z axis moves parallel to the spindle centre line and determines the position along the length of the workpiece (Fig. 12.8). The relationship of these machine motions can be shown as a graph, Fig. 12.9. The up and down motions, or X axis, corresponds to the vertical line of the graph while the Z axis corresponds to the horizontal line. Again for CNC purposes the origin of the graph is commonly called the part origin or part zero and establishes the point of reference for motion commands

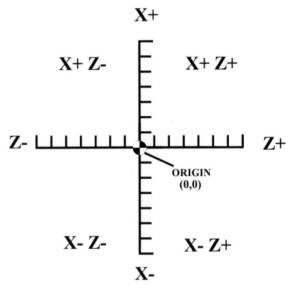

Figure 12.9 Graph of X, Z axes

in a CNC program and thus all movements are programmed from a common location.

Imagine standing in front of a lathe with the tool position at top right. If X0 is the centre line of rotation, then all moves along that axis to produce a diameter will be away from you in a plus direction, i.e. X+. All Z moves will be to your left in a minus direction, i.e. Z−. All moves will be in the top left X+Z− quadrant. It should be noted that X values relate to the diameter being produced, not the radius.

When programming a turning operation X zero (X0) is always the centre of rotation. Z zero (Z0) is usually the front face of the workpiece as shown in Fig. 12.10.

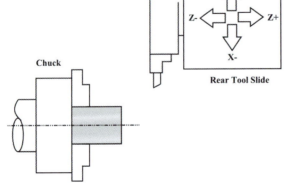

Figure 12.8 Lathe machine axes

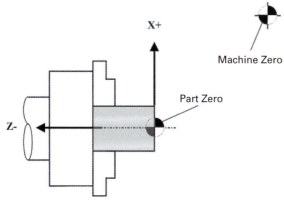

Figure 12.10 Part zero for turning

Although some CNC lathes have their cutting tools positioned at the front, as in a conventional lathe, turning centres, with automatic tool changers, have the tool positioned at the rear of the machine, usually at an angle, referred to as a slant bed. This arrangement enables the metal cuttings, or swarf, to clear and fall to the base of the machine. Turning tools at the rear can be mounted upside down, with forward spindle rotation so the relationship between front-mounted and rear-mounted tools remains the same. An example of a CNC slant bed turning centre is shown in Fig. 12.11.

12.2.1 Dimensioning system

The dimension system can be absolute or incremental as described for milling; incremental where the move is from the previous position and absolute where the move is the distance from a fixed datum (part zero).

The lathe differs from milling in this function. In a word address programming system, a lathe uses X and Z to represent a move in absolute and U and W to represent an incremental move on the corresponding axes. (If you remember, milling used the codes G90 and G91.)

A lathe has the unique capability to do absolute and incremental at the same time. This could be useful in producing a taper, chamfer or a series of grooves. For example a block could be written:

G01 X50 W-15 (X axis move in absolute and the Z axis in incremental)

or

G01 U5 Z-20 (X axis move in incremental and the Z axis in absolute).

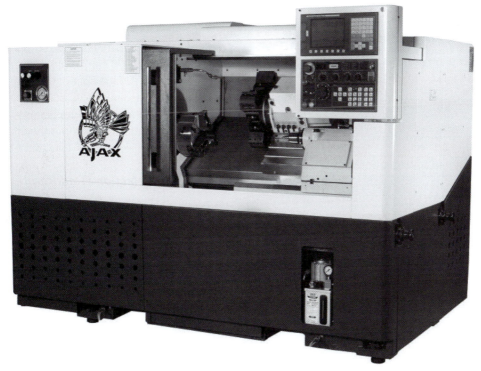

Figure 12.11 CNC slant bed turning centre

12

12.2.2 Part programming

Part programming for a lathe follows the same word address system as described for milling with the same use of words and blocks, read, interpreted and executed sequentially, by the machine control system. Common word addresses for the lathe include:

Code	Function
N	sequence number
G	preparatory function
X, Z	linear absolute axis co-ordinate
U, W	linear incremental axis co-ordinate on X and Z respectively
R	radius designation
F	feed rate in mm/revolution
S	spindle speed in rev/minute
T	tool selection 4 digits (1 and 2 are tool number, 3 and 4 are tool offsets)
M	miscellaneous function

12.2.3 Explanation of words

12.2.3.1 Sequence number (N address)

Same description as milling.

12.2.3.2 Preparatory function (G code)

Same description as milling. Common codes include:

Code	Function
G00	rapid traverse
G01	linear interpolation (straight line feed move)
G02	circular interpolation, clockwise (CW)
G03	circular interpolation, counterclockwise (CCW)
G20	imperial data input (inches)
G21	metric unit input (millimetres)
G28	rapid traverse to home position
G40	tool nose radius compensation cancel
G41	tool nose radius compensation, left
G42	tool nose radius compensation, right
G96	constant surface speed control
G97	constant surface speed control cancel

12.2.3.3 Co-ordinate word (X, Z address)

A co-ordinate word to specify the tool move to position in absolute mode.

12.2.3.4 Co-ordinate word (U, W address)

A co-ordinate word to specify the tool move in incremental mode.

12.2.3.5 Radius designation (R address)

This word designates the size of a corner radius, e.g.

N025 G01 X20 R2 F0.3 (move to 20mm diameter at a feed rate of 0.3mm/rev through a 2mm radius)

N030 Z-30 (to 30mm length)

12.2.3.6 Feed rate (F address)

A cutting tool feed rate is programmed as an F address except for rapid traverse. Rapid traverse is set by the manufacturer and is programmed with a G00 code. The feed rate is mm/revolution, e.g. F0.4 is 0.4mm/rev.

12.2.3.7 Spindle speed (S address)

Spindle speed is commanded by an S address and is always revolutions per minute (rev/min), e.g. S2000 is 2000 rev/min. Provided the spindle is already running, G96 can be programmed to select a constant surface speed in metres/minute e.g. S500. The machine would automatically alter the spindle speed relative to the diameter being machined to maintain the programmed cutting speed.

12.2.3.8 Tool selection (T address)

The T address selects a tool and offset while initiating the tool change process. The format is T0202 where the first two digits are the tool turret position and the last two digits select the tool offset.

12.2.3.9 Miscellaneous function (M address)

Same description as milling. Common codes include:

Code	Function
M00	program stop (stops program – will start at same position if cycle start is pressed)
M03	spindle ON clockwise (when viewed from rear of headstock)
M04	spindle ON counterclockwise
M05	spindle STOP
M08	coolant ON
M09	coolant OFF
M30	program end, reset for another start

Before a program can be run, in the same way as milling, the machine control system needs to know where the machine axes are positioned and establish their relationship to the workpiece, i.e. the machine co-ordinate system and the machine control system must be synchronised.

The origin of the machine co-ordinate system is called 'machine home'. This position is established by the machine tool manufacturer.

When a CNC machine tool is first turned on, it does not know where the axes are positioned in the workspace. The machine zero or home position is found by the 'power up/restart' sequence initiated by the operator. This 'power up/restart' sequence drives both axes slowly to their extreme positive locations until their limit switches are reached. This signals to the control system that the home position of each axis has been reached. Once both axes have stopped moving the machine is said to be 'homed'. The machine co-ordinates are synchronised with the machine control system.

The position of the part zero can now be determined relative to the home position. X zero (X0) is always the centre line of the part. Normally the front face of the part is designated as Z zero (Z0).

When setting up the machine, the operator needs to determine the distance that different tools are from the home position to part zero. Each tool is manually touched off the face and diameter and their distance from the home position is saved as values of X and Z in the tool offset page of the machine control system.

Having established the correct relative positions, the program can be run and the machining operations carried out.

Modern CNC control systems have specially designed functions to simplify the programming process mainly in avoiding repetitive or more complex commands and reducing the length of the program. Most of these functions are specific to a particular system, are beyond this introductory stage, and are therefore not covered in this chapter. These functions include drilling, boring and tapping cycles, roughing and finishing cycles and grooving and thread-cutting cycles.

The following example of a simple program is intended to show a number of machining operations using a single cutting tool and is intended to show the general arrangement only. It does not relate to any particular machine tool or any particular control system.

12.2.4 Example program of part shown in Fig 21.12

The program is written with the following points in mind:

▶ The workpiece has been rough turned.
▶ The program is a single finishing cut ending at P4.
▶ Part zero is spindle centre line and front face of workpiece.
▶ Start point of the tool is X200 Z100.

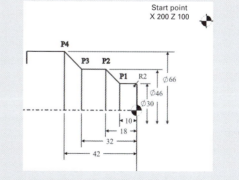

Figure 12.12 Turning example

12

12.2.4.1 Program

N005 T0202 (tool 2 offset number 2)

N010 G21 S750 M03 (metric units, spindle on clockwise at 750 rev/min)

N015 G00 G42 X0 Z1.5 M08 (rapid traverse to centre, 1.5 mm off face, coolant on, compensation on)

N020 G01 Z0 F0.4 (feed to front face at 0.4 mm/rev)

N025 X30 R2 (feed to diameter 30 mm around 2 mm radius)

N030 Z-10 (feed to P1)

N035 X46 Z-18 (feed to P2)

N040 Z-32 (feed to P3)

N045 X66 Z-42 (feed to P4)

N050 M09 (coolant off)

N055 G40 G00 X200 Z100 M05 (compensation off, rapid traverse to start point, spindle stop)

N060 T0200 (cancel tool offset)

N065 M30 (end of program)

Although manufacturing industry accepts standardised systems, machine tool manufacturers will have arrangements, specific to their own requirements. A machine tool can be fitted with any one of a number of control systems either developed in-house or purchased from a third party, which may differ in some way depending on the provider. As stated earlier in this chapter, this introduction is provided as a guide to general principles and you may find in practice that the machine tools and control systems with which you may become involved will differ in some of their details.

Review questions

1. Name two common types of CNC machine.
2. Describe the difference between incremental and absolute dimensioning.
3. To what do the initials CNC refer?
4. What is meant by a machine 'HOME' position and why is it important?
5. Describe what is meant by a word address format for CNC part programming.
6. State the six degrees of freedom and how they are designated.
7. State the difference between modal and non-modal codes.
8. By means of a sketch, show and designate the axes movements on two types of CNC machine.
9. What are 'G' codes and what are they used to define in CNC part programming?
10. What are 'M' codes and what are they used to define in CNC part programming?

12

CHAPTER 13

Joining methods

Some method of joining parts together is used throughout industry, to form either a complete product or an assembly. The method used depends on the application of the finished product and whether the parts have to be dismantled for maintenance or replacement during service.

There are five methods by which parts may be joined:

▶ mechanical fasteners – screws, bolts, nuts, rivets;
▶ soldering;
▶ brazing;
▶ welding;
▶ adhesive bonding.

Mechanical fasteners are most widely used in applications where the parts may need to be dismantled for repair or replacement. This type of joint is known as non-permanent. The exception would be the use of rivets, which have to be destroyed to dismantle the parts and so form a permanent joint. Welding and adhesives are used for permanent joints which do not need to be dismantled – any attempt to do so would result in damage to or destruction of the joints and parts.

Although soldered and brazed joints are considered permanent, they can be dismantled by heating for repair and replacement.

13.1 Mechanical fasteners

Mechanical fasteners can be made from many materials but most bolts, nuts and washers are made from carbon steel, alloy steel, or stainless steel depending upon their industrial use. Carbon steel is the cheapest and most common for general use.

To prevent corrosion, mechanical fasteners may be plated or coated in some way, again depending on the application. The most common surface treatments are zinc, nickel and cadmium. Phosphate coatings are also used but have limited corrosion resistance.

13.1.1 Machine screws

These are used for assembly into previously tapped holes and are manufactured in brass, steel, stainless steel and plastics (usually nylon) and threaded their complete length. Various head shapes are available, as shown in Fig. 13.1.

Depending on the style, thread diameters are generally available up to 10 mm, with lengths up to 50 mm. For light loading conditions where space is limited, a headless variety known as a grub screw is available. A typical application would be to retain a knob or collar on a shaft.

Although Fig. 13.1 shows head types with slotted head drives, these screws are available with a

Workshop Processes, Practices and Materials, Fifth Edition. 9781138784727.
© 2015 Bruce J. Black. Published by Taylor & Francis. All rights reserved.

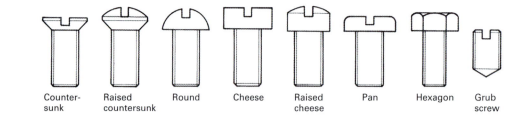

Figure 13.1 Types of screw head

variety of head drives as outlined on page 39 with Phillips, Pozidriv and Torx the most common.

13.1.2 Socket screws

Manufactured in high-grade alloy steel with rolled threads, this type of screw is used for higher strength applications than machine screws. Three head shapes are available, all of which contain a hexagon socket for tightening and loosening using a hexagon key, Fig. 13.2.

Headless screws of this type – known as socket set screws – are available with different shapes of point. These are used like grub screws, where space is limited, but for higher strength applications. Different points are used either to bite into the metal surface to prevent loosening or, in the case of a dog point, to tighten without damage to the work, Fig. 13.3.

13.1.3 Self-tapping screws

Self-tapping screws are used for fast-assembly work. They also offer good resistance to loosening through vibration. These screws are specially hardened and produce their own threads as they are screwed into a prepared pilot hole, thus eliminating the need for a separate tapping operation.

There are two types:

▶ the thread-forming type, which produces its mating thread by displacing the work material and is used on softer ductile materials, Fig. 13.4(a);
▶ the thread-cutting type, which produces its mating thread by cutting in the same way as a tap. This type has grooves or flutes to produce the cutting action, Fig. 13.4(b), and is used on hard brittle materials, especially where thin-wall sections exist, as this type produces less bursting force.

More rapid assembly can be achieved by self-piercing-and-tapping screws. These have a special piercing point and a twin-start thread, Fig. 13.4(c). Used in conjunction with a special gun, they will pierce their own pilot hole in the sheet metal (up to 18 SWG (1.2 mm) steel) or other thin materials and are then screwed home in a single operation.

13.1.4 Bolts

Bolts are used in conjunction with a nut for heavier applications than screws. Unlike screws, bolts are

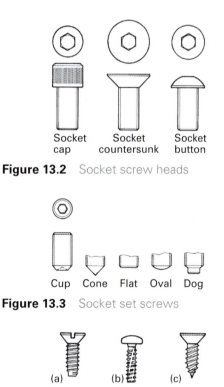

Figure 13.2 Socket screw heads

Cup Cone Flat Oval Dog

Figure 13.3 Socket set screws

(a) (b) (c)

Figure 13.4 Self-tapping screws
(a) thread-forming (b) thread-cutting
(c) self-piercing-and-tapping

13

threaded for only part of their length, usually twice the thread diameter.

Bright hexagon-head bolts are used in engineering up to 36 mm diameter by 150 mm long. Larger sizes are available in high-tensile materials for use in structural work.

13.1.5 Nuts

Standard hexagon nuts are used with bolts to fasten parts together. Where parts require to be removed frequently and hand tightness is sufficient, wing nuts are used, Fig. 13.5(a). If a decorative appearance is required, a dome or acorn nut can be fitted, Fig. 13.5(b).

Where thin sheets are to be joined and access is available from only one side, rivet bushes or rivet nuts are used. These provide an adequate length and strength of thread which is fixed and therefore allows ease of assembly, Fig. 13.5(c). Available in thread sizes up to 12 mm for lighter applications, blind nuts of the type shown in Fig. 13.5(d) can also be used. The nut is enclosed in a plastics body which is pressed into a predrilled hole. A screw inserted into the nut pulls it up and, in so doing, expands and traps the plastics body.

Spring-steel fasteners are available which as well as holding also provide a locking action. If access is available from both sides, a flat nut can be used, Fig. 13.5(e), or from one side a J-type nut can be used, Fig. 13.5(f). In their natural state these nuts are arched, but they are pulled flat when the screw is tightened.

13.1.6 Washers

Washers distribute the tightening load over a wider area than does a bolt head, screw head or nut. They also keep the surface of the work from being damaged by the fastener.

Plain flat washers spread the load and prevent damage to the work surface. Flanged nuts are available with a plain flange at one end which acts as an integrated, non-slipping washer. The flange face may be serrated to provide a locking action but can only be used where scratching of the work surface is acceptable (Fig. 13.6).

Washers which provide a locking action are discussed in Section 13.3.

Figure 13.6 Flanged nuts

13.1.7 Spring tension pins

These pins are made from spring steel wrapped round to form a slotted tube, Fig. 13.7. The outside diameter is produced larger than the standard-size drilled hole into which it is to be inserted. When inserted in the hole, the spring tension ensures that the pin remains securely in position and cannot work loose. A chamfer at each end of the pin enables it to be easily inserted in the hole, where it can be driven home using a hammer.

Pins of this type are now being used to replace solid hinge pins, split pins, rivets and screws, eliminating the need for reaming, tapping, counterboring and countersinking. They are

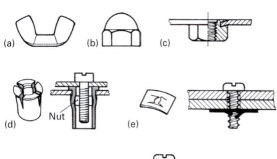

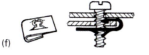

Figure 13.5 Types of nut (a) wing (b) dome or acorn (c) rivet bush or nut (d) blind (e) flat spring steel fastener and (f) J type fastener

Figure 13.7 Spring tension pin

187

available in a range of diameters from 1 mm to 12 mm and lengths from 4 mm to 100 mm.

13.2 Screw threads

Since 1965, British industry has been urged to adopt the ISO (International Organization for Standardization) metric thread as a first-choice thread system, with the ISO inch (unified) thread as the second choice. The British Standard Whitworth (BSW), British Standard Fine (BSF), and British Association (BA) threads would then become obsolete. The British Standard Pipe (BSP) thread is to be retained. The changeover has been extremely slow in taking place, and all these threads are available and still in use.

13.2.1 ISO metric thread

This thread, based on a 60° triangular form, provides a range of coarse and a range of fine pitches (Appendix 1). The threads are designated by the letter M followed by the diameter and pitch in millimetres, e.g. M16 × 2.0. The absence of a pitch means that a coarse thread is specified, e.g. M16 indicates an M16 × 2.0 coarse pitch.

13.2.2 Unified thread

This thread is also based on a 60° form, but with a rounded crest and root. A coarse series of pitches (UNC) is provided from $\frac{1}{4}$ inch to 4 inch diameter, and a fine series of pitches (UNF) from $\frac{1}{4}$ inch to $1\frac{1}{2}$ inch diameter. The threads are designated by the diameter of thread in inches followed by the number of threads per inch (t.p.i.) and whether coarse or fine series, e.g. $\frac{1}{4}$ 20 UNC, $\frac{1}{4}$ 28 UNF (Appendix 2).

13.2.3 British Standard Whitworth (BSW) thread

This thread is based on a 55° vee-thread form, rounded at the crest and root, covering a range of thread diameters from $\frac{1}{8}$ inch to 6 inches. The threads are designated by the diameter in inches followed by the thread series, e.g. $\frac{3}{8}$ BSW (Appendix 3). It is not usual to include the number of t.p.i.

13.2.4 British Standard Fine (BSF) thread

This thread has exactly the same thread form as BSW but with finer pitches, covering a range from $\frac{3}{16}$ inch to $4\frac{1}{4}$ inches. The threads are designated by the diameter in inches followed by the thread series, e.g. $\frac{3}{8}$ BSF (Appendix 3).

13.2.5 British Standard Pipe (BSP)

This thread has exactly the same thread form as BSW and covers a range from $\frac{1}{8}$ inch to 6 inch diameter.

The BSP parallel threads, known as 'fastening' threads, are designated BSPF, the size referring to the bore size of the pipe on which the thread is cut, e.g. 1 inch BSPF has an outside diameter of 1.309 (Appendix 3).

Where pressure-tight joints are required, BSP taper threads have to be used. These have the same form and number of threads, but the thread is tapered at 1 in 16.

13.2.6 British Association (BA) thread

Threads in this range have extremely fine pitches and are used for applications less than $\frac{1}{4}$ inch diameter in preference to BSW or BSF. They are designated by a number, e.g. 4 BA (Appendix 4).

In general engineering, preference is given to the even-numbered BA sizes, i.e. 0, 2, 4, 6, 8 and 10 BA.

13.3 Locking devices

13.3.1 Self-locking screws and bolts

These screws and bolts eliminate the need for nuts and washers and so provide cost savings. Common types incorporate a nylon insert, either as a small plug or as a strip along the length of the thread, Fig. 13.8(a). A development of this principle is the application of a layer of nylon over a patch of thread. As the screw is engaged in the mating part, the nylon is compressed and completely fills the space between thread forms. This provides an interference which will resist

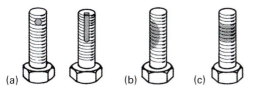

Figure 13.8 Self-locking screws and bolts (a) nylon insert (b) nylon layer (c) adhesive film

rotation of the screw, Fig. 13.8(b). A more recent development is the application to an area of thread of a chemical adhesive which is completely dry to the touch, Fig. 13.8(c). The liquid adhesive is encapsulated within the film. When the two threaded parts are assembled, the micro-capsules of adhesive are broken, releasing the adhesive, which hardens and provides a reliably sealed and locked thread.

13.3.2 Locking nuts

The simplest method of locking a nut in position is by applying a lock nut. Lock nuts are a little over half the thickness of a standard nut. When used in conjunction with a standard nut and tightened, the lock nut is pushed against the thread flanks and locked, Fig. 13.9(a).

Slotted and castle nuts are used in conjunction with wire or a split pin through a hole in the bolt to prevent the nut from working loose, Fig. 13.9(b).

Self-locking nuts are available which are easy to assemble and do not require a hole in the bolt or the use of a split pin. One type, known as a 'Nyloc' nut, Fig. 13.9(c), incorporates a nylon insert round the inner top end of the nut. As the nut is screwed on, the nylon yields and forms a thread, creating high friction and resistance to loosening.

A second type, known as an 'Aerotight' stiff nut, Fig. 13.9(d), has two arms formed on top of the nut. These arms, which are threaded, are deflected inwards and downwards. When the

nut is screwed on, the arms are forced into their original position and the resistance of these arms gives a good grip on the thread, preventing it from working loose.

A third type, known as a 'Philidas' self-locking nut, Fig. 13.9(e), has a reduced diameter above the hexagon. Two slots are cut opposite each other in the reduced diameter and the metal above the slots is pushed down, which upsets the thread pitch. When screwed in position, the thread is gripped by the upset portion, preventing the nut from working loose.

A fourth type, known as a torque lock nut, Fig. 13.9(f), has the top part of the nut deformed to an elliptical shape which grips the thread as the nut is applied. This ensures close contact between the threads, preventing the nut from working loose.

13.3.3 Threadlocking

As well as the pre-applied adhesives already mentioned, threadlocking products can be applied separately to secure any threaded fastener against vibration and shock loads. Applied as easy flowing liquids which fill the gap between mating threads, they cure into a strong insoluble plastic. The joint formed is shock, leak, corrosion and vibration proof and depending on the grade used, can be undone using hand tools and by the application of heat.

Liquid threadlock is available in a variety of strengths to suit different applications. High strength produces a permanent joint and is intended for applications which do not need to be dismantled, e.g. permanently locking studs on engine blocks and pump housings. Use of localised heat will be necessary if dismantling is required.

Medium strength is effective on all types of metal fastener and prevents loosening on vibrating parts such as pumps, motor mounting bolts and gearboxes and is recommended for use

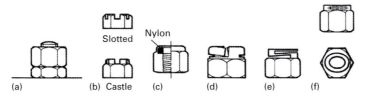

Figure 13.9 Locking nuts (a) standard lock nut (b) slotted and castle nuts (c) 'Nyloc' nut (d) 'Aerotight' nut (e) 'Philidas' self-locking nut (f) torque lock nut

13

where dismantling with hand tools is required for servicing and where parts are contaminated with oil. Parts can be loosened using normal hand tools without damaging the threads.

Low strength is recommended for use on adjusting screws, countersink head screws and set screws on collars and pulleys and is also used on low-strength metals such as aluminium and brass which could break during dismantling. Screws can be loosened using normal hand tools without breaking or damaging the threads.

13.3.4 Locking washers

A locking washer is inserted under the head of a screw, bolt or nut to prevent it working loose during service.

A tab washer may be used, similar to a plain washer with the addition of a tab which is bent up on the hexagon face of the nut, screw or bolt to prevent it working loose, Fig. 13.10(a).

Helical-spring locking washers are commonly used as locking devices and are available for threads up to 24 mm diameter. They may be of square or rectangular section in a single coil, with the ends of the coil raised in opposite directions. These ends form sharp points which dig into the surfaces. In addition, the spring is flattened during the tightening of the screw, bolt or nut, which gives constant tension during use, Fig. 13.10(b).

Shake-proof washers are used for thread sizes up to 16 mm and can have external or internal teeth. The teeth are twisted out of flat so that the washer bites into the surfaces as it is compressed during tightening, Fig. 13.10(c).

Where rigid permanent fixing is required on shafts, a range of spring fixing washers which eliminate the use of threads and nuts is available. One type is shown in Fig. 13.10(d). As it is pushed on to the shaft, the 'prongs' are deformed and

bite into the shaft and cannot be removed without destroying it. This type is available up to 25 mm diameter and can be used on all types of material, including plastics.

13.3.5 Circlips

Circlips are used to lock a variety of engineering features. An external circlip, Fig. 13.11(a), usually fitted in a groove in a shaft, prevents the shaft from moving in an axial direction or prevents an item fitted to the end of a shaft from coming loose, e.g. a bearing or pulley. Similarly, an internal circlip, Fig. 13.11(b), can be used in a groove to prevent an item such as a bearing from coming out of a recess in a housing and will withstand high axial and shock loading.

Circlips are available in sizes from 3 mm to 400 mm and larger and are manufactured from high-carbon spring steel. Lugs with holes are provided for rapid fitting and removal using circlip pliers. Smaller shafts can use a variation of the circlip, known as an E-type circlip, or retaining ring, Fig. 13.11(c). These provide a large shoulder on a relatively small diameter, e.g. rotating pulleys can act against the shoulder.

Wire rings or snap rings, Fig. 13.11(d), can be used as a cost-effective replacement for the traditional type of circlip in both internal and external applications such as the assembly of needle bearings and needle cages and sealing rings.

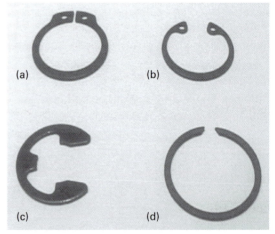

Figure 13.11 Circlips

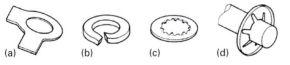

Figure 13.10 Locking washers (a) tab (b) helical spring (c) shake proof (d) spring fixing

13.4 Riveting

13.4.1 Solid and tubular rivets

Riveting as a means of fastening is used because of its speed, simplicity, dependability and low cost. Light riveting, used for general assembly work up to about 6 mm diameter, is carried out in industry using high-speed rivet-setting machines having cycle times as short as $\frac{1}{3}$ second. Rivets are used on assemblies where parts do not normally have to be dismantled, i.e. permanent joints. They may also be used as pivots, electrical contacts and connectors, spacers or supports.

The cost of riveting is lower than that of most other methods of fastening, due to the absence of plain washers, locking washers, nuts or split pins; also, the use of self-piercing rivets eliminates the need to predrill holes. Rivets are available in steel, brass, copper and aluminium in a variety of types. The more standard head types used are shown in Fig. 13.12.

Solid rivets are strong but require high forces to form the end. In riveting, forming the end is known as clinching. Solid rivets are used in applications where the high forces used in clinching will not damage the work being fastened. Tubular rivets are designed for application where lighter clinching forces are used.

Short-hole tubular rivets, Fig. 13.13(a), have the advantage of a solid rivet with easier clinching. These have a parallel hole and can be used for components of varying thicknesses.

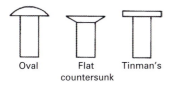

Oval   Flat   Tinman's
        countersunk

Figure 13.12 Solid-rivet head types

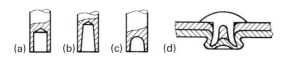

(a)   (b)   (c)   (d)

Figure 13.13 Tubular rivets (a) short-hole (b) double-taper semi-tubular (c) bull-nose semi-tubular (d) self-piercing

Double-taper semi-tubular rivets, Fig. 13.13(b), have a taper hole to minimise shank expansion during clinching and are used to join brittle materials.

Bull-nose semi-tubular rivets, Fig. 13.13(c), are used where maximum strength is required. The rivet shank is intended to expand during the setting operation in order to fill the predrilled hole in the work. This ensures a very strong joint.

Self-piercing rivets, Fig. 13.13(d), have been specially developed to pierce thicker metal and clinch in the same operation. For metal up to 4.7 mm thick, the self-piercing rivet is made from special steel and is heat-treated to give the required hardness for piercing and ductility for clinching. One use of this type of rivet is automatic riveting in the production of garage doors.

13.4.2 Blind rivets

Blind rivets, also known as pop rivets, are rivets which can be set when access is limited to one side of the assembly. However, they are also widely used where both sides of the assembly are accessible.

Used to join sheet metal, blind rivets are readily available in sizes up to around 5 mm diameter and 12 mm long in aluminium, steel and monel. Plated steel rivets are used where low cost, relatively high strength and no special corrosion-resistance is required; aluminium for greater resistance to atmospheric and chemical corrosion; and monel for high strength and high resistance to corrosion.

Blind rivets consist of a headed hollow body inside of which is assembled a centre pin or mandrel. The rivet is set by inserting the mandrel in a tool having a means of gripping, and the rivet is inserted into a predrilled hole in the assembly. Operation of the tool causes its gripping jaws to draw the mandrel into the rivet, with the result that the head of the mandrel forms a head on the rivet on the blind side of the assembly, at the same time pulling the metal sheets together. When the joint is tight, the mandrel breaks at a predetermined load, Fig. 13.14. The broken-off portion of the mandrel is then ejected from the tool.

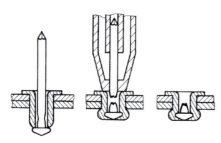

Figure 13.14 Blind rivet

13.5 Soft soldering

Soft soldering is a process of joining metal parts by heating and running a low-melting-point alloy between the two surfaces being joined. When cooling takes place, the alloy solidifies, resulting in a secure joint. For most purposes, the low-melting-point solder alloy is a composition of tin and lead. Solder melts at a temperature less than 300°C, which is far below the melting temperature of the metal being joined, and produces a low-strength joint.

Soldering is used not only to make a mechanical bond between surfaces but also to provide a leak-proof seal when liquids have to be contained. It can also be used to provide a permanent electrical connection. To provide a secure joint, it is essential that the joint surfaces are perfectly clean and free from rust, grease or any other substance likely to prevent good metal-to-metal union. The most likely cause of a bad joint, assuming the surface is clean, is the thin oxide film which is present on all metals. Oxide films can be removed using emery cloth or a flux.

13.5.1 Fluxes

Fluxes used in soft soldering are either active or passive.

An active flux chemically removes the oxide film, has an acid base, and is highly corrosive. These fluxes are usually hydrochloric acid in which zinc has been dissolved to form zinc chloride, known as 'killed spirits'. Any joint prepared using an active flux must be thoroughly washed in warm water when soldering is completed, to remove any flux residue. For this reason an active flux is not suitable for electrical applications.

A passive flux is used after the oxide film has been removed using emery cloth, to prevent the

oxide film reforming. Passive fluxes are usually resin-based. Cored solder, containing a resin-based flux, is usually sufficient for electrical work.

13.5.2 Heating

The type of heat source used for soldering depends largely on the size of the parts to be joined. The greatest problem is usually heat loss to the surrounding area by conduction through the metal. The temperature in the joint area must be high enough to melt the solder and allow it to flow and combine with the surfaces to be joined.

Where the conduction of heat is likely to cause damage, e.g. to electrical insulation or electronic components, a piece of bent copper can be placed in contact with the conductor, positioned between the joint and the insulation or component. The copper readily absorbs the heat before it can do any damage and is known as a 'heat sink'.

For small parts, an electric soldering iron is quite sufficient. The soldering iron, as well as melting the solder and heating the work, acts as a reservoir for the solder to deposit an even amount in the required position. Larger parts can be heated using a gas/air or butane torch or by placing the work on a hot-plate.

When the area of the surfaces being joined is large, solder cannot be satisfactorily run between them. In such a case, the surfaces should first be 'tinned', i.e. each surface should be separately coated with solder. The two parts are then assembled and the parts are reheated until the solder melts to make the joint.

13.5.3 Joint design

Solder is not as strong as other metals. When a mechanical joint is required, the joint design should provide as much additional strength as possible and not rely on the solder strength alone. A higher mechanical strength can be obtained by an interlocking joint, the solder providing an additional leak-proof seal as shown in Fig. 13.15(a). This method is used in the production of cans of drink and food stuffs.

Additional mechanical strength can be obtained in electrical joints by winding the wire round a pillar or bending the wire through a hole in a tab connector, as shown in Fig. 13.15 (b) and (c).

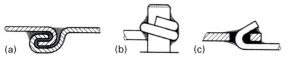

Figure 13.15. Soldering-joint design
(a) interlocking (b) winding round a pillar
(c) bending through a tab connector

13.6 Solders

Soft solders are alloys of tin and lead. All the plain tin/lead solders become solid at 183 °C. The temperature at which they become completely liquid depends on the composition, the temperature increasing as the lead content increases.

Some solders pass through a considerable pasty stage from being completely liquid to becoming solid. Reference to Fig. 13.16 shows a 20% tin/80% lead solder, completely liquid at 276 °C and solid at 183 °C. The solder passes through a pasty stage in the transformation from liquid to solid. This feature is useful in some plumbing applications or where work is coated by dipping in a bath of molten solder.

A solder which contains 60% tin/40% lead has a lower melting temperature and a very small temperature interval between being completely liquid and becoming solid, 188 °C to 183 °C. This is preferred, especially in electrical work, where a higher melting temperature and longer cooling period could result in damage to insulation or components. Where soldering is carried out by machine and too fine a control of temperature cannot be maintained, the 50/50, 40/60 ranges of solders are used.

Table 13.1 shows the temperatures, strength and uses of a range of compositions of plain tin/lead solder alloys.

13.7 Brazing

Brazing is defined as a process of joining metals in which, during or after heating, molten filler metal is drawn by capillary action into the space between closely adjacent surfaces of the parts being joined. In general, the melting point of the filler metal is above 450 °C but always below the melting temperature of the metals being joined.

Brazing is used to join any combination of similar or dissimilar metals and results in a high-strength joint of good reliability.

To form a strong joint, the surfaces must be free of any rust, grease or oxide film.

13.7.1 Brazing alloys

Brazing alloys are available in a wide variety of forms, including rod, strip, wire, foil and powder.

The choice of brazing alloy depends upon the materials being joined and the temperature at which the brazed parts are to operate. Brazing brasses are widely used with hand-torch heating for joining ferrous-metal parts. The common

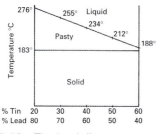

Figure 13.16 Tin–lead diagram

Table 13.1 Tin/lead solders

% Tin	% Lead	Solidus* (°C)	Liquidus† (°C)	Strength (MN/m²)	Uses
60	40	183	188	58	General purpose, especially electrical works
50	50	183	212	46	Machine solders
40	60	183	234	41	
30	70	183	255	38	Plumbers' solders and dipping baths
20	80	183	276	37	

*Solidus – the temperature at which the alloy has completely solidified.
†Liquidus – the temperature at which the alloy is completely liquid.

13

composition of brazing brass is shown in Table 13.2.

Silver brazing alloys have excellent brazing properties and are the most widely used for joining most ferrous and non-ferrous materials with the exception of those based on aluminium, zinc and magnesium. Silver brazing alloys have lower melting points than the brazing brasses and are capable of penetrating narrow joint gaps. The composition of two typical silver brazing alloys is shown in Table 13.3. Silver brazing is sometimes referred to as silver soldering or hard soldering and should not be confused with soft soldering as described in Section 13.5.

Although the addition of cadmium has particular advantages, fumes given off can have serious health effects. As a result, safer, cadmium-free alloys are available and should be used wherever practicable. Exposure to alloys containing cadmium must be adequately controlled as covered by COSHH Regulations.

Aluminium–silicon alloys are used for the brazing of aluminium and aluminium alloys. The composition of a commonly used aluminium brazing alloy is shown in Table 13.4.

13.7.2 Fluxes

The function of a flux is to dissolve or remove the surface oxide film from the metal to be joined and any oxides formed during heating and promote the flow and wetting of the brazing alloy. The ideal flux should be active at a temperature below the solidus and remain active at a temperature above the liquidus. No single flux can achieve this over the range of temperatures used in brazing and so a flux has to be chosen to suit a particular temperature range. The use of cadmium-containing alloys is not permitted in the manufacture of food- and drink-handling equipment and medical instruments.

A range of proprietary fluxes are available to suit different applications usually as a pre-mixed paste or as a powder which can be made into a paste when required by stirring in water until the mixture has the consistency of thick cream. Ideally the flux should be applied to both joint surfaces before assembly.

These fluxes must be removed after brazing. This can be done by quenching the work in hot or cold water shortly after the brazing alloy has solidified. If this is not practical, the flux can be removed by chipping, filing, scraping or steel-wire brushing.

13.7.3 Heating

Any heat source capable of raising the temperature above the liquidus of the selected brazing alloy can be used. Many types of controlled automatic heating are used in industry, but for workshop purposes the hand torch is the

Table 13.2 Brazing brass

Composition %		Melting range (°C)	
Copper	Zinc	Solidus	Liquidus
60	40	885	890

Table 13.3 Silver brazing alloys

Composition %				Melting range (°C)	
Silver	Copper	Zinc	Tin	Solidus	Liquidus
40	30	28	2	660	720
55	21	22	2	630	660

Table 13.4 Aluminium brazing alloy

Composition %			Melting range (°C)	
Aluminium	Silicon	Copper	Solidus	Liquidus
86	10	4	520	585

13

most widely used. A hand torch has the advantage of being flexible in use, but it requires a skilled operator to produce consistent results.

A wide variety of gas mixtures can be used, the most common being

▶ oxyacetylene;
▶ oxypropane;
▶ compressed air + coal gas;
▶ compressed air + natural gas.

When using a hand torch, care must be taken to achieve an even distribution of heat, especially when using oxyacetylene with its intensely hot localised flame.

13.7.4 Joint design

The strength of a brazed joint relies on the capillary action of the brazing alloy between the faces being joined.

Wherever possible, joints should be designed so that loads applied in service act on the joint as shear stresses rather than as tensile stresses. This means that lap joints are preferred to butt joints, Fig. 13.17. The recommended length of overlap on lap joints is between three and four

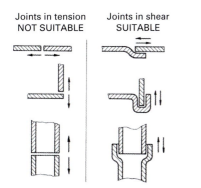

Figure 13.17 Brazing – joint design

times the thickness of the thinnest component in the assembly.

To take full advantage of the capillary action, it is essential that there is a sufficient gap or clearance between the faces being joined, to allow penetration of the brazing alloy. Depending on the metals being joined, the joint gap should be between 0.04 mm and 0.20 mm.

13.8 Welding

Welding differs from soldering and brazing in that no alloy is used to join the metals together. In welding the metals being joined are locally melted and when solidified produce a solid mass, giving a joint strength as strong as the metals being joined. A filler rod is sometimes used to make up for material loss during welding, to fill any gaps between the joint surfaces and to produce a fillet. A flux is required with some metals and with some welding methods to remove the oxide film and provide a shield to prevent oxides from re-forming.

The definition of a weld is: 'a union between pieces of metal at faces rendered plastic or liquid by heat or by pressure or by both'. This can be realised by:

▶ Fusion welding – where the metal is melted to make the joint with no pressure involved.
▶ Resistance welding – where both heat and pressure are applied.
▶ Pressure welding – where pressure only is applied, e.g. to a rotating part where the heat is developed through friction, as in friction welding.

For the purpose of this book we will deal with fusion welding.

Fusion welding processes are distinguished by the methods of producing the heat, arc and gas.

13.8.1 Arc welding

An electric arc is produced by passing an electric current between two electrodes separated by a small gap. In arc welding, one electrode is the welding rod or wire, the other is the metal plate being joined.

The electrodes are connected to the electrical supply, one to the positive terminal and one to the negative. The arc is started by touching them and withdrawing the welding rod about 3 or 4 mm from the plate. When the two electrodes touch, a current flows, and, as they are withdrawn, the current continues to flow in the form of a spark. The resulting high temperature is sufficient to melt the metal being joined. The circuit is shown in Fig. 13.18(a).

When the electrode also melts and deposits metal on the work, it is said to be consumable.

13

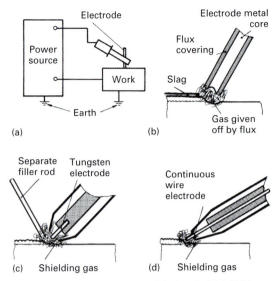

Figure 13.18 Arc welding (a) circuit (b) MMA Welding (c) TAGS welding (d) MAGS welding

Electrodes made from tungsten which conduct current but do not melt are known as non-consumable.

The most common arc-welding methods are manual metal arc welding, tungsten arc gas-shielded welding and metal arc gas-shielded welding.

13.8.1.1 Manual metal arc (MMA) welding, Fig. 13.18(b)

In this process the arc is struck between a flux-covered consumable electrode and the work. This method is the most widely used form of arc welding and is used on all materials with the exception of aluminium. The flux produces gas which shields the surface of the molten metal and leaves behind a slag which protects the hot metal from the atmosphere while cooling and has to be chipped off when cool.

13.8.1.2 Tungsten arc gas-shielded (TAGS) welding, Fig. 13.18(c)

In this process the arc is struck between a non-consumable tungsten electrode and the workpiece. The tungsten electrode is held in a special gun through which argon gas flows to shield the electrode and molten metal from atmospheric contamination – the process is often referred to as TIG or argon arc welding.

Additional filler metal can be applied separately as rod or wire. The argon shield enables aluminium, magnesium alloys and a wide range of ferrous metals to be welded without the use of a flux. This method is used primarily for welding sheet metal and small parts and produces a high-quality weld.

13.8.1.3 Metal arc gas-shielded (MAGS) welding, Fig. 13.18(d)

In this process the arc is struck between a continuous consumable wire electrode fed through a special gun. A shielding gas – argon, carbon dioxide (CO_2), oxygen or a mixture of these – is also fed through the gun to shield the arc and molten metal from contamination. Using different filler wires and types of gas, this method is suitable for welding aluminium, magnesium alloys, plain-carbon and low-alloy steels, stainless and heat-resisting steels, copper and bronze. This process is often referred to as MIG or MAG welding.

Using carbon-dioxide shielding gas for plain-carbon and low-alloy steels, this method is referred to as CO_2 welding.

13.8.2 Electron beam welding (EBW)

This is a fusion welding process in which the joint to be welded is bombarded with a finely focused beam of high-velocity electrons. As the electrons hit the workpiece, their energy is converted to heat. The heat penetrates deeply, producing parallel-sided welds and making it possible to weld thick workpieces, with a typical maximum of 50 mm. Because the beam is tightly focused, the heat-affected zone is small, resulting in low thermal distortion. This gives the ability to weld close to heat-sensitive areas and the capability of welding in otherwise inaccessible locations. Weld face preparation requires a high degree of accuracy between the two mating surfaces.

Almost all metals can be welded; the most common of these are aluminium, copper, carbon steels, stainless steels, titanium and refractory metals. The process can also be used to weld dissimilar metal combinations. It is used in the aerospace and car industries.

The beam generation and welding process is carried out in a vacuum and so there is a restriction on workpiece size. There is also a time delay while the vacuum chamber is being evacuated.

A mode of EBW is available called non-vacuum or out-of-vacuum since it is performed at atmospheric pressure. The maximum material thickness is around 50 mm and it allows for workpieces of any size to be welded since the size of the welding chamber is no longer an issue.

13.8.3 Laser beam welding (LBW)

This is a welding technique used to join materials together using a laser as an energy source. The laser is focused and directed to a very small point where it is absorbed into the materials being joined and converted to heat energy which melts and fuses the materials together. They can produce deep narrow welds with low heat input and so cause minimal distortion. They can be automated, producing aesthetically pleasing joints at fast rates.

There are two types of laser commonly used, solid-state and gas.

Solid-state lasers use a synthetic crystal of yttrium aluminium garnet (YAG) doped with neodymium and referred to as Nd:YAG. For a pulsed welding laser, the single crystal, shaped as a rod, is surrounded by a pump cavity which also contains the flash lamp. When flashed, pulses of light lasting about 2 milliseconds are emitted by the laser, which are delivered to the weld area using fibre optics. Cooling is provided to both the flash lamp and laser rod by flooding the entire pump cavity with flowing water. Power outputs can be in excess of 6 kW.

Gas lasers use CO_2 as a medium. They use high-voltage, low-current power sources to supply the energy needed to excite the CO_2 gas mixture. A rigid lens and mirror system is used to deliver the beam. Power outputs for gas lasers can be much higher than solid-state lasers, reaching 25 kW.

LBW is a versatile process and can be used to weld a variety of materials including carbon steels, stainless steel, titanium, aluminium, nickel alloys and plastics. Lasers are often used in high-volume production applications as they have high welding speeds and a level of automation which allows them to be used in numerically controlled machines and robots. They are also used to weld dissimilar metal combinations.

This process is used in a wide variety of industries including car, aerospace, shipbuilding and electronics.

Lasers are also used in laser cutting and laser marking in an extensive range of applications including manufacture of medical devices, car industry, electronics and jewellery.

13.8.4 Gas welding

Gas welding is usually known as oxy-acetylene welding, after the gases commonly used in the welding process, oxygen and acetylene.

Oxy-acetylene welding uses the combustion of the two gases, oxygen and acetylene, to create a source of intense heat, providing a flame temperature of around 3000/3200 °C. This heat is sufficient to melt the surface of the metals being joined, which run together and fuse to provide the weld. Additional material in the form of a filler rod may sometimes be required.

The gases are supplied at a high pressure in steel cylinders which are made to rigid specifications. Low-pressure systems are also available. Oxygen is supplied in cylinders painted black and provided with a right-hand threaded valve. Acetylene is supplied in cylinders painted maroon and provided with a left-hand threaded valve. The use of opposite-hand threads prevents incorrect connections being made. For ease of recognition, all left-hand nuts on acetylene fittings have grooved corners.

Pressure regulators are fitted to the top of each cylinder, to reduce the high pressure to a usable working pressure of between 0.14 and 0.83 bar. These regulators carry two pressure gauges, one indicating the gas pressure to the torch and the other indicating the pressure of the contents of the cylinder. Regulators are also identified by colour: red for acetylene and blue for oxygen.

The welding torch or blowpipe consists of a body, the gas mixer and interchangeable copper nozzle, two valves for the control of the oxygen and acetylene, and two connections for their supply. The body serves as a handle so that the operator

13

can hold and direct the flame. The oxygen and acetylene are mixed in the mixing chamber and then pass to the nozzle, where they are ignited to form the flame. The interchangeable nozzles each have a single orifice or hole and are available in a variety of sizes determined by the hole diameter. As the hole size increases, greater amounts of the gases pass through and are burned to supply a greater amount of heat. The choice of nozzle size is determined by the thickness of the work, thicker work requiring a greater amount of heat and consequently a larger nozzle size. Welding torch manufacturers supply charts of recommended sizes for various metal thicknesses and the corresponding gas pressure is to be used.

Synthetic rubber hose, with fabric reinforcement, is used to supply gas from the regulator to the torch and is colour coded: red for acetylene and blue for oxygen. It is essential that the hoses are correctly fitted. The connection for attaching the hose to the torch contains a check valve to prevent the flow of gas back towards the regulator. Between the hose and regulator a flashback arrestor is fitted to minimise the risk of a flame travelling back into the cylinder and causing an explosion.

A schematic of the complete welding set is shown in Fig. 13.19.

The oxy-acetylene welding process can be applied in steels up to and exceeding 25 mm thick, but is mainly used on gauges up to 16 SWG (1.6 mm), where the heat input needs to be flexible, and for welding die-castings and brazing aluminium where the heat must be maintained within a critical range.

13.8.5 Flame settings

During oxy-acetylene welding it is the flame which does the work and is therefore most important. The welding equipment already described merely serves to maintain and control the flame. It is the operator who sets the controls to produce a flame of the proper size, shape and condition to operate with maximum efficiency to suit the particular conditions.

To start the process, open the acetylene control valve on the torch and light the gas with a suitable spark lighter. Adjust the flame so that it burns without smoke or sooty deposit. Open the oxygen control valve on the torch and adjust to give a well-defined inner cone as described in neutral flame. When welding is finished, extinguish the flame by first closing the acetylene control valve on the torch followed by closing the oxygen control valve on the torch.

By adjusting the amounts of oxygen and acetylene it is possible to achieve three types of flame condition: neutral flame, carburising flame and oxidising flame.

13.8.5.1 Neutral flame

The neutral flame, Fig. 13.20(a), consists of approximately equal amounts of oxygen and acetylene being burned. Having adjusted the acetylene as described, the oxygen is adjusted to give a clearly defined light-blue inner cone, the length of which is two to three times its width. This indicates equal amounts of the gases are being used, combustion is complete and the flame is 'neutral'. This flame is the one most extensively used, having the advantage that it adds nothing to the metal being joined and takes nothing away – once the metal has fused, it is chemically the same as before welding. It is typically used for welding applications in steel, cast iron, copper and aluminium. The hottest part of the flame is approximately 3 mm forward of the inner cone which when welding should be held close to but clear of the molten pool.

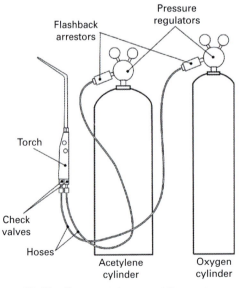

Figure 13.19 Oxy-acetylene welding set

13.8.5.2 Oxidising flame

The oxidising flame, Fig. 13.20(b), has an excess of oxygen being burned. Start with a neutral flame and slightly increase the quantity of oxygen. The inner core will take on a sharper pointed shape and the flame is more fierce with a slight hissing sound. This type of flame is used when welding brasses and should be avoided when welding steels.

13.8.5.3 Carburising flame

The carburising flame, Fig. 13.20(c), has an excess of acetylene being burned. Start with a neutral flame and slightly increase the quantity of acetylene. The inner cone will become surrounded by a white feathery plume which varies in length according to the amount of excess acetylene. This flame is used when hardfacing – a process in which a layer of hard metal is deposited on the surface of a soft metal to give localised resistance to wear. This type of flame should be avoided when welding steels.

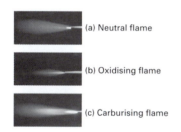

(a) Neutral flame

(b) Oxidising flame

(c) Carburising flame

Figure 13.20 Oxy-acetylene flame settings

13.8.6 Safety in the use of oxy-acetylene welding

Common causes of accidents result from:
▶ fires started by flames, sparks and hot material;
▶ explosion (working on vessels which contain or have contained flammable substances);
▶ working in confined spaces (leading to oxygen depletion and risk of suffocation);
▶ falling gas cylinders;
▶ gas leaks from hoses, valves and other equipment;
▶ health issues from inhalation of harmful fumes and noise (particularly during weld preparation);

▶ misuse of oxygen (if used with incompatible materials, oxygen reacts explosively with oil and grease).

Remember that you have a duty under the various health and safety regulations already covered in Chapter 1. To avoid the risk of accident:

▶ Never use oxy-acetylene equipment unless you have undergone the required training.
▶ Always wear goggles with the correct filter.
▶ Always wear required PPE (e.g. gauntlets, apron, leg protection and boots).
▶ Handle a lighted torch with great care.
▶ Always turn off torch when not in use.
▶ Shut off gas supply after use.
▶ Remove any combustible material from the work area.
▶ Always work in a safe location away from other people.
▶ Always work in a well-ventilated area.
▶ Use guards to prevent hot particles passing through openings.
▶ Clamp workpieces; never hold by hand.
▶ Keep hoses away from working area to prevent contact with flames, sparks and hot spatter.
▶ Carry out regular checks for gas leaks.
▶ Never allow oil or grease to contact oxygen valves or fittings.
▶ Maintain a constant fire watch.
▶ Keep fire extinguishers nearby and know how to use them.
▶ Avoid cylinders coming in contact with heat.
▶ Secure cylinders by chains during storage, transport and use.
▶ Store and transport cylinders in an upright position.
▶ Store flammable and non-flammable gases separately.
▶ Store full and empty cylinders separately.

13.9 Adhesives

An adhesive is a non-metallic material used to join two or more materials together. Adhesive bonding is the modern term for gluing, and the technique is used to join metals to themselves and also a wide variety of metallic and non-metallic materials including thermoplastic and thermosetting plastics, metal, glass, ceramics, rubber, concrete and brick. Many types of adhesive are available,

13

their applications varying from use in surgery to heal internal wounds and organs to joining structural members in aircraft. When selecting adhesive bonding as a joining method, a number of factors should be considered such as:

▶ Materials being joined, e.g. similar/dissimilar, metallic/non-metallic and thick/thin.
▶ Joint and loading conditions: different types of load arise depending on the joint geometry and the direction of loading. These are classified as tensile, shear, cleavage or peel (see Fig. 13.21). In general, shear loading is more desirable than tensile, cleavage or peel.
▶ Operating environment, e.g. wet, humidity or temperature range that the joint will experience.
▶ Manufacturing facilities, e.g. methods of adhesive application, cure rate, health and safety issues (when solvents are involved).

to remove dust, dirt, grease, oils and even finger marks.

▶ Surface roughening – to remove unwanted layers and generate a rough surface texture, e.g. by shot blasting.
▶ Chemical treatments – by immersion in a solution which will etch or dissolve part of the surface, so that the surface becomes chemically active, e.g. acid etching and anodising.
▶ Primers – applied by dipping, brushing or spraying to chemically alter or protect the surface. The primer acts as a medium which can bond readily to the adhesive.

The choice of joint design will depend on the nature of the structure being created. As already indicated, joint strength is higher under shear loading and so it is desirable to choose a joint geometry to take advantage of this as shown in the lap joint in Fig. 13.22(a). This lap joint creates a large load-bearing area and, in shear, results in a strong joint. The butt joint in Fig. 13.22(b) has a small area of contact and is in tension, giving a weak joint, and is an example of poor joint design.

Having selected the adhesive and designed the joint, the bonding process basically consists of thoroughly cleaning the surfaces to be joined, preparing and applying the adhesive and finally assembling the parts. The complete process may be carried out at room temperature or at a higher temperature, with or without accelerators, with or without the application of pressure and for a period depending on the time required for the adhesive to set (or cure). Some adhesives known as radiation or light-cured adhesives can be cured quickly when exposed to radiation, usually ultraviolet (UV) light.

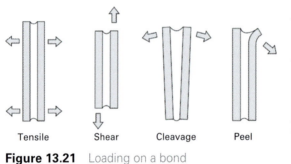

| Tensile | Shear | Cleavage | Peel |

Figure 13.21 Loading on a bond

As with all bonding operations appropriate surface preparation and joint design is essential to obtain a good bond. Surface preparation is achieved by carrying out a suitable surface pre-treatment such as:

▶ Cleaning/degreasing – solvent cleaning or detergent wash by wiping, dipping or spraying

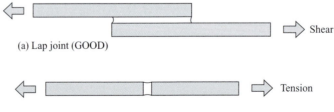

(a) Lap joint (GOOD) — Shear

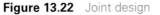

(b) Butt joint (BAD) — Tension

Figure 13.22 Joint design

There are advantages and limitations in using adhesives, some of which are listed below.

13.9.1 Advantages

▶ A variety of materials can be bonded. These may be similar or dissimilar; thick or thin; metallic or not metallic.
▶ Thin, delicate and heat-sensitive parts, which heat methods of joining would distort or destroy, can be bonded.
▶ When used to replace mechanical methods of joining:
  i) hole drilling is eliminated – reducing time,
                      – reducing costs,
                      – avoiding weakness in the region of the hole;
  ii) loads are distributed over total joint area;
  iii) a weight saving is made;
  iv) outer surface is smooth and free from bolt, screw and rivet heads;
  v) easier assembly – several components can be joined in one operation.
▶ The adhesive layer provides:
  i) a good seal against moisture and leaks;
  ii) good thermal and electrical insulation;
  iii) some flexibility in joint giving good damping properties. Useful in reducing sound and vibration.

13.9.2 Limitations

▶ Problems may exist:
  i) in preparing joint surfaces;
  ii) in storing, preparing and applying the adhesive;
  iii) in the time required for curing.
▶ Adhesive is not as strong as metal.
▶ Maximum bonding strength is not usually produced instantly.
▶ Assembled joint may need to be supported for at least part of the time to allow bond strength to build up.
▶ Bonded structures are mostly difficult to dismantle for repair or replacement.

▶ In most cases, temperature limitations in service are below those of other joining methods.
▶ There are health, safety and fire hazards when using solvent-based fluids.

13.9.3 Types of adhesive

There are many different types of adhesive available today, covering a wide range of applications, and they usually fall into two categories: natural and synthetic.

Natural adhesives derived from animals, caseins from protein isolated from milk, and starch, a carbohydrate extracted from vegetable plants, for example are used for bonding paper and card in the packaging industries.

Engineering applications require greater strengths of bond and usually come under the heading of structural adhesives. With the emergence of the chemical industry and synthetic polymers the range of synthetic adhesives (and sealants) has increased dramatically.

It is now more usual to classify the synthetic adhesives, based on the methods and way the adhesive cures, i.e. whether the bonding involves a physical or chemical mechanism.

13.9.4 Physical hardening adhesives

These are adhesives which, on application, are already present in their final chemical state. Two common types are:

13.9.4.1 Hot melt adhesive

These are thermoplastic polymers that are tough and solid at room temperature but very liquid at elevated temperatures.

They are used for fast assembly of structures designed to be only lightly loaded. Hot melts can be in the form of blocks, rods, granules, powder and film. The adhesive is heated and applied to one surface as a melt. Joining is then carried out immediately after application and the adhesive cools and develops strength by its consequent solidification. The quicker the surfaces are joined, the better for the bond. Because the bond strength is reached in seconds, the need for clamps of fixtures is eliminated. These adhesives can be used for joints which may subsequently need to be detached and/or re-attached due to their thermoplastic structure. In use, however, the bonded joint must not be heated up to its melting temperature.

Hot melts are widely used in the packaging, printing, shoemaking and wood processing industry as well as car manufacture and electronics.

13.9.4.2 Contact adhesive

These are mixtures of soluble elastomers and resins in the form of a solution in an organic solvent or as a dispersion in water. In principle, the solvents evaporate and the adhesive solidifies.

The adhesive is applied to both of the surfaces to be joined and the solvent is allowed to almost evaporate before the surfaces are joined. Joining is carried out under as high a pressure as possible and as soon as the surfaces contact, the adhesive polymers diffuse into each other and create a strong bond. It is not the duration of pressure which is important but rather the initial pressure.

Because the coated surfaces bond immediately on contact, great care needs to be taken in positioning prior to contact, since re-positioning is nearly impossible. Organic solvent-based adhesives dry faster than water based, but do have health and safety implications requiring careful handling.

Although water is a more environmentally friendly solvent used in some contact adhesives, it has not been able to replace organic solvents in all applications because of the possible effects of the moisture present. The drying time is slower than organic solvents but can be shortened by heating or increasing the air movement.

Contact adhesives are used in the woodworking and construction industries and in car manufacture.

13.9.5 Chemically curing adhesives

These are reactive materials that require chemical reaction to convert them from liquid to solid. Once cured these adhesives generally provide high-strength bonds which vary from flexible to rigid and resist temperature, humidity and many chemicals.

The chemical reactions that form the solid adhesive have to be blocked for a sufficient period to allow the adhesive to get to its final destination, i.e. the bonding joint. Some adhesives, which after mixing with their reaction partners spontaneously react at room temperature, are sold as two-component adhesives. These are present as 'resin' and 'hardener' in separate containers and are hence physically apart. They are only mixed together a short time before application.

With single-component adhesives, the adhesive components are pre-mixed in their final proportions. They are, however, chemically blocked. As long as they are not subjected to the specific conditions which activate the hardener, they will not bond. They require either high temperature or substances or media (light and humidity) from their surroundings to initiate the curing mechanism. The containers in which this type of adhesive is stored are carefully chosen to prevent any undesirable reaction.

13.9.5.1 Cyanoacrylates

These single-component adhesives cure through reaction with moisture held on the surfaces to be bonded. They are usually very thin liquids meaning that only gap widths of around 0.1 mm can be bridged. Some products have gap-filling properties up to 0.25 mm. They usually solidify in seconds and are commonly called 'superglues'. Although hand-tight joints are realised in seconds, their final strength is, however, only reached after several hours. Curing results from moisture found on most surfaces, which is usually slightly alkaline, and it is this combination which creates the bond. Any acidic surface such as wood will tend to inhibit the cure and in extreme cases will prevent curing altogether. They are very economical in use as it requires only a few drops of adhesive to provide strong joints.

These adhesives are thermoplastic when cured and therefore have a limited temperature capability and chemical resistance. They are suitable for bonding all types of glass, most plastics and metals and are, in general, used for bonding small components.

Great care must be taken when handling, as conditions are usually ideal for bonding skin to itself.

13.9.5.2 Anaerobic adhesives

These are also single-component adhesives but cure under the absence of oxygen (hence the name anaerobic – 'living in the absence of free oxygen'). Anaerobic adhesives are often known as 'locking compounds', being used to secure, seal and retain turned, threaded and similar close-fitting cylindrical parts. After application of the adhesive there is an oxygen-free environment because of the close fit of the joint. In order to then start the curing reaction, contact with a metal is also required. They are available in a number of strength grades. Since hardening only takes place in the absence of oxygen, any adhesive outside the joint will not cure and can be wiped off after assembly.

So that the adhesive does not cure prematurely it must remain in contact with oxygen. Hence it is kept in air-permeable plastic bottles which are only half filled and which, prior to filling, are flushed with oxygen.

Anaerobic adhesives are thermosets and resulting bonds have high strength and high resistance to heat. The joints are, however, brittle and not suitable for any joint likely to flex.

Besides their bonding function, anaerobic adhesives are often simultaneously used for their sealing properties because they are resistant to oils, solvents and moisture. These properties make this type of adhesive suitable for mounting engines in the car industry. Other applications include thread locking nuts, bolts, screws and studs to prevent loosening due to vibration, eliminating the use of locking washers and other locking devices. They are also used to reliably retain cylindrical-fitted parts such as bearings, pulleys, couplings, rotors and gears in the car and machine tool industry.

13.9.5.3 Toughened acrylic adhesives

These adhesives were developed from earlier work on single-component anaerobic adhesives. A toughened adhesive has small rubber-like particles dispersed throughout the adhesive which improves some of the properties. Toughened acrylics are relatively fast curing and offer high strength and toughness and have more flexibility than the common epoxies.

Two types are available:

13.9.5.3.1 Two-component, two-step

These adhesives do not require mixing prior to use and are cured using a liquid activator. The adhesive is applied to one surface, the more absorbent (more porous) surface, in beads or spread in a thin layer using a flat edge. The activator is applied to the other surface, the less absorbent (less porous) surface. The two parts are then brought together and clamped or held in position to achieve best results. Full bond strength will typically be achieved within 24-hour cure time.

They give excellent adhesion to all metals, glass and composite materials.

13.9.5.3.2 Two-part

These adhesives require mixing, usually a mix ratio of 1:1 resin to accelerator. They give an excellent strength of bond with good impact and durability and bond well to many metals, ceramics, wood and most plastics. The resin and accelerator are supplied in a dual cartridge and used in an applicator with a mixing nozzle. Small quantities can be mixed thoroughly by hand. For maximum bond strength the adhesive is applied evenly to both surfaces to be joined. The surfaces are then brought together as quickly as possible and allowed to cure at 16 °C or above until completely firm. The application of heat (between 49 °C and 66 °C) will speed up curing.

The parts should be held secure and not allowed to move during curing.

13.9.5.4 Epoxy resins

Epoxy-based systems are the most widely used structural adhesives. They are encountered everywhere in car manufacture, aircraft, building and construction industries, metal fabrication and the home. They are available as single-part and two-part.

Single-part epoxy adhesives are available as liquid, paste or film and require heat to cure. The resin and hardener (or catalyst) are pre-mixed but curing does not occur because the hardener is inactive at room temperature. It only becomes reactive as the temperature is raised, usually above 100 °C, forming a hard thermoset polymer which will not remelt on further heating.

13

Two-part epoxies start to react at room temperature once the two components have been mixed together. They are supplied as resin and hardener in separate containers. The mechanism for curing requires exact quantities of resin and hardener and their thorough mixing together. The final cure at room temperature can range from a few minutes to a few hours. Curing time can be shortened by the application of heat which also results in an increase in strength and stability of the bond.

Epoxy adhesives can be used to bond a wide variety of materials with high strength. In some cases they have been used to replace traditional joining methods such as nuts and bolts, rivets, welding, soldering and brazing.

Additives can be used to give additional properties, e.g. in microelectronics silver powder is added to give electrically conducting adhesives, while aluminium oxide powder is added to give heat-conducting properties.

13.10 Electrical connections

Any electrical connection must securely anchor all the wires of the conductor and not place any appreciable mechanical stress on the terminal.

13.10.1 Mechanical connections

The simplest form of mechanical connection is where the wire is looped round clear of the diameter of a screw and is firmly clamped by a washer under the screw head or, alternatively, by means of a washer and nut, Fig. 13.23(a) and (b). Another form is the brass pillar terminal found in plugs, sockets and lampholders, which has a hole drilled through it in which the wire is securely held by means of a brass screw, Fig. 13.23(c).

Connection to a socket can be made by squeezing or crimping directly to the wire. Alternatively, the wire can be soldered into the socket, Fig. 13.23(d).

Wrapped joints are used to connect wires to terminal posts, Fig. 13.23(e). The terminal post, called a wrapping post, is square or rectangular in section. Several turns of the wire are twisted under pressure round the post, using a special wrapping tool. The electrical connection is made by the wrapped wire digging into the corners of the post.

13.10.2 Soldered connections

Solder alone should not be relied upon to make a secure connection. The wire should be bent or wrapped round to give good mechanical strength and then be soldered to give the required electrical connection. Sufficient solder should be applied to enable the wire to be seen through the solder, and with just enough heat to allow the solder to flow freely round the connection (see Section 13.5).

Soldered connections are made directly to a circuit or to tags of various design, one of which is shown in Fig. 13.23(f).

13.10.3 Vehicle connections

Electrical connections on vehicles may need to be easily disconnected for replacement of components. One such type is the 'bullet' connector, shown in Fig. 13.23(g), where the ends are soldered or crimped to the wire and then pushed firmly into the insulated connector. Quick-connect types as shown in Fig. 13.23(h), usually crimped to the wire, are pushed together to form the connection. An insulating sleeve can then be pushed over the completed connection.

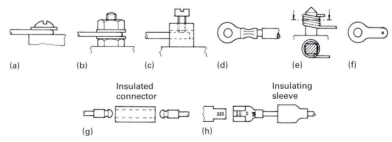

Figure 13.23 Electrical connections (a) washer (b) nut (c) pillar terminal (d) socket (e) wrapped joint (f) tag (g) bullet connector (h) quick-connect connector

13.11 Relative merits of joining methods

	Heat requirement	Type of joint	Heat conductivity
Soldering	Less than 300 °C	Permanent – can be dismantled using heat	Not above 183 °C
Brazing	Above 450 °C	Permanent – can be dismantled using heat	Not above the solidus of the brazing alloy used
Welding	Melting point of metals being joined	Permanent	Up to melting point of metals joined
Adhesives	None (except the hot-melt films)	Permanent	Insulator
	Electrical conductivity	Strength of joint	Type of material joined
Soldering	Conductor	Low	Similar and dissimilar metals
Brazing	Conductor	Medium	Similar and dissimilar metals
Welding	Conductor	High	Similar metals
Adhesives	Insulator	Low to high	Similar and dissimilar metals and non-metals

Review questions

1. What is the composition of a general-purpose soft solder?
2. State four advantages of using adhesives instead of mechanical or heat joining methods.
3. How does welding differ from soft soldering or brazing?
4. Name four methods used to join parts.
5. State the purpose of using a flux during a brazing operation.
6. Name three types of thread systems.
7. To what process does MAGS refer and where would it be used?
8. Name four types of locknut.
9. What considerations should be made in designing a soft soldered joint?
10. State two types of adhesive used in metal joining.
11. Name the two gases associated with gas welding.
12. State three types of flame setting used in gas welding. Which is most commonly used and why?

13

CHAPTER 14

Materials

A wide range of materials is used in engineering, and it is important to be aware of the ways in which these are applied and of the properties which make them suitable for these applications.

Properties of materials can be divided into two groups: physical and mechanical. Physical properties are those properties of a material which do not require the material to be deformed or destroyed in order to determine the value of the property. Mechanical properties indicate a material's reaction to the application of forces. These properties require deformation or destruction tests in order to determine their value. The value of these properties can be altered by subjecting the material to heat treatment and cold or hot working.

14.1 Physical properties

14.1.1 Coefficient of linear expansion

This is a measure of the amount by which the length of a material increases when the material is heated through a one-degree rise in temperature. Thus

$$\text{increase in length} = \text{original length} \times \text{temperature rise} \times \text{coefficient of linear expansion}$$

Thus, if the coefficient of linear expansion of copper is 0.000017 per °C (written 0.000017/°C or 17×10^{-6}/°C) then for each degree rise in temperature a length of copper will expand by 0.000 017 of its original length; e.g. a 100 mm long copper rod will expand $0.000017 \times 100 = 0.0017$ mm for each degree rise in temperature. If this 100 mm long copper rod is heated through 20 °C, then the amount of expansion will be $100 \times 0.000017 \times 20 = 0.034$ mm.

Different metals expand or contract by different amounts for a given temperature change; e.g. aluminium expands at a greater rate than cast iron. That different metals have different values for the coefficient of linear expansion can be useful on some occasions while on other occasions it can be a disadvantage.

A typical application of advantage is in the construction of a thermostat. This device makes use of two strips of different materials clamped together, the different expansion rates when heated causing the strip to bend and so make or break an electrical contact, Fig. 14.1.

The disadvantages are many and have to be allowed for during design. For instance, the clearance between the aluminium piston and the cast-iron cylinder block in motor-vehicle engines

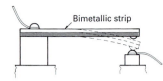

Figure 14.1 Bimetallic strip thermostat

will be less when the engine is hot than when it is cold.

14.1.2 Specific heat capacity

The specific heat capacity of a material is the amount of heat energy (in joules) required to raise the temperature of unit mass (1 kg) of the material by unit rise in temperature (1°C).

The quantity of heat energy required to raise the temperature of a piece of material depends upon the type of material and its mass. Thus equal masses of two different materials will require different amounts of heat energy to raise their temperature by the same amount; e.g. 1 kg of water with a specific heat capacity of 4200 J/(kg °C) will require 4200 joules of heat energy to raise its temperature by 1°C. Similarly, 1 kg of copper of specific heat capacity 386 J/(kg °C) will require 386 joules to raise its temperature by 1°C.

A knowledge of this property is required when dealing with heating or cooling operations. Various liquids are used when cooling after heat-treatment operations. The liquid used must be able to absorb the heat energy from the block of metal, and this depends upon the type of liquid and its mass. In metal-cutting operations, the coolant used must be of a type and delivered in sufficient volume to remove heat from the tool and cutting area without itself becoming too hot.

14.1.3 Density

Equal volumes of different materials have different masses. The mass in a given volume is a measure of the density.

The density of a material is the mass per unit volume and is given by the equation

$$\text{density} = \frac{\text{mass of material}}{\text{volume occupied by the material}}$$

Since mass is measured in kilograms and volume in cubic metres, the unit of density is kilograms per cubic metre, written kg/m³. For example, the density of aluminium is 2700 kg/m³ and that of lead is 11 300 kg/m³. This means that a volume of one cubic metre of aluminium has a mass of 2700 kg and the same volume of lead has a mass of 11 300 kg.

This property must be considered where the mass has to be restricted. In the production of aircraft, for instance, special materials have been developed which are as strong as steel but are only a fraction of its density.

14.1.4 Melting point

This is the temperature at which a material changes from the solid to the liquid state. This may be an important consideration in some material applications; e.g. it is important to know the melting point of a solder if it is used on a joint which may be subjected to temperatures approaching the solder's melting point. The many plastics now available must be used within their temperature limits. Equipment used in hot-working processes – such as furnaces, casting machines and forging dies – must be designed to withstand their high working temperatures.

14.1.5 Thermal conductivity

When one end of a metal bar is heated, the heat will be conducted along the length of the bar. The rate at which the heat is conducted depends on the bar material, some materials being better conductors than others. This heat-conducting ability of a metal is measured by the thermal conductivity.

Thermal conductivity is given as a rate of transfer of heat energy, measured in J/(m s °C) – i.e. the number of joules of energy transferred per second through one metre for each degree Celsius rise of temperature. Since one joule per second equals one watt, thermal conductivity can also be quoted in W/(m °C).

A good conductor such as copper has a high thermal conductivity and is used where heat has to be readily transferred, e.g. in a soldering iron or a motor-car radiator. Bad conductors, such as the non-metallic materials, have a low thermal conductivity and are used where heat has to be retained, e.g. lagging materials on hot-water tanks and pipes.

14.1.6 Electrical resistivity

Some materials allow electricity to pass through them very easily and are electrical conductors. These include carbon and most of the metals, such as aluminium, copper, brass and silver. Other materials offer a high resistance to the flow of electricity and are bad electrical conductors, known as insulators – these include non-metallic materials such as plastics, rubber, mica, ceramics and glass.

The resistance of an electrical conductor is measured in ohms and depends on the dimensions of the conductor as well as the material from which it is made. It is fairly easy to see that conductors of similar shape but made of different materials may have different resistances.

In order to compare the resistance effect of different conductor materials, a standard size and shape of conductor is considered. The standard shape chosen is a cube whose sides are one metre. The resistance of this metre cube of material is known as the resistivity of the material, measured in ohm metres. Thus an electrical-conducting material will have a low resistivity while an insulator will have a high resistivity.

The resistance to the flow of electricity can be found from the following equation, knowing the area and length of the conductor and its resistivity:

$$R = \frac{\rho l}{a}$$

where R = resistance, in ohms

l = length of conductor, in metres

a = cross-sectional area of conductor, in square metres

ρ = resistivity of the conductor material, in ohm metres

14.1.7 Example 14.1

What is the resistance of an electrical conductor 1 mm diameter and 20 metres long whose resistivity is 2.5×10^{-8} ohm metres?

$$\text{area of conductor} = \frac{\pi \times 1^2}{4} = 0.7854 \text{ mm}^2$$

Since the other values are expressed in metres,

$$\text{area in m}^2 = \frac{0.7854}{1000 \times 1000} = 0.7854 \times 10^{-6} \text{ m}^2$$

Resistance is given by $\frac{\rho l}{a}$ which is

$$\frac{2.5 \times 10^{-8} \times 20}{0.7854 \times 10^{-6}} = 0.637 \text{ ohms}$$

14.2 Mechanical properties

14.2.1 Hardness

A material which is hard is able to resist wear, scratching, indentation and machining. Its hardness is also a measure of its ability to cut other materials. Hard materials are required for cutting tools and for parts where wear must be kept to a minimum.

14.2.2 Brittleness

A brittle material will break easily when given a sudden blow. This property is associated with hardness, since hard materials will often be brittle. Brittle materials cannot be used in the working parts of power presses, which are subjected to sudden blows.

14.2.3 Strength

A strong material is able to withstand loads without breaking. Loads may be applied in tension, compression or shear, and a material's resistance to these loads is a measure of its tensile strength, compressive strength and shear strength. Connecting rods in internal-combustion engines must be strong in tension and compression, while the gudgeon pin must be strong in shear, Fig. 14.2.

14.2.4 Ductility

A ductile material can be reduced in cross-section without breaking. In wire-drawing, for instance,

Tension   Compression   Shear

Figure 14.2 Tension, compression and shear

14

the material is reduced in diameter by pulling it through a circular die. The material must be capable of flowing through the reduced diameter of the die and at the same time withstand the pulling force.

14.2.5 Malleability

A malleable material can be rolled or hammered permanently into a different shape without fracturing. This property is required when forging, where the shape of the metal is changed by hammering. Lead is a malleable material, as it can easily be shaped by hammering, but is not ductile since it is not strong enough to withstand a load if attempted to be drawn into wire. Heat may be used to make a material more malleable.

14.2.6 Elasticity

A material which is elastic will return to its original dimensions after being subjected to a load. If loaded above a point known as the elastic limit, the material will not return to its original dimensions and will be permanently deformed when the load is removed. Elasticity is essential in materials used in the manufacture of springs.

14.2.7 Toughness

A material is tough if it is capable of absorbing a great deal of energy before it fractures. A tough material will withstand repeated flexing or bending before it begins to crack or break. The working parts of power presses must be tough to withstand the repeated blows in pressing operations.

As previously stated, mechanical properties indicate a material's reaction to the application of forces and their values are determined through a series of standard tests. These tests are carried out on small specimens of the material under test until deformed or destroyed. Such tests are referred to as destructive tests. These tests include tensile, compressive, torsion, bend and impact testing and are used to determine the mechanical properties already outlined.

14.3 Comparison of properties

The mechanical and physical properties of common plastics and metallic materials are compared in Table 14.1.

14.4 Non-destructive testing (NDT)

Tests of a different nature and purpose are used to examine manufactured components and assemblies for internal flaws and faults and surface cracks and defects without destroying the component. These tests are known as non-destructive tests (NDT), since the component is not physically damaged as a result of the test and therefore remains 'fit for purpose'. These tests are carried out to check possible defects produced during machining, welding, casting and heat treatment and are also carried out on 'in-service' component's, e.g. jet engine turbine blades and aircraft components.

Non-destructive evaluation (NDE) is a term often used together with NDT. However, NDE is used to describe measurements that are more quantitative in nature, i.e. detection of the defect is not enough, quantitative information is required regarding flaw size, shape and orientation, information which could be used to determine the component's suitability for use.

No single NDT method will work for all flaw detection applications. Each method has advantages and disadvantages when compared to each other. There are many types of test but the following six are the most often used.

1. Visual testing – this involves an inspector using the naked eye or magnifying glass. This can only be used for locating surface cracks or defects large enough to be seen by these methods and has to be carefully carried out to be successful. The use of mirrors, and fibrescopes, borescopes and inspection cameras incorporating fibre optics with integral light source, allows access to areas otherwise inaccessible.
2. Penetrant testing – this method uses a penetrant dye applied to a pre-cleaned surface. The liquid penetrant which has high surface wetting characteristics is pulled into surface

Table 14.1 Comparison of mechanical and physical properties of common plastics and metallic materials

	Density (kg/m³)	Tensile strength (N/mm²)	Coefficient of linear expansion (10⁻⁶/°C)	Specific heat capacity (J/(kg °C))	Thermal conductivity (W/(m °C))	Melting point (°C)	Resistivity (Ωm)
Mild steel	7800	505	15	463	47	1495	16×10^{-8}
Grey iron	7000–7300	150–400	11	265–460	44–52	1100	10×10^{-8}
Malleable iron	7300–7400	280–690	11	520	40–49	1100	–
S.G. iron	7100–7200	370–800	11	460	32–36	1100	–
Copper	8900	216	17	386	385	1083	1.7×10^{-8}
70/30 brass	8530	320	20	379	117	935	6.2×10^{-8}
Phosphor	8820	400	18	379	70	1000	9.5×10^{-8}
Aluminium	2700	80	24	965	240	660	2.6×10^{-8}
Aluminium alloy	2790	250	22	965	150	600	4×10^{-8}
Zinc alloy	6700	280	27	418	113	400	5.9×10^{-8}
Lead	11300	15	29	126	35	327	21×10^{-8}
Platinum	21450	350	9	136	69	1773	11×10^{-8}
Silver	10 500	15	19	235	419	960	1.6×10^{-8}
Gold	19 300	120	14	132	296	1063	2.4×10^{-8}
						Softening point (°C)	
Polyethylene LD	925	7–16	160–180	2300	0.34	85–87	10^{14}
Polyethylene HD	950	21–38	110–130	2220	0.46–0.52	120–130	
P.V.C.	1390	58	50	840–2100	0.14	82	10^{14}
Polystyrene	1055	34–84	60–80	1340	0.11–0.14	82–103	10^{11}
ABS	1100	17–62	60–130	1380–1680	0.062–0.36	85	1.2×10^{13}
Acrylic	1200	48–76	50–90	1470	0.17–0.25	80–98	10^{12}
Polypropylene	900	29–38	110	1930	0.14	150	10^{14}
Nylon 6.6	1140	48–84	100–150	1680	0.22–0.24	75	$0.45 - 4 \times 10^{12}$

14

cracks by capillary action. Time is allowed for the penetrant to seep into defects. Excess penetrant is then carefully removed and the surface allowed to dry. A developer is then applied to the surface which acts like a blotter, pulling the trapped penetrant out of any imperfection open to the surface. The dyes which are coloured, show up as a 'bleed' on the surface of the developer and can be easily seen. Fluorescent dyes are used in conjunction with ultraviolet light which increases the sensitivity, allowing imperfections to be more easily seen and avoids the need for a developer. Penetrant testing is used to locate cracks and other defects that break the surface of the material such as fatigue, quench and grinding cracks, porosity and pin holes in welds and can be used to inspect large areas. The main disadvantages are that it will only detect surface defects, surface preparation is critical and post-cleaning is necessary to remove the chemicals.

3. Magnetic particle testing – a magnetic field is set up in the component which must be of a ferromagnetic material, i.e. a material that can be magnetised. This field can be set up using portable or stationary equipment. The surface is then dusted with iron particles finely coated with a dye pigment and which can be dry or suspended in a liquid. Surface or near-surface cracks or voids distort the magnetic field and concentrate iron particles near the imperfections and so appear as a visual indication of the flaw. This method can only be used with ferromagnetic materials and is used to detect surface or near-surface defects in castings, forgings and welded parts. Demagnetisation and post-cleaning is usually necessary.

4. Ultrasonic testing – uses high frequency sound waves which are sent into the material by the use of a transducer. The sound waves travel through the material and are reflected from the inside bottom surface and picked up by the probe which also acts as a receiver. The reflected wave signal, or echo, is transformed into an electrical signal and displayed on a screen. The time taken from signal generation to the echo return is proportional to the thickness of the material. If there is a discontinuity in the wave path (such as a crack) the pulse is interrupted and will be reflected back from the flaw surface. Since this echo returns in a shorter time, this will show the distance of the defect from the surface. Material thickness and changes in material properties can also be measured. This testing method is used to locate surface and sub-surface defects in materials such as metals, plastics and wood.

5. Eddy current testing – alternating electrical current is passed through a coil producing a magnetic field. This source is in the form of a probe where a range of sizes are available to suit the required application. If another electrical conductor, i.e. the component under test, is brought into close proximity a current will be induced in this second conductor known as eddy currents since they flow in a circular path. Interruptions in the flow of these eddy currents caused by imperfections, dimension changes, or changes in the material's conductive properties can be detected and monitored on an instrument display. This method is sensitive in detecting surface and near-surface defects giving immediate results. The equipment is very portable and can be used to inspect complex shapes and sizes. The limitations are that only conductive materials can be inspected, the surface must be accessible to the probe and reference standards are needed for set-up. This method can be used to perform a variety of inspections and measurements. These include:
   ▶ crack detection;
   ▶ material thickness measurement, e.g. tubes;
   ▶ coating thickness measurement, e.g. paints and plastics;
   ▶ conductivity measurements for:
      ▶ material identification;
      ▶ heat damage detection;
      ▶ case depth determination;
      ▶ heat treatment monitoring.

6. Radiographic testing – involves the use of penetrating X-rays or gamma rays to inspect components for defects. The source of radiation is either from an X-ray generator

14

or in the case of gamma rays, a radioactive isotope. The radiation is directed through the component under test onto film or other imaging media which is sensitive to the radiation. The thickness and density of the material which the rays have to penetrate affect the amount of radiation reaching the film or other source. Any variation in radiation produces an image which shows the internal features of the test component including any defect contained within. As well as producing an image on film similar to an X-ray of broken bones, technological advances have now resulted in an image produced electronically which can be viewed immediately on a computer screen similar to an airport security system. Radiography is used to inspect almost any material for surface and sub-surface defects and can also be used to locate and measure internal features and to measure thickness. The disadvantages include extensive operator training and skill requirement, relatively expensive equipment and the potential radiation hazard.

14.5 Plain-carbon steel

Plain-carbon steels are essentially alloys of iron and carbon together with varying amounts of other elements such as manganese, sulphur, silicon and phosphorus. These additional elements are found in the raw materials used in the steel-making process and are present as impurities. Both sulphur and phosphorus are extremely harmful and cause brittleness in the steel – they are therefore kept to a minimum. The effect of these is offset by the presence of manganese. The carbon content varies up to about 1.4%, and it is this carbon which makes the steel harder and tougher and able to respond to the various heat-treatment processes.

Low-carbon steels cover a range of steels with carbon content up to 0.3%. These cannot be hardened by direct heating and quenching, but can be case-hardened. Steels containing 0.2% to 0.25% carbon, referred to as mild steels, are used in lightly stressed applications and can be readily machined and welded. They are used for general engineering purposes as bar, plate, sheet and strip and for cold-forming operations. The tensile strength of rolled section for 0.2% carbon content is $300 \, \text{N/mm}^2$, and for 0.25% carbon content increases to $430 \, \text{N/mm}^2$.

Medium-carbon steels cover a range of steels with a carbon content above 0.3% up to 0.6%. They can be hardened by direct heating and quenching, and tempered to improve the mechanical properties. Steels containing 0.4% carbon are used where higher stressing and toughness is required for forgings, levers, shafts and axles. In the normalised condition the tensile strength is around $540 \, \text{N/mm}^2$, and in the hardened and tempered condition it can increase to around $700 \, \text{N/mm}^2$.

Those steels with 0.6% carbon have a higher tensile strength, $700 \, \text{N/mm}^2$ in the normalised condition, and can be hardened and tempered up to $850 \, \text{N/mm}^2$. They are used where wear properties are of greater importance than toughness, for sprockets, machine-tool parts and springs.

High-carbon steels cover the range of steels with a carbon content between 0.6% and 1.4%. Those containing 0.6% to 0.9% carbon are widely used in the hardened and tempered condition for laminated and wire springs and in the manufacture of spring collets. Steels having a carbon content above 0.9% are used in the hardened and tempered condition for hand and cutting tools where hardness is important, e.g. cold chisels, punches, files and woodworking tools.

14.6 Heat treatment of plain-carbon steel

Heat treatment is a process in which a metal, in its solid state, is subjected to one or more cycles of heating and cooling in order to obtain certain desired mechanical properties. The mechanical properties listed in Section 14.2 can be altered by changing the size, shape and structure of the grains from which the material is made up.

When a plain-carbon steel is heated through a sufficient temperature range, there is a particular temperature at which the internal grain structure begins to change. This temperature, known as the

14

Thed document

lower critical temperature, is about 700°C and is the same for all plain-carbon steels, Fig. 14.3.

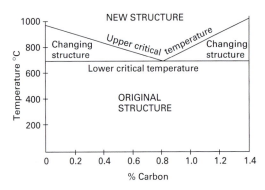

Figure 14.3. Relationship of critical temperature and carbon content

When the steel is heated still further, the structural changes continue until a second temperature is reached where the change in the internal structure of the steel is complete. This temperature is known as the upper critical temperature and varies for plain-carbon steel according to the percentage carbon content, Fig. 14.3.

The influence of the carbon content is so great that the heat treatment and subsequent employment of the steel is determined by this factor.

The temperature range between the lower and upper critical temperatures is known as the critical range.

If the steel, at a point above its upper critical temperature, is plunged into a cold liquid – a process known as quenching – the result will be to permanently fix this new structure, i.e. the structure is suddenly 'frozen' before it can change back to its original state.

If the steel is heated to above its upper critical temperature and instead of being quenched is allowed to cool slowly, structural changes take place in the reverse order to those during heating. When cold, the steel will have returned to its normal structure.

It follows from this that, if the rate of cooling is varied, considerable changes in structure and therefore variations in mechanical properties can be obtained.

It should be noted that the steel is still in a solid state during the heating and cooling cycles. The changes which take place are internal structural

changes only – the steel never reaches its melting point.

Heat-treatment operations in industry are carried out in correctly controlled furnaces, the most common processes being annealing, normalising, hardening and tempering.

14.6.1 Annealing

This process is carried out to soften the steel so that it may be machined or so that additional cold-working operations such as pressing and bending can be carried out.

The process involves heating the steel to a temperature depending upon its carbon content, Fig. 14.4, holding it at this temperature for a period of time depending upon the thickness of the steel, so that the whole mass reaches the correct temperature (known as 'soaking'), and finally allowing the steel to cool as slowly as possible. This slow rate of cooling is achieved by switching off the furnace, allowing the steel in the furnace and the furnace itself to cool at the same slow rate.

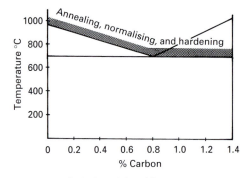

Figure 14.4. Relationship of heat-treatment temperature and carbon content

The result is a steel having large grains in its structure which is soft, ductile, low in strength and easily shaped by machining, pressing and bending.

14.6.2 Normalising

This heat-treatment process is carried out to give the steel its 'normal' structure. For example, a steel which has been forged has a grain structure which has been distorted due to the hot working. Such a steel requires normalising, to return the grains to their normal undistorted structure to be in the best condition for use.

The process differs from annealing only in the rate of cooling. The steel is heated to the required temperature, depending again upon the carbon content, Fig. 14.4, and is allowed to soak. The steel is then removed from the furnace and is allowed to cool in still air. This gives a faster rate of cooling than annealing, resulting in a steel with smaller grains which is stronger but less ductile than an annealed steel.

14.6.3 Hardening

In contrast to annealing, this heat-treatment process is designed to produce a steel that is hard. Steel will vary in hardness depending upon the carbon content, the hardness increasing as the carbon content increases, Fig. 14.5. As the hardness increases so will the brittleness, which must be borne in mind when deciding what the material is to be used for.

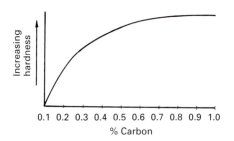

Figure 14.5. Hardness obtainable with a quenched plain-carbon steel

Hardening is carried out by raising the temperature, again depending on the carbon content, Fig. 14.4, in the same way as for annealing and normalising, and allowing the steel to soak. The difference is again in the rate of cooling. This time the steel is removed from the furnace and is cooled very quickly, or quenched, by immersion in a suitable liquid such as water or oil.

Plain-carbon steels having a carbon content below 0.3% cannot be effectively hardened this way, due to the small amount of carbon present. These low-carbon steels can be hardened by a method known as case-hardening. The steel is heated to above the upper critical temperature while in contact with a carbon-rich material. Carbon is absorbed into the surface of the steel, raising the carbon content at the surface to around 0.9%, a process known as *carburising*. The steel can

then be hardened as a 0.9% steel as previously described.

Carburising can be carried out using solid or liquid materials or a gas. Pack carburising is carried out by packing the steel in a solid carbon-rich material similar to charcoal in a steel box with the lid sealed using fireclay. The box is put in a furnace and the temperature is raised to between 900°C and 950°C and kept at this level for a period depending upon the required depth of penetration of the carbon. At the end of this period, the furnace is switched off and the box and its contents are allowed to cool slowly. Thus the steel now consists of a core with the original low carbon content, usually 0.15%, and an outer case of high carbon content to the required thickness.

Subsequent heat treatment is carried out which produces a tough core (0.15% carbon) and a hard wear-resisting outer case (0.9% carbon). This method is used for shafts and spindles which, as well as requiring a hard wear-resistant outer surface, also require toughness to resist bending and torsional loads.

14.6.4 Tempering

This heat-treatment process is carried out after a steel has been hardened. Steel in a hard state is brittle and if used could easily break. Tempering removes some of the hardness, making the steel less brittle and more tough. This is done by reheating the hardened steel to a temperature between 200°C and 450°C, to bring about the desired structural change. The steel can then be quenched or cooled slowly, since it is the temperature to which the steel is raised which brings about the necessary structural change, not the rate of cooling. In general, the higher the temperature to which the hardened steel is raised the tougher the steel will be, with a corresponding reduction in hardness and brittleness, Fig. 14.6.

A component such as a scriber which requires to be hard and will not be subjected to shock loads would be tempered by reheating to a lower temperature. The head of a hammer which requires to be tough to withstand shock loads, with no great degree of hardness, would be tempered by reheating to a higher temperature.

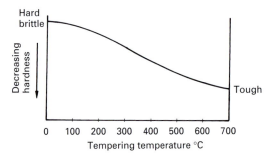

Figure 14.6. Relationship of tempering temperature and hardness

Industrial tempering is carried out in correctly controlled furnaces, but it is a curious phenomenon that the colours caused by surface oxidation as steel is heated correspond fairly closely to the temperature within the tempering range, shown in Table 14.2.

If, after hardening, a steel is rubbed with emery cloth to give a bright surface and is then reheated, the bright surface will take on the oxide colours – pale straw at the lower end, approximately 230°C, through to blue as the temperature is increased to approximately 300°C. These colours cannot of course be relied upon to give accurate results, but they may be a useful aid in the workshop.

More accurate results can be obtained by the use of thermal crayons. A crayon of an appropriate colour is selected for the desired temperature and is rubbed on the surface of the cold metal. The metal is then heated and the crayon deposit changes to a different colour, in accordance with a colour chart provided, when the correct temperature is reached.

14.7 Cast iron

There are many types of cast iron available, covering a wide range of mechanical and physical properties. All of them can be easily cast into a variety of simple or complex shapes. Like steel, cast iron is an alloy of iron and carbon, but with the carbon content increased to between 2% and 4%.

14.7.1 Grey cast iron

Grey cast iron is widely used for general engineering applications, due to its cheapness and ease of casting. It is available in a variety of different grades, from those that are relatively soft and of low strength to relatively high-strength harder materials. Tensile strengths range from 150 to 400 N/mm².

Grey cast irons contain carbon in the form of graphite flakes distributed throughout, which create a weak structure, Fig. 14.7(a). Other elements present include:

▶ silicon, which aids in the formation of the free graphite and may be present up to about 3%;
▶ phosphorus, which helps make the cast iron more fluid and may be present up to 1.5%;
▶ sulphur, present as an impurity from the furnace – too much sulphur tends to produce unsound castings so the sulphur content is kept to a minimum, about 0.1%;
▶ manganese, which toughens and strengthens the iron, partly because it combines with the unwanted sulphur, and is present up to about 1.5%.

(a)   (b)   (c)

Figure 14.7 Cast irons (a) grey (b) malleable (c) spheroidal-graphite

Grey irons are brittle materials, have a high compressive strength between three and four times the tensile strength, and can be easily cast to shape and easily machined.

There is no limit to the size of grey-iron castings. Although not a corrosion-resistant material, grey iron does have a useful resistance to chemicals, water, gas and steam, and for this reason is used for valves, pipes and fittings with these substances.

Table 14.2 Tempering temperatures and oxide colours

Temperature (°C)	230	240	250	260	270	280	300
Oxide-film colour	Pale straw	Dark straw	Brown	Brownish purple	Purple	Dark purple	Blue

Grey iron has a good wear-resistance and is used for castings in machine tools containing slideways where continuous lubrication is not always possible. The free graphite on the surfaces acts as a temporary lubricant and, when used up, creates minute pockets which then act as small reservoirs for the machine lubricant.

The reasonable mechanical strength and good thermal conductivity make grey iron a suitable material for cylinder heads, and brake and clutch applications in motor vehicles.

14.7.2 Malleable cast iron

All grades of malleable iron start as white-iron castings, free from graphite. The castings are then made tough and machineable by an annealing process. Due to the limited capacity of heat-treatment equipment, malleable iron castings are usually light in section and seldom weigh more than 50 kg.

Malleable irons are used in place of grey irons because of their increased tensile, impact, and fatigue strengths.

The three main types of malleable iron are: *whiteheart*, *blackheart* and *pearlitic*. These have different properties due to the different heat treatments given to the initial white iron resulting in rosettes or clusters of graphite, Fig. 14.7(b).

Whiteheart-malleable-iron castings are used where there is a need for small ductile components, particularly with thin sections. Uses include pipe fittings, frame sockets, steering-column housings and engine bearers on motor cycles, and parts for agricultural and textile machinery. Tensile strengths range from 340 to 410 N/mm².

Blackheart-malleable-iron castings have somewhat better tensile strengths than grey iron but much higher ductility and good machineability. They can be used under shock-loading conditions and are widely used in the car industry for small components such as door hinges and brackets where wear-resistance is not a primary requirement. Tensile strengths range from 290 to 340 N/mm².

Pearlitic-malleable-iron castings are used where high strength is required together with a good wear-resistance. There are many applications, particularly in the car industry, which include axle and differential housings and gears. Tensile strengths range from 440 to 690 N/mm².

14.7.3 Spheroidal-graphite cast iron

Spheroidal-graphite cast iron combines the strength, toughness and ductility of steel with the ease of casting of grey cast iron. Usually abbreviated to SG iron, it is also known as 'ductile' or 'nodular' iron.

In SG iron, the graphite is present as spheroids or nodules, Fig. 14.7(c), which are induced by adding magnesium before casting. This gives values of mechanical properties higher than with most other cast irons.

There is no practical limit to the section size or mass of SG iron castings, which leads to their use in a wider variety of applications than the malleable irons.

In many instances SG iron is used as a substitute for steel and replaces forgings, the advantages of the cast shape and reduced machining making it economical. Uses include pipes and fittings for gas, oil, water, sewage and chemicals. In motor vehicles the uses of SG iron include cylinder blocks, crankshafts, connecting rods, exhaust manifolds, water pumps, timing gears, gearbox casings, steering boxes and many other items. In agriculture the uses include a variety of tractor components, e.g. front-axle, steering and suspension units, transmission housings, and drive gears, as well as equipment such as plough shares, cultivator discs and mowing-machine parts.

SG iron is widely used in civil, mining and power engineering, construction, the steel industry and in the manufacture of machine tools.

The tensile strength varies over a wide range from 370 to 800 N/mm².

14.8 Copper and its alloys

14.8.1 Copper

Copper is a soft ductile material which increases in hardness and strength when cold-worked, i.e. in bending, spinning and drawing. The main advantages of copper are its high thermal and electrical conductivities and excellent

14

corrosion-resistance to chemicals, water and the atmosphere.

The electrical and thermal conductivities of high-purity copper are greater than those of any other metal except silver. Many grades of commercial purity are available.

14.8.1.1 Tough-pitch copper

Tough-pitch copper, of 99.85% purity, is used in chemical and general engineering applications where the highest conductivity is not required. Where the highest conductivity is required for conductors and electrical components, a highly refined grade known as tough-pitch high-conductivity copper, of 99.9% purity, is used.

14.8.1.2 Phosphorus deoxidised arsenical copper

Phosphorus deoxidised arsenical copper, of 99.2% purity with 0.3% to 0.5% arsenic, is widely used for copper tube and general engineering applications where brazing and welding are required.

14.8.1.3 Brass

Brass is essentially an alloy of copper and zinc, but may also contain small amounts of other alloying elements to improve strength, corrosion-resistance and machining characteristics.

14.8.1.4 70/30 brass

This alloy contains 70% copper and 30% zinc and is highly ductile. It is often referred to as cartridge brass, due to its use in the manufacture of ammunition. Because of its ductility it can be used in pressing, spinning and drawing operations.

The addition of 1% tin at the expense of zinc gives better corrosion-resistance, and the resulting material, known as Admiralty brass, is used for condenser tubes.

14.8.1.5 60/40 brass

This alloy contains 60% copper and 40% zinc and is used where the material is to be hot-worked. It is ideal for producing hot stampings and extruded bars, rods and sections.

To produce a brass with excellent high-speed machining properties, known as free-cutting brass, 2% to 3% lead is added.

14.8.1.6 High-tensile brass

This alloy is basically a 60/40 brass with additional alloying elements such as tin, iron, manganese and aluminium. The effect of these alloying elements is to increase the strength and corrosion-resistance.

This material can be cast, forged and extruded. Uses include high-pressure valves, components for pumps, and marine propellors; non-sparking tools for the gas, oil and explosives industries; and in the manufacture of nuts, bolts and studs.

14.8.2 Bronze

Bronze is essentially an alloy of copper and tin, but may also contain additional elements such as zinc and phosphorus. Bronzes containing copper, tin and phosphorus are known as phosphor bronze, while those containing copper, tin and zinc are known as gunmetal. A wide range of bronzes with additional alloying elements are available for a wide range of applications. These materials include aluminium bronze, nickel aluminium bronze and manganese bronze.

14.8.2.1 Phosphor bronze

Phosphor bronze is available in a wrought condition as sheet, strip, plate, rod, bar, wire and tube and also as cast materials.

As a wrought alloy in the form of wire and strip containing 5% tin and up to 0.4% phosphorus with the remainder copper, the alloy can be heat-treated. In the heat-treated condition it has good elastic properties as well as corrosion-resistance and is used for springs, diaphragms, clutch discs, fasteners and lock-washers.

As a cast material containing a minimum of 10% tin and 0.5% phosphorus with the remainder copper, it is used for bearing applications.

14.8.2.2 Gunmetal

The most common material of this type is Admiralty gunmetal, containing 88% copper, 10% tin and 2% zinc. This alloy casts extremely well

14

and has a high resistance to corrosion with good mechanical properties.

Gunmetal castings are largely used for naval purposes – for valves, pump bodies and fittings used with water and steam.

Up to 5% lead may be added to improve pressure tightness, ease of casting and machining, and antifriction properties.

14.9 Aluminium and its alloys

14.9.1 Advantages

Pure aluminium is light, soft, ductile, corrosion-resistant and highly conductive to heat and electricity. It is used in its pure form where strength is not of major importance, e.g. as foil for packaging. Alloying with other elements such as copper, magnesium, manganese, silicon and zinc increases the strength and hardness and enables the material to be heat-treated to give additional properties.

One of the most important characteristics of aluminium and its alloys is the thin oxide film which forms on their surfaces when exposed to the atmosphere. If the oxide film is broken it will reform quickly, and this gives these materials excellent corrosion-resistance. This oxide film can be artificially thickened to give added protection – a process known as anodising – and can be easily coloured to provide a highly decorative appearance.

14.9.2 Forms of supply

Aluminium and its alloys are available in a wide range of shapes and forms.

▶ *Foil* – in thicknesses from 0.2 mm to 0.005 mm, usually from aluminium of 99% purity.
▶ *Sheet* – available in standard sizes 1000 mm × 2000 mm and 1250 mm × 2500 mm in thicknesses from 0.5 mm to 3.0 mm.
▶ *Strip* – available in widths of 500 mm, 1000 mm and 1250 mm in thicknesses from 0.25 mm to 2.0 mm.

Both sheet and strip are available with the surface prepainted in a range of colours, used in the manufacture of caravans.

▶ *Plate* – defined as having a minimum thickness of 3 mm, available in a range of sizes.

▶ *Bars* – defined as round, square, rectangular and polygon solid section in sizes greater than 6 mm and supplied in straight lengths.
▶ *Extruded sections* – it is possible to produce an enormous variety of shapes by extrusion. The cost of the extrusion die is high, and, to be economical, large quantities of section have to be produced. Although some standard extruded sections are available, such as angles, the majority are made to a customer's own requirements.
▶ *Tube* – may be extruded, drawn or seam-welded from strip in a variety of sizes.
▶ *Wire* – available in sizes up to 10 mm diameter, wire is used in the production of rivets, nails, screws, bolts and welding rods, in metal spraying, and for electrical cables.
▶ *Forgings* – a number of alloys forge very well. The process was originally developed for aircraft components where high-strength and low-weight properties were required.
▶ *Castings* – where intricate shapes are required, the range of cast alloys can be used. These may be cast by sand, gravity-die and pressure-die-casting methods and with other more specialised processes.

14.9.3 Applications

The uses of aluminium and its alloys are virtually limitless, covering the fields of transport; electrical, structural, civil and general engineering; household items; packaging in the chemical and food industries and many more.

Applications in transport include cladding and floor sections of commercial vehicles, prepainted sheet for caravans, the superstructure of ships and hovercraft and a variety of components in aircraft.

Electrical engineering uses include large and small cables, foil, and strip windings for coils and transformers.

Structural and civil engineering uses include roofing and structural applications, door and window frames and a variety of decorative items.

General engineering uses include watches, photographic equipment, machinery for textiles, printing and components for machine tools.

Household items using aluminium include electrical appliances, pans and furniture; and packaging uses foil for food, drink, tobacco and chemicals.

14.10 Die-casting alloys

The process of die-casting uses a split metal mould into which molten metal is:

▶ poured under the action of gravity (gravity die-casting); or
▶ forced in under pressure (pressure die-casting).

When the metal solidifies, the split die is opened and the casting is ejected.

The common metals used in die-casting are zinc- and aluminium-based alloys.

Zinc alloy containing 4% aluminium with the remainder zinc is most widely used for the production of castings required in large quantities, accurately, and with a good surface appearance. This alloy is cast at temperatures around 400°C and is very fluid in the molten state, which leads to its use for intricate shapes and thin sections.

Zinc alloys can be easily electroplated or painted to give a decorative or protective finish. Uses include the mass production of car parts such as door handles, lock parts, carburettors, fuel pumps and lamp units, as well as camera parts, power tools, clock parts and domestic appliances.

Aluminium alloys, typically containing 3.5% copper and 8.5% silicon, with the remainder aluminium, are less tough than zinc alloy but have the advantage of low density. Aluminium alloys require casting temperatures of around 650°C. Their many uses include car crankcases, gearboxes, timing-case covers and components for electrical and office equipment.

14.11 Lead

Lead in its pure state is very soft and has low mechanical strength. In this form it is used widely in the chemical industry, due to its high corrosion-resistance. Because of its low mechanical strength it is applied in the form of a lining to other, stronger materials. It is also used for radiation shielding. When alloyed with antimony, the strength and hardness is increased and it is used for the production of lead bricks for nuclear

shielding. Two of its largest uses are as cable sheaths for power cables and in the manufacture of connectors and grids in lead-acid batteries. Lead-tin alloys give a range of soft solders.

14.12 Contact metals

Contacts are used where there is a need to make and break electrical connections. They must be reliable when working under a variety of conditions such as heat, cold, humidity, vibration, dust and corrosive atmospheres, and must give a long service life against mechanical wear, heat, fatigue, metal transfer and corrosion.

14.12.1 Platinum

Pure platinum has two important properties as a contact material: it has a high melting point, 1770°C, and is therefore able to resist the effects of the electrical current arcing across the faces; also it has a high resistance to corrosion.

Pure platinum is relatively soft and is used in light-duty applications on sensitive relays and instruments. When alloyed with iridium, greater hardness and mechanical properties can be obtained, leading to its use in medium-duty applications.

14.12.2 Silver

Silver has the highest thermal and electrical conductivities of any metal and is resistant to corrosion. Pure silver is used on contacts in the form of an electroplated film. It is used in a variety of light- and medium-duty applications in telephone relays, sliding contacts, thermostats and voltage regulators. When alloyed with copper, the electrical conductivity is lowered and the hardness and mechanical properties are increased.

14.12.3 Gold

The high electrical conductivity of gold, surpassed only by silver and copper, together with its resistance to tarnishing, makes it suitable as a contact material. Alloying with copper and silver gives good resistance to wear and is used for sliding contacts in light-duty applications.

A 5% nickel/gold alloy is widely used.

14.12.4 Tungsten

Tungsten has the highest melting point of any known metal: 3380°C. It is alloyed with other metals in powder form by a process known as sintering. Alloying with silver or copper gives varying degrees of conductivity, hardness and wear-resistance.

14.13 Bearing materials

Plain bearings are the oldest form of bearing. They may be manufactured completely from a single bearing material or composition of materials, or, alternatively, a thin layer of bearing material can be attached to a backing which is usually of a stronger material.

Where a metal bearing material is used with a metal shaft, metal-to-metal contact must be avoided to prevent seizure of the two metals. This is done by providing a film of lubricant between the two surfaces. Where it is not possible to provide a film of lubricant, a non-metallic bearing material will have to be used.

Ideally a bearing material should possess the following properties:

▶ good thermal conductivity, to carry heat away from the bearing;
▶ sufficient strength to carry the loading of the shaft or sliding part without permanent deformation;
▶ resistance to corrosion by lubricants or the atmosphere;
▶ the ability to operate over a range of temperatures (melting point and coefficient of expansion are important factors);
▶ the ability to deform slightly, to compensate for small misalignments or surface irregularities;
▶ the ability to allow dirt, grit or filings to embed in the surface rather than pick up and seize on the shaft or sliding part;
▶ the ability to resist wear.

14.13.1 Phosphor bronze

Cast phosphor-bronze bearings possess high strength to support heavy loads at low speeds and have excellent corrosion-resistance. The bearing and shaft must be accurately aligned.

Lubrication is required and must be plentiful and reliable – oil grooves or indents are usually provided in the bore of the bearing to give good distribution over the complete surface.

Porous bronze bearings are available, made from powdered metal. The powder is pressed into shape leaving air spaces or pores. When the bearing is subsequently immersed in oil, these pores absorb oil. In operation, the oil seeps from the pores to provide an oil film which prevents metal-to-metal contact. Alternatively, graphite can be incorporated in the initial powder to provide a self-lubricating property.

Porous bronze bearings can be used at high speeds with light loading and offer good corrosion-resistance. They are designed to last the life of the assembly without additional lubrication and are used in domestic appliances, starter motors and car parts.

14.13.2 Cast iron

Cast iron is not used as a bearing material for rotating parts: it has high load characteristics and is used mainly for sliding surfaces such as machine-tool parts, e.g. milling-machine slides, lathe beds, etc.

Because of the free graphite it contains, cast iron has certain self-lubricating properties but requires additional lubrication as there is a tendency for the material to pick up and quickly seize.

14.13.3 PTFE

This is a thermoplastic material, the correct name for which is polytetrafluoroethylene, though it is usually referred to as PTFE.

It possesses a wide temperature range of use from −200°C to +250°C, is resistant to most chemicals, and can be easily machined. Its main advantage as a bearing material is its low coefficient of friction.

However, PTFE has big disadvantages as a bearing material – it has marginal wear-resistance, poor thermal conductivity and a high coefficient of expansion. These disadvantages can be reduced by the inclusion of additives such as glass fibre, graphite, bronze and molybdenum disulphide;

14

alternatively, a film of PTFE on a steel backing provides strength, wear-resistance, good heat conductivity and low thermal expansion.

It is used for bearings required to operate without lubricant and as static bearing pads for bridge expansion joints.

14.13.4 Nylon

Nylon is a thermoplastic material which can operate under light to medium loads, is tough, abrasion-resistant, resists chemical attack, is non-toxic and can be used continuously at temperatures up to about 90°C.

Like other thermoplastic materials, nylon has a high coefficient of expansion and low thermal conductivity. It also readily absorbs moisture. The inclusion of additives such as glass fibre, graphite and molybdenum disulphide reduces these disadvantages and increases the wear-resistance and operating temperature.

Uses of nylon bearings include bushes in domestic appliances; bushes, thread guides, rollers and slides in textile machinery; ball joints and bearings in suspension and steering systems in cars.

14.13.5 Polyimide

Thermosetting polyimides are one of the highest performing engineering plastics. They have high temperature resistance and can perform up to 315°C with only negligible loss of mechanical properties. Polyimide is relatively easy to machine, has good chemical resistance, is dimensionally stable and has good thermal and electrical properties. It has a high wear resistance and this together with the high temperature resistance leads to its use as a bearing material in the aerospace and automotive industries particularly in high-temperature environments where weight saving is an advantage. It can be used to replace metal and ceramic parts. The frictional properties of polyimide materials can be further enhanced by the use of additives such as graphite and molybdenum disulphide.

14.13.6 PEEK

This is a thermoplastic material, the correct name for which is polyetheretherketone. PEEK is also regarded as a high-performing polymer, offering excellent chemical resistance, very low moisture absorption and good wear, abrasion and electrical resistance. It can be used continuously to 250°C in hot water and steam without permanent loss in physical properties. It is used to replace metal parts in the aerospace, automotive, oil and gas industries and is available reinforced with glass and carbon fibre.

14.13.7 Graphite

Graphite is used as an additive for bearings required to operate without lubricant. It is added to PTFE and nylon and gives a lower coefficient of friction than is possessed by the materials themselves.

The antifriction properties of graphite can be illustrated by sliding a block of metal across a surface plate. Now sprinkle some shavings from a pencil lead containing graphite on the surface plate and again slide the block across the surface. It will now slide very much more easily.

14.14 Metal protection

Many thousands of pounds are lost each year as a result of corrosion. The consequences of corrosion are many and varied and the safe, reliable and efficient operation of equipment and structures is often more serious than the simple loss of the mass of metal. Some of the harmful effects of corrosion are:

▶ reduction of metal thicknesses leading to loss of mechanical strength and structural failures or breakdown;
▶ hazards or injuries to people arising from structural failure or breakdown, e.g. bridges, cars and aircraft;
▶ reduced value of goods due to the deterioration of appearance, e.g. rusty car;
▶ contamination of fluids in vessels or pipes;
▶ perforation of vessels and pipes allowing contents to escape and create a hazard or damage surroundings, e.g. leaking oil tank damaging the floor and toxic fluids escaping;
▶ loss of surface properties, e.g. frictional and bearing properties, fluid flow in a pipe and electrical contacts;

14

- mechanical damage to shafts, valves and pumps;
- added complexity and cost of design and manufacture to allow for convenient replacement of corroded parts.

Much of the corrosion problems can be prevented easily by applying known corrosion-control techniques.

Corrosion will occur and can be controlled. It is therefore essential that engineering technicians are aware of the causes of corrosion and of the methods available for its control.

One of the most common methods used in industry for the control of corrosion is to apply a protective coating to material surfaces. The type of protective coating and the method of application depend on a wide variety of factors which include operating conditions, the immediate environment, whether indoors or outdoors, the likely risk of damage, and appearance.

14.15 Corrosion

Corrosion is an electrochemical process in which a metal reacts with its surroundings to form an oxide or compound similar to that of the ore from which it was extracted.

Metals vary greatly in their corrosion-resistance – chromium and titanium have good resistance, while steel readily corrodes. The oxide film formed on chromium and titanium closely adheres to the surface and protects the metal from further oxidation. In the case of steel, the oxide film in the form of rust is loose, allows moisture to be retained, and promotes further corrosion. If corrosion is allowed to continue, the steel will eventually be completely consumed, i.e. the metal will have returned to the condition of the ore from which it was extracted.

The electrochemical process is caused by a series of cells known as corrosion cells. A cell is made up of:

- an *anode* – an area where corrosion takes place;
- an *electrolyte* – an electrically conducting solution;
- a *cathode* – which completes the cell and is not consumed in the corrosion process.

The anode and cathode may be two surfaces in contact with each other or they may occur on the same surface due to small variations in the metal structure, Fig. 14.8.

The electrolyte is usually water – present as moisture, rain or sea water – which may also contain elements of dust and gases which accelerate the corrosion process. The metal may be in constant contact with the electrolyte, e.g. underground structures and liquids in pipes, tanks and various vessels; alternatively, the metal may be indoors subjected to differing degrees of humidity or dampness or outdoors in all weather conditions. The rate of corrosion is influenced by the electrical conductivity of the electrolyte; i.e. high rate in salt solutions, low rate in high-purity water.

Corrosion can be controlled by using a material having good resistance. Unfortunately, materials which have good resistance to corrosion are more expensive than those with poor resistance. It is seldom possible or even desirable to replace all materials with high-cost corrosion-resistant ones, and other methods have to be considered.

As already stated, corrosion can take place only if an electrolyte is present. The obvious method of controlling corrosion, then, is to prevent electrolyte from contacting the metal surfaces; i.e. exclude the environment from the metal. One

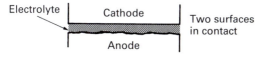
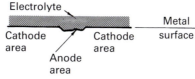

Figure 14.8. Corrosion cells

method of doing this is to provide a protective coating on the metal surfaces, the method chosen depending on the type of metal to be protected, the environment in which the metal will operate, and the coating material. The coating may consist of:

▶ another metal, e.g. zinc coating on steel;
▶ a protective coating derived from the material, e.g. aluminium oxide on 'anodised' aluminium;
▶ organic coatings, e.g. resins, plastics, paints, enamels, oils and greases.

14.16 Protective coatings

14.16.1 Electroplating

Electroplated coatings are commonly applied to meet a variety of service requirements and also for decorative purposes. Service requirements include corrosion resistance, wear resistance and contact with chemicals and foodstuffs. The coatings described here are those used chiefly for their corrosion-resistant properties.

14.16.1.1 Chromium plating

Chrome plate is a widely used electroplated coating and its appearance can be bright, satin, matt or black. It combines resistance to corrosion, wear and heat, use in contact with foodstuffs and decorative qualities. By itself it does not give a high level of protection against corrosion. To obtain a high level of corrosion protection, the chromium is deposited over a coating of nickel. The deposit of nickel, which itself has good corrosion resistance, gives protection to the base metal and is itself protected from surface oxidation by the chromium. As a decorative finish, chrome plate can also be applied to plastics, usually acrylonitrile-butadiene-styrene (ABS). Hard chrome deposits, typically 150–500 μm, are used to build up rollers, hydraulic rams, valves, etc., as a hard, abrasive-resistant wearing surface which is subsequently ground to give a highly accurate surface finish.

14.16.1.2 Nickel plating

Nickel deposits are most widely used as a base for chromium plating; the thicker the deposit (typically 20 μm), the greater the corrosion resistance.

Nickel plating is used in engineering where wear resistance, hardness and corrosion resistance are required, e.g. oil valves, rotors, drive shafts and in printed circuit board manufacturer. It is also used for its decorative properties, e.g. doorknobs.

14.16.1.3 Cadmium plating

Cadmium plating is used principally to give a corrosion-resistant coating to steels although it can be used on brass and aluminium. It provides sacrificial protection to the underlying steel, e.g. slight damage to the coating causes no loss of protection to the steel as the cadmium, being electrochemically more active, 'sacrifices' itself at a very slow rate.

The normal thickness of coating is 5–25 μm and has low frictional properties, making it an ideal surface on fasteners which reduces the tightening torque and prevents jamming. It also provides an effective barrier to prevent bimetallic reaction between steel fasteners and aluminium, e.g. where parts are fixed to an aluminium framework. Cadmium-plated surfaces can be easily soldered without the use of corrosive fluxes.

After plating, a chromate conversion coating is usually applied which gives the coating its iridescent green/brown appearance and adds to its corrosion resistance.

Cadmium-plated parts are safe to use and handle normally. Under certain conditions, cadmium can present a health hazard (covered by the COSHH Regulations). Use fume extraction and respirators as directed when welding or heating above its melting point of 320°C as cadmium gives off cadmium oxide fumes which are highly toxic. Hazards also exist in the handling of cadmium metal in fine powder form as they produce dust very easily. Cadmium is strictly controlled and can only be used in specific applications. Its main applications are in aircraft, mining, military and defence, offshore, nuclear and electrical industries.

14.16.1.4 Tin plating

Tin is used mainly in the manufacture of tinplate for food cans usually with a coating thickness of between 0.4 and 2 μm. The material is bright, easily soldered and readily formed. Tin coatings are applied to vessels and equipment used in the food industry, e.g. baking tins, where the minimum coating thickness is around 30 μm.

Coatings of tin/zinc alloy, containing around 25% zinc, give good protection in some environments and are used in hydraulic components. These coatings are used on electrical components for their corrosion resistance and ease of soldering.

14.16.1.5 Zinc plating

Zinc plating is ideally suited to mass production and is usually zinc or zinc alloy, i.e. zinc iron and zinc nickel. Zinc iron and zinc nickel give increased corrosion protection over zinc. They are widely used for fasteners such as nuts, bolts, screws and for metal stampings in the car industry and can be coloured, e.g. black, blue and iridescent.

Zinc flake non-electrolytic systems are used for the highest corrosion protection as a base coat and top coat and provide a variety of additional properties and are available in a range of colours including black, silver, blue, red, yellow and green.

A hot dip process of applying zinc to mild steel, known as galvanising, provides a strong protective coating. The material, after undergoing a series of critical cleaning operations, is immersed in molten zinc at 450°C until it reaches the same temperature as the zinc and is then gradually withdrawn allowing excess zinc to drain off. The process will safeguard mild steel against corrosion in an external UK environment for between 40 and 70 years and for over 100 years in indoor locations.

14.16.2 Anodising

Anodising is used extensively for the production of protective and decorative films on aluminium and its alloys. In the anodising process, the naturally occurring film of aluminium oxide is artificially thickened by an electrolytic process to give increased corrosion resistance. This is done by making the article the anode, immersed in a weak solution of sulphuric, chromic, oxalic or phosphoric acid depending on product requirements. The thickness of film depends on the electrical current and the process used. The natural film colour is grey and thicknesses of 1–50 μm are normal. Anodising provides a decorative finish where the surface can be dyed to give a wide range of attractive colours.

Hard anodising is a process where the anodising parameters are changed, leading to a much thicker, harder, abrasion- and corrosion-resistant film. The film formed is dark in colour, so black is the only sensible film colouring.

Hard anodised composite coatings, containing polymer particles, give enhanced lubrication and wear-resistant properties. The polymer particles are permanently locked within the anodised layer giving non-stick properties and low friction, resulting in reduced wear. These coatings provide the ideal surface for slideways and bearing surfaces in a wide range of industries including aerospace, automotive, pneumatics, marine and electronics.

14.16.3 Plasma electrolytic oxidation (PEO)

Plasma electrolytic oxide coating under the proprietary name 'keronite' involves a high-energy plasma discharge around a component, immersed in a proprietary electrolyte, that provides for surface oxidation, resulting in a superhard ceramic layer on light alloys of aluminium, magnesium and titanium. The process takes place basically at room temperature typically between 15°C and 20°C with different electrolytes being used for the different metals. The process resembles anodising because it uses an electrical supply and bath of electrolyte but is significantly different as it produces harder and thicker layers, typically 50–70 μm but 150 μm is possible. The components need to be degreased prior to being coated but do not require surface etching as the discharge of the electrolyte sufficiently cleans the surface. After treatment the component is washed in warm water for several minutes. The proprietary electrolytic solution is non-hazardous as it contains no heavy metals and no toxic or aggressive chemicals.

As a ceramic, 'keronite' is resistant to most chemicals and therefore performs extremely well in most corrosive environments. Because of the hardness of the layer, treated aluminium alloys can now be used as direct replacements for steel with corresponding weight savings.

14

Applications are still being assessed by many industries but include:

▶ Automotive – pistons, piston crowns, cylinder liners and magnesium wheels.
▶ Consumer goods – magnesium bicycle frames, MP3 players, laptops and binoculars.
▶ Textile machinery – lightweight components combined with hard, wear-resistant surface.
▶ Plastics moulding – extending life of moulds.

These are in addition to military and defence, aerospace, oil and gas, nuclear, chemical, marine and electronics industries.

Primary features of the process include:

▶ superb corrosion resistance;
▶ extremely hard;
▶ thick coatings up to $150\,\mu m$;
▶ predictable coverage and dimensional control;
▶ high wear resistance;
▶ effective undercoat for paints and polymers;
▶ superb heat resistance.

14.16.4 Chemical coatings

Metal surfaces can be treated chemically with an appropriate solution which provides a coating with limited protection against corrosion. These coatings also provide an excellent base for paint. The combined effect of the chemical and paint coatings gives an extremely high degree of corrosion resistance.

14.16.4.1 Chemical blacking

This is a non-electrolytic process providing a very adherent black oxide coating with a jet black finish, usually applied to steels but can be used with other materials. Processes are available that can be carried out at room temperature (20°C) or at temperatures around 141°C. The process involves a series of cleaning and rinsing operations before immersing in a bath containing the blacking solution for up to 10 minutes. Further rinsing is carried out before a final immersion in dewatering oil, heavy oil or wax to further enhance corrosion resistance.

14.16.4.2 Phosphating

This process is applied mainly to steel components which are sprayed or dipped in a phosphoric acid solution containing iron, zinc or manganese phosphate. The most common processes use zinc or manganese phosphate and are employed in the treatment of car bodies and domestic appliances before painting. Where long-term corrosion protection is required, the phosphate coating must be sealed with oil lacquer or paint.

14.16.4.3 Chromating

A number of proprietary chromate filming treatments are available usually for aluminium, magnesium, cadmium and zinc alloys. The process involves a series of cleaning and rinsing operations, and then immersing in a strong acid chromate solution, followed by more rinsing. The resulting film is coloured depending on the base metal, being golden yellow on aluminium. The film contains soluble chromate which acts as corrosion inhibitor, and although the film provides a modest corrosion resistance, its main purpose is to provide a suitable surface for sealing resins, painting or powder coating.

A process for zinc alloy consists of immersing in a sodium dichromate solution with other additives, at room temperature, followed by rinsing and drying to produce a dull yellow zinc chromate coating.

14.16.5 Autodeposition

Autodeposition is a waterborne process of applying a layer of anti-corrosion paint to a metal using a chemical reaction rather than by using electrical energy. The coatings can be used as standalone coatings or as a primer for a subsequent top coat. Because of the chemical reaction, it is possible to evenly coat hidden or recessed areas where corrosion typically starts. Coatings are available in acrylic, polyvinylidine chloride (PVDC) and epoxy-based polymers. A thermoset epoxy–acrylic-based hybrid coating is available for use as a single coat or primer in black, off-white and shades of grey.

The process consists of four basic steps:

1. Clean the metal.
2. Apply the coating.
3. Rinse off any untreated material.
4. Oven dry.

The coating bath contains a mildly acidic emulsion of the proprietary chemicals at 22°C. When the metal component (must be ferrous metal) is immersed in the solution, it is attacked by the chemicals which de-stabilise the emulsion locally at the metal surface, causing the paint layer to be deposited. Initially the deposition rate is quite rapid but slows down as the film increases in thickness. Mixers can be added to the bath which will affect the deposition rate (30 μm can be deposited in 3 minutes). Typical film thickness is 12–25 μm in a deposit time of between 30 and 100 seconds.

The process lends itself to automated production lines and is widely used in the car industry for frames, chassis, seat frames and suspension parts and in construction, agricultural machinery, domestic appliances, electric motors and compressors.

Benefits include:

▶ fewer steps than conventional painting;
▶ reduced processing costs;
▶ simple process equipment;
▶ waterborne process/non-electrical;
▶ flexible rate of throughput;
▶ enhanced corrosion resistance;
▶ less waste.

14.16.6 Electrocoating

Also known as E-coat or electrodeposition, this is a method of using electrical energy to deposit paint on a metal surface. The process is based on the fundamental physics principle that opposites attract. The metal parts are charged with direct current and immersed in a bath that has oppositely charged paint particles in suspension (the paint acts like a magnet to the metal part to be coated). This gives the ability to coat even the most complex parts and assembled products, so that even corners, edges and recessed areas are completely protected. The direct current is controlled to allow the paint film to build up to the desired thickness where it then acts as an insulator to stop the deposition process. Typical film thickness ranges from 18 to 30 μm.

Two types of system are available: anodic and cathodic. The cathodic electrocoating system is the most common with the part to be coated given a negative charge, attracting positively charged paint particles. Cathodic coatings give high-performance coatings with excellent corrosion resistance, toughness and exterior durability.

Prior to dipping, the parts are first cleaned and pre-treated with a phosphate conversion coat to prepare the part for electrocoating. Iron and zinc phosphate are commonly used.

The electrocoat bath consists of 80–90% de-ionised water and 10–20% paint solids. The water acts as the carrier for the paint solids which are constantly agitated. Paint solids consist of resin and pigment. The resin, which is an epoxy or acrylic-based thermoset, is the backbone of the final paint film and provides the corrosion protection, durability and toughness of the finish. Pigments are used to provide the colour and gloss.

As the parts are removed from the bath, paint solids cling to the surface and are rinsed off. The rinse is then filtered and the solids returned to the bath keeping waste products to a minimum.

After final rinsing, the part enters the bake oven which cross-links and cures the paint film to ensure that the maximum performance properties are achieved. Bake temperatures range from 85°C to 204°C depending on the paint used.

E-coat is used in thousands of everyday products. One of the major users is the automotive industry where the vast majority of car bodies are primed using this process. The electrocoat primer is what protects your car from rusting and allows for the extended corrosion warranties available today. Other uses in the automotive industry include steering gear parts, engine blocks, seat components, radiators, mirror brackets and trailers.

E-coat is widely used in appliances such as washing machines and dryer drums, office and patio furniture, garden equipment such as lawnmowers, tractors and wheelbarrows, and sports equipment such as golf carts and snowmobiles.

The benefits of E-coat include:

▶ ability to evenly coat complex surfaces;
▶ elimination of drips and runs;
▶ high material usage and low wastage because of rinsing and filtration;
▶ low levels of hazardous air pollutants;

▶ readily lending itself to automation;
▶ prior assembly possible due to ability to cover evenly;
▶ precise control of the film thickness.

14.16.7 Powder coating

Powder coating is the technique of applying dried paint to a part as a free-flowing dry powder. In conventional liquid paint the solid pigments are in suspension in the solvent which must evaporate to produce a solid paint coating. Dry powder does not contain a solvent.

Two types of powder are available:

▶ *Thermoplastic powders* that will melt on heating and are usually polyvinyl chloride (PVC), polyethylene and polypropylene. They require higher temperature to melt and flow at between 250°C and 400°C depending on the coating required. Thermoplastics have good chemical, detergent and water resistance and are used in dishwasher baskets, frozen food shelving, furniture and some car parts.

▶ *Thermosetting powders* will not remelt on heating. During the heating or curing process, a chemical cross-linking reaction is triggered at the curing temperature and it is this chemical reaction which gives the powder coating many of its desirable properties. Thermoset powders are much more widely used than thermoplastics. The primary resins used in the formulation of thermoset powders are epoxies, polyesters and acrylics. Some powders use more than one resin and are referred to as hybrids, e.g. epoxy–polyester. Curing varies depending on the resin used, but is typically 180°C for 10 minutes. Speciality powders are available, formulated to give a wide range of texture, structure, hammer and antique finishes. Thermosetting powders provide toughness, chemical and corrosion resistance, and electrical insulation and are used in an extensive range of applications including domestic appliances, garden tools, radiators, office furniture, instrument casings, vehicle components, window frames, electric motors and alternators.

The basis of any good coating is preparation. Removal of oils, greases and oxides is essential prior to the coating process which can be done in a series of cleaning, rinsing and etching operations. As previously outlined, chemical pre-treatments such as phosphating and chromating improve bonding of the powder to the metal. Another method of preparing the surface is by abrasive blasting (also known as sandblasting or shot blasting) which not only cleans the surface, but also provides a good bond for the powder. Powder coating will not protect against corrosion when applied to ungalvanised exterior steelwork but will do so on galvanised steel where the powder-coated surface can be guaranteed for as long as 25 years.

In powder coating the powder paint can be applied by either of the following two techniques:

▶ *Spraying:* Here the powder paint is electrostatically charged and sprayed on to the part which is at earth, or ground potential. The spray gun imparts a charge on the powder which is attracted to and wraps around the grounded part. The powder will remain attached to the part as long as some of the electrostatic charge remains on the powder. To obtain the final solid, tough, abrasion- and corrosion-resistant coating, the parts are heated to temperatures in the range 160°C to 210°C depending on the powder. The curing can be done by convection ovens, infrared, a combination of both or by ultraviolet (UV) radiation, which in the case of an automated production line is located at the final stages. Typical coating thickness is around 50–65 μm.

▶ *Fluidised bed:* This is a simple dipping process which can be conventional or electrostatic. The conventional fluidised bed is a tank containing the plastic powder with a porous bottom plate where low-pressure air is applied uniformly across the plate. The rising air surrounds and suspends the fine plastic powder particles so that the powder/air mixture resembles a boiling liquid or fluid, hence 'fluidised bed'. Products, preheated to above the melting temperature of the powder, are dipped in the fluidised bed where the powder melts and fuses into a continuous coating. This method is used to apply heavy coats (75–250 μm) uniformly to complex-shaped parts. An electrostatically fluidised bed has a high voltage d.c. grid above the bottom plate which charges the powder particles. As the air renders the powder fluid, the parts, which are not preheated, are dipped into the bed and attract the powder in the same way as spraying. This is followed by a curing process. Small products such as electrical components can be coated uniformly and quickly.

Benefits of powder coating versus conventional liquid paints are as follows:

▶ Eliminates the use of solvents.
▶ High film thickness can be achieved with a single coat.
▶ Powder coatings can produce much thicker coatings without running or sagging.
▶ Virtually all over-sprayed powder can be recovered for reuse.
▶ Powder coating is readily automated, enabling high volume processing.
▶ Powder coating production lines produce less hazardous waste.
▶ Capital equipment and operating costs are generally less.
▶ Powder coatings generally have fewer appearance differences between horizontal and vertical coated surfaces.
▶ A wide range of speciality effects can be accomplished.
▶ Reduced rejects – damaged coatings can be blown off before heat is applied.
▶ Powder coatings require no pre-mixing, viscosity adjustments or stirring.

14.16.8 Coil coating

Coil coating is an advanced technique for applying organic coatings as a liquid or film to sheet metal in a continuous process. This process is also referred to as pre-coated, pre-painted or pre-finished. The technology is based on the simple fact that it is easier to coat a flat surface, than paint individual irregular shapes after they have been formed.

Coil coating can be carried out on:

▶ aluminium and its alloys;
▶ hot dip galvanised steel – where a layer of zinc has been applied by a hot dip process;
▶ electrogalvanised steel – where a layer of zinc has been applied by an electrolytic process;
▶ cold-rolled steel.

The coil of metal, weighing from 5 to 6 tonnes for aluminium and up to 20 tonnes for steel, is delivered to the start of the coating line. The new coil is attached to the previous coil by a metal stitching process without slowing down or stopping the line, allowing the process to be continuous at speeds up to 220 m/minute. As the coil is unwound, it passes into a tension leveller which flattens the metal strip before it passes into the pre-treatment section where it is chemically cleaned and a chemical treatment, usually zinc phosphate or chromate, is applied to provide a good bond between the coating and the metal. The surfaces are then dried before the metal strip passes through the primer roller coat which applies a primer to one or to both sides. The strip then passes through a curing oven which heats the coating to remove solvents and cure the coating to achieve its required properties. The time spent passing through the oven can vary from 15 to 60 seconds and the thickness of the primer coat is usually between 5 and 35 μm.

The strip then passes through the second set of rollers which apply the finish coat, again to one side or to both, as required, to provide the final colour and appearance of the final coated system before passing through the second curing oven. The final coat thickness is between 15 and 200 μm depending on the paint used and the final application of the pre-coated metal.

At this stage an embosser or laminator can be fitted to either produce an embossed pattern on the surface of the coating or apply a plastic film onto the coated surface. A final water quench is carried out before being inspected and re-coiled.

The fully coated coil has a consistent quality with excellent formability, corrosion and weather resistance and can be delivered to the customer in this form or slit into narrower coils or cut into sheets ready for use. A multitude of colours, gloss levels, textures and patterns can be achieved with exceptional consistency from batch to batch.

Types of liquid coatings used include, among others, acrylics, epoxies, PVC plastisols, polyesters and polyurethanes. Coatings must be easy to apply, have a very short cure time and be flexible enough to enable the coated metal to be bent or formed without cracking or suffering loss of adhesion of the paint film.

14

Plastic film or laminate applied at the final stages provides high flexibility and suitability for deep drawing, while others are resistant to rain, sun, heat, fire, staining, abrasion and chemicals. Pre-coated metals are available with additional functional features such as anti-bacterial, anti-fingerprint, anti-dirt, easy clean, non-stick and heat reflective.

There are hundreds of applications for pre-coated steel and aluminium, the largest user being the construction industry taking advantage of the architectural properties. These include cladding, infill panels, partitioning, rainwater goods, roofing tiles and suspended ceilings.

In the automotive industry there are a variety of individual applications covering a range of products including vans, caravans, trailer bodies and panels for cars, buses and coaches, instrument dials and car licence plates. Van, trailer and caravan bodies are ideal products with large surface areas combined with some forming requirements. The coating is generally polyesters or polyurethanes which provide attractive, smooth, glossy or matt finishes together with superior abrasion resistance to resist impact damage. On the reverse side of the pre-coated panel, a suitable coating is applied which is compatible with foams and adhesives used with insulated body construction.

Domestic and consumer products are further applications where the coating systems have been developed to be highly flexible and tough to resist constant handling and cleaning, resistant to staining and attack by a whole range of domestic products which may come in contact with the coating, and be capable of operating in humid and potentially corrosive environments, e.g.:

▶ washing machines – wet and corrosive;
▶ refrigerator – frequent handling;
▶ cookers – heat;
▶ microwave – humidity.

Coatings on the reverse side also perform important functions such as enhancing corrosion resistance in washing machines and aiding adhesion of foam backings in refrigerators.

Office furniture and filing cabinets are further applications while teletronics products such as CD player cases, video recorders, televisions, DVD players and decoder boxes have coating systems which are attractive, capable of working in tough environments and remain functional.

14.16.9 Paints

A wide range of paints is available for industrial use and their primary function is to give added corrosion resistance to the metal surface as well as providing for decorative appearance. With the increase in the use of powder paints this type of paint is often referred to as 'wet' or 'liquid' paint.

Paint consists of a mixture of pigments which gives body and colour, and a resin or binder which is the actual film-forming component and acts as a 'glue' to hold the pigment together and stick them to the surface. Binders include synthetic resins such as acrylics, polyurethanes, polyesters and epoxies and can be a combination of resins, e.g. epoxy/acrylic and polyurethane/acrylic. To adjust the curing properties and reduce the viscosity so that the paint can be easily applied, a solvent or carrier is used. These evaporate after application and do not form part of the paint film. In waterborne paints, the carrier is water. With solvent-borne paints, also called oil-based paints, the carrier is a solvent such as acetone, turpentine, naphtha, toluene, xylene and white spirit.

The most significant trend over recent years is the growth of waterborne paints and the decline in solvent-borne paints in view of environmental and health and safety requirements to reduce the emission of volatile organic compounds (VOCs) produced by organic solvents.

Note that drying and curing are different processes. Drying generally refers to evaporation of the solvent or carrier. Curing (or hardening) of the coating takes place by chemical reaction to cross-link (or polymerise) the binder resins. These are generally the two-pack coatings which polymerise by chemical reaction, from mixing the resin and a hardener in the correct ratio. In UV curing paints, the solvent is evaporated first and hardening is then initiated by the UV light.

As in all cases of protective coating systems, proper surface preparation is essential.

The success of any paint coating is totally dependent on the correct and thorough preparation of the surface prior to coating.

It is therefore essential as a first stage to degrease, i.e. remove all oil, grease and surface contaminants. This can be done using proprietary degreasing fluids or vapour degreasing plants. For larger items, emulsion cleaners with rust inhibitor properties are used. All water-based cleaners should be thoroughly rinsed off. Previously painted or rusty surfaces can be prepared by shot blasting.

To provide the maximum protection, the paint system should ideally consist of three types of coating:

▶ Primer – applied to the clean surface and designed to adhere strongly and prevent corrosion. These can be chromate or phosphate coatings, as previously described, or etch–primer solutions and anti-corrosive primers, containing zinc phosphate or iron oxide, and can be single or two pack. Some waterborne primers are available.

▶ Undercoat – applied as an intermediate film to provide a basis for the final colour.

▶ Finish or top coat – the choice of this paint will depend on the degree of protection and the decorative effect required. Finishes can be smooth, gloss, semi-gloss, matt, eggshell or textured and are available in a wide range of colours.

14.16.10 Epoxy paints

These are two-pack (base and hardener) cold-cure solvent-borne coatings for use with steel, aluminium and light alloys and applied by spraying on suitably pre-treated surfaces. The two parts are well mixed together in the ratio of 2:1 or 3:1 by volume, base to hardener, as recommended by the manufacturer. Only sufficient material should be mixed to meet the immediate requirements, as the mixture has a limited time, known as pot life, while remaining in a usable condition. The pot life varies depending on the manufacturer and can range from 8 to 24 hours under normal working conditions (20°C) and should not be used after the stated time.

Epoxy coatings provide an extremely tough and abrasion-resistant finish with outstanding resistance to a wide range of chemicals, solvents and cutting fluids. They are available in gloss or matt finish that can be used as an undercoat

or finish coat. Although they are cold cure, i.e. in air at ambient temperature of 20°C, this can be speeded up by heating at up to 120°C, after allowing an approximately 10-minute 'flash off' period to allow the solvents to evaporate. Single-pack epoxy stoving finishes are available where the coating, after spraying, is allowed a 10-minute 'flash off' period before stoving for 30 minutes at 140°C in a convection oven.

14.16.11 Polyurethane paints

These two-pack cold-cure solvent-borne coatings have excellent durability and abrasion resistance. They are designed for spray applications on steel, aluminium and light alloy surfaces which have been suitably pre-treated. Base and hardener are mixed in 3:1 or 5:1 ratios as recommended by the manufacturer and the pot life is around 2 hours under normal working conditions (20°C) and should not be used after the stated time. Polyurethane/acrylic resin combination is also available. Two-pack polyurethanes are widely used in vehicle body repair and can contain isocyanate which is a basic constituent in the production of polyurethane and is the most common cause of industrial asthma and it is also a skin irritant. It should be easily identified by a label on the container which should say 'contains isocyanates, harmful by inhalation and in contact with skin'.

Waterborne two-pack polyurethane finishes are available.

It has good resistance to diesel oil and petrol and applications include agricultural equipment, construction plant and industrial machinery.

14.16.12 Acrylic paints

These are two-pack cold-cure coatings and can be a blend of acrylic and epoxy resins for use with ferrous and non-ferrous metals as well as plastics. Designed to be applied by spraying to suitably pre-treated surfaces they are available as an isocyanate-free alternative to two-pack polyurethane systems.

Again, the two parts, base and hardener, are well mixed in the ratio recommended by the manufacturer, e.g. 4.5:1 base to hardener. They have a typical pot life of around 12 hours and should not be used after the stated time. They are fast drying, highly durable and impact resistant with a high gloss finish and applications include agriculture and construction equipment, commercial vehicles, plant and machinery and street furniture.

Low-temperature acrylic stoving finishes are available. After spraying on a suitable pre-treated surface the components are allowed a 10-minute 'flash off' period to allow the solvents to evaporate before stoving. The component is then baked, or stoved, usually in a convection oven for 30 minutes at 120°C depending on the required finished properties.

Waterborne acrylic is available as primers and finish coats which are quick drying and have a low odour and low VOCs. The application equipment is easy to clean.

14.16.13 Chlorinated rubber paints

These single-pack protective coatings, based on chlorinated rubber, are fast drying and give very good water, chemical and corrosion resistance. They are suitable for coating steel and concrete surfaces and are recommended for applications subject to chemically laden atmospheres or in a coastal environment and can be applied by spray, roller or brush. Typical uses include bridges, chemical plants, marine structures, ships and structural steelwork.

Review questions

1. What is annealing and why is it carried out?
2. What is the definition of plain carbon steel?
3. Name four physical properties of materials.
4. What is the main characteristic of aluminium?
5. Name six mechanical properties of materials.
6. Describe the process of hardening a piece of high-carbon steel.
7. State four properties required of a bearing material.
8. State the make-up of a corrosion cell.
9. Describe the basic difference between steel and cast iron.
10. Name the two main constituents of brass.
11. Name two types of electroplating process and state where each would be used.
12. Give four advantages of using powder coating.
13. Briefly describe the coil coating process.
14. Name a plastics material and indicate the properties which would make it suitable as a bearing material.

CHAPTER 15

Plastics

Most plastics used today are man-made and are described as synthetic materials, i.e. they are made by the process of building up from simple chemical substances.

Some plastics are soft and flexible, others hard and brittle, and many are strong and tough. Some have good thermal and electrical properties while others are poor in these respects. Crystal clear plastics are available, while others can be produced in an extremely wide range of colours. Most plastics can be easily shaped using heat or pressure or both.

All plastics products are made from the essential polymer mixed with a complex blend of materials known collectively as 'additives'. Without additives, plastics would not work, but with them included, plastics can be made safer, cleaner, tougher and more colourful. Additives include:

▶ *Antimicrobials* – help prevent microbiological attack which can cause infection and disease. Microbes can also be responsible for undesirable effects such as product degradation, discolouration and food contamination.

▶ *Antioxidants* – oxygen is the major cause of polymer degradation and antioxidants are included to slow down this degradation.

▶ *Antistatic agents* – added to minimise the natural tendency of plastics to accumulate a static charge which can result in dust accumulation.

▶ *Biodegradable plasticisers* – used in applications such as rubbish bags, food packaging and check-out bags whereby the material breaks down in a pre-programmed manner under defined conditions.

▶ *Fillers* – natural cheaper substances such as chalk, talc and clay are incorporated to improve strength and lower the cost of the raw materials by increasing their overall bulk.

▶ *Flame retardants* – to prevent ignition or spread of flame in plastics materials.

▶ *Fragrances* – used in the preparation of perfumed plastics articles. Deodorisers can be included to absorb more unwanted odours.

▶ *Heat stabilisers* – to prevent decomposition of the polymer during processing.

▶ *Impact modifiers* – enable plastics products to absorb shocks and resist impact without cracking.

▶ *Pigments* – tiny particles used to create colour.

▶ *Plasticisers* – used to make the plastics softer and more flexible.

▶ *UV stabilisers* – ultraviolet radiation is destructive to polymer materials. These additives 'absorb' the UV radiation.

Plastics materials have a wide range of properties which make them invaluable to industry. These

Workshop Processes, Practices and Materials, Fifth Edition. 9781138784727.

include electrical and heat insulation, being lightweight, durable, recyclable and energy efficient in their production and providing freedom of design.

The largest user of plastics is the packaging industry, producing bottles, crates, food containers and films, taking advantage of their flexibility and light weight, transparency, durability and cost effectiveness. They are safe and hygienic and practically unbreakable.

The construction industry is the second largest user, producing pipes, cladding, insulation and seals and gaskets for doors, windows, roofing and linings.

The transport industry takes advantage of the lightness of plastics, reducing the weight of cars, boats, trains and aircraft, cutting fuel consumption and operating costs. Composites are now widely used in military jets and helicopters as well as body and wing skins, nacelles, flaps and a wide range of interior fittings in commercial passenger jets. The latest Boeing 787 passenger jet has been dubbed 'Boeing's plastic dream machine' as it is built mostly of composites. Plastics composites make up 100% of the skin and 50% of all materials in the aircraft. Fewer parts are required, so there is less to bolt together and maintenance costs are also reduced. Any weight saving will mean less fuel is used.

Electrical and electronic industries also benefit from the use of plastics, making electrical goods safer, lighter, more attractive, quieter and more durable. Applications are endless but include televisions, DVD and CD players, washing machines, refrigerators, kettles, toasters, vacuum cleaners, lawnmowers, telephones, computers, printers, etc.

15.1 Thermoplastics and thermosetting plastics

With a few exceptions, plastics are compounds of carbon with one or more of the five elements hydrogen, oxygen, nitrogen, chlorine and fluorine. These compounds form a diverse group of different materials, each with its own characteristic properties and uses. All plastics materials are based on large molecules which are made by joining together large numbers of smaller molecules. The small molecules, known as monomers, are derived from natural gas and crude oil and are subjected to suitable conditions to join up and form long-chain-molecular products known as polymers. The process of joining the molecules together is known as polymerisation, and the names of the plastics made in this way frequently contain the prefix 'poly'. For example, the monomer ethylene is polymerised to form very-long-chain molecules of the polymer polyethylene (PE; also called 'polythene').

Plastics are mostly solid and stable at ordinary temperatures, and at some stage of their manufacture they are 'plastic', i.e. soft and capable of being shaped. The shaping process is done by the application of heat and pressure, and it is the behaviour of the material when heated that distinguishes between the two classes of plastics: thermoplastics and thermosetting plastics.

Plastics made up of molecules arranged in long chain-like structures which are separate from each other, soften when heated, and become solid again when cooled. By further heating and cooling, the material can be made to take a different shape, and this process can be repeated again and again. Plastics having this property are known as thermoplastics and examples include polyethylene (PE) (polythene), polyvinyl chloride (PVC), polystyrene (PS), acrylonitrile-butadiene-styrene (ABS), acrylics, polypropylene (PP), nylon and polytetrafluoroethylene (PTFE).

Other plastics, although they soften when heated the first time and can be shaped, become stiff and hard on further heating and cannot be softened again. During the heating process, a chemical reaction takes place which cross-links the long chain-like structures, thus joining them firmly and permanently together – a process known as curing. Plastics of this type are known as thermosetting plastics, or thermosets, and examples include phenolics and melamine formaldehyde.

15.2 Types of plastics

15.2.1 Thermosetting plastics

15.2.1.1 Phenolics

Phenolic plastics are based on phenol formaldehyde resins and are often referred to as

'Bakelite'. The material has a limited colour range and is supplied only in dark colours, principally brown and black.

Phenolics are good electrical insulators and show good resistance to water, acid and most solvents. They are rigid and have low thermal conductivity. The normal operating temperature limit for phenolic mouldings is 150 °C, but grades are available which will operate at up to 200 °C for limited periods. Typical uses include handles and electrical plugs and sockets.

Phenol and urea formaldehyde resins are used in the casting industry to produce cores and moulds and although formaldehydes have not been proven to cause cancer in humans, they can, when heated, irritate eyes, skin and respiratory tract.

Phenolic laminates are widely used and are made by applying heat and pressure to layers of paper, linen, cotton or glass cloth impregnated with phenolic resin. These laminates are strong, stiff, have high impact and compressive strength and are easy to machine. They are available in a variety of sheet, rod and tube sizes and are used for gears, insulating washers and printed circuit and terminal boards. They are known by various trade names such as 'Tufnol', 'Paxolin', 'Richlite' and 'Novotext'.

15.2.1.2 Melamine formaldehyde

Usually shortened to melamine, melamine formaldehyde is often used in kitchen utensils and tableware and in electrical insulating parts. It is strong and glossy and has good resistance to heat, chemicals, moisture, electricity and scratching, has excellent moulding properties and is available in a range of vibrant colours.

Melamine resin is the main constituent of high-pressure laminates with a heat-resistant and wipe-clean surface, widely known by the trade name 'Formica'.

Melamine foam, a special form of melamine, is used as an insulating and soundproofing material and as a cleaning abrasive.

15.2.1.3 Polyimide

Thermosetting polyimides are one of the highest performing engineering plastics, exhibiting superior performance in applications requiring low wear and long life in severe environments. Polyimide materials feature:

▶ high temperature resistance (can perform up to 315 °C with only negligible loss of mechanical properties);
▶ high wear resistance;
▶ low thermal expansion;
▶ dimensionally stable;
▶ good thermal and electrical insulation;
▶ resistance to radiation;
▶ good chemical resistance;
▶ relatively easy to machine.

Polyimide materials can be used as unfilled base polymers or have graphite, PTFE and molybdenum disulphide added to further enhance the low frictional properties and can be used to replace metal and ceramic parts. Material is available as moulded stock shapes, e.g. rod, rings and plate, which can be machined to the final dimensionally accurate product. Some grades can be processed by compression moulding while other grades can be processed to produce a finished product by direct forming.

Direct forming is a process which uses powder metallurgy techniques to produce the final part. The polymer, in powder form, is loaded into a die and compressed to produce a solid compact. The part known as 'green part' is then moved to an oven where it is heated up (or sintered) at a high temperature. After a few hours, sintering is complete and the parts have their full range of properties.

Thermosetting polyimides are used in a variety of industries including aircraft, aerospace, automotive, electrical and engineering, for a wide range of applications, which include rotary seals, bearing, thrust washers and bushings, and in many cases as an alternative to metal parts particularly in high-temperature environments and where weight saving is advantageous.

In jet engines they are used as wear pads and strips, thrust washers, bushings and bearings to save weight and take advantage of wear resistance and low thermal expansion.

Other aerospace uses include control linkage components, door mechanisms and bushings to take advantage of the wear properties.

They are used in vehicles and other industrial equipment requiring higher temperature capabilities, wear resistance and dimensional stability.

Thermoplastic polyimides are available with similar high-temperature and high-performance properties as the thermosets but can be formed using typical melt-processing techniques such as injection moulding. The materials can be used as unfilled base polymer, glass or carbon fibre filled, or PTFE and graphite added. As with the injection moulding process, complex-shaped parts can be produced.

15.2.1.4 Polymer composites

Polymer composites are plastics within which fibres are embedded. The plastic is known as the matrix (resin) and the fibres dispersed within it are known as the reinforcement. Thermosetting matrix materials include polyester, vinyl ester and epoxy resins. For higher temperature and extreme environments, bismaleimide, polyimide and phenolic resins are used. Composites can be used to replace metal parts but care must be taken during design. Most engineering materials have similar properties in any direction (called isotropic) where composites have not. This can, however, be offset by arranging the reinforcement layers in varying directions.

Reinforcement materials are most commonly glass, carbon (graphite) and aramid fibres. The most common are glass and carbon. Glass fibres are cheaper than carbon and come in various forms, making them suitable for many applications. Carbon fibres have higher stiffness and strength than glass fibres, making them suitable for lightweight structures that require stiffness. Aramid fibres are typically used in high-value products where high energy absorption is needed.

15.2.1.5 Glass fibres

This is the commonest reinforcing material. Glass fibre reinforced plastic (GRP), more commonly referred to as 'fibreglass', has many useful properties such as high tensile and compressive strength, hard smooth surfaces unreactive to chemicals, fire resistance, insulator of heat, electricity and sound, impervious to water, easily

moulded and coloured, long lasting and low maintenance and can be manufactured into all manner of products. It can be translucent, opaque or coloured, flat or shaped, thin or thick and there is virtually no limit to the size of the part that can be made (boat hulls over 60 m long have been made). The most common matrix is polyester resin. Phenolic resins have a high resistance to ignition and do not support combustion, and are used where fire resistance, low fume toxicity and low smoke emission are major requirements, e.g. in transport systems, defence, aerospace and public buildings.

GRP mouldings can be produced by hand laminating, spray depositing, resin transfer injection and hot and cold press. GRP requires some sort of former or tool from which it is moulded and can be quite simple and cheap, made from wood, plaster, plastics or GRP itself. The most common method of producing a part is hand laminating, often referred to as hand lay-up.

The mould is first cleaned and a few coats of wax release agent applied (similar to car polish). A thickened resin, containing the required coloured pigment and known as a 'gelcoat', is applied to the surface of the mould. The gelcoat gives the final smooth, protective, durable, coloured surface to the part. This is allowed to cure and is followed by layers of chopped strand mat together with the laminating resin and repeated until the required thickness is achieved. The resin is mixed with a hardening agent (or catalyst) in a precise amount which allows the resin to harden (or cure). The product is formed by the hardening of the resin and the glass into an integrated finished moulding. When cured, the finished product is released from the mould and any excess trimmed off and finished as required. In the manufacture of signs, graphics can be encapsulated within the laminate for both protection and backlit displays.

Spray depositing can be used to speed up the moulding process. Glass fibre rovings (rope-like form) are fed into a chopper on a special spray gun and the resultant strands are blown into a stream of liquid resin and catalyst in the correct ratio. Moulds are pre-treated as before, but instead of hand lay-up, the mixture is sprayed onto the mould, resulting in a composite which has a random array of glass fibres.

The applications for GRP are endless and cover a wide range of industries including:

▶ automotive – cab panels, doors, engine covers, exterior body and instrument panels;
▶ construction – canopies, doors, light panels, water tanks and signs;
▶ defence – armoured vehicles, ships, trucks and submarines;
▶ electrical – custom-designed enclosures;
▶ engineering – machine guards and covers;
▶ transport – train body ends, external panels and boat hulls;
▶ water industry – tanks.

15.2.1.6 Carbon fibres

Carbon (sometimes referred to as graphite) fibre is the reinforcement material of choice for 'advanced' composites. Carbon fibres have a higher fatigue resistance than glass or aramid.

Carbon fibre properties depend on the structure of the carbon used and are typically defined as standard, intermediate and high modulus fibres. Several thousand fibres are twisted together to form a yarn which may be used by itself or woven into a fabric. The yarn or fabric is combined with a resin, usually epoxy, and wound or moulded to shape to form a wide variety of products.

Pre-impregnated sheet materials called 'prepregs' are available, where the carbon fibres are impregnated with reactive resins, usually modified epoxy resins. These can be worked on, layed-up and moulded to shape with heat applied to cure the resins. They are also available with glass fibre/phenol and aramid fibre/phenol combinations.

Uses of carbon fibre range from parts for aircraft and on the space shuttle to tennis rackets and golf clubs.

In the aerospace industry uses include fuselage parts, both military and passenger, wings and control surfaces, engine cowls and landing gear doors.

In sport and leisure uses include boat hulls, propeller shafts, masts, rudders, bicycle frames, golf clubs and tennis rackets.

15.2.1.7 Aramid fibres

Aramid fibres have the highest strength to weight ratio compared to the other fibres but broadly similar tensile strength to glass fibre. There are a number of manufacturers but 'Kevlar' and 'Nomex' are the familiar brand names.

Applications for Kevlar include aircraft interiors, ducting, helicopter rotor blades, radomes and avionics.

A major use of Kevlar is in rigid and soft body armour protective applications. Kevlar fabric is bullet and fragment resistant, lightweight, flexible and comfortable, has excellent thermal properties, is resistant to cuts and chemicals and is flame resistant and self-extinguishing. The uses include bulletproof vests, chainsaw leg protection and military uses for helmets and armoured vehicles, cargo containers, armour shields and cockpit doors.

Kevlar and Nomex fabrics are used in the construction of flame-resistant protective clothing used in petroleum and petrochemical operations, by utility workers, NASA astronauts, racing drivers and their crews, the military and any industry where there is a chance of flash fire exposure or electric arc flash or blast.

These glass, carbon and aramid fibres can be produced as continuous profiles by a process called pultrusion. The pultrusion process starts by pulling continuous fibres, together with reinforcing mats and fabrics through a resin impregnation system. All the materials are coated with a resin and passed through preforming guides, before passing through a heated die where the resin is cured. The cross-section of the die determines the finished size and shape, which can be solid round and square, round and square tube and various sections such as angle, channel and I-shapes. At the end of the process the sections are cut to length. The resins are usually polyester, vinylester, epoxy or phenolic. The resulting composite is strong and stiff with high corrosion resistance. The weight is 80% less than steel and 30% less than aluminium and its high strength to weight ratio makes it ideal for structural components in transport, aerospace, building and civil engineering applications. It has good electrical and thermal insulation and fire-retarding properties and is non-magnetic.

Pultrusions can be fabricated by machining (drilling, sawing, etc.), bolting, riveting and adhesive bonding. Applications include structural beams, handrails, floor supports and ladders.

15.2.2 Thermoplastics

There are many thermoplastics available today and the following describe a few of those more widely used.

15.2.2.1 Polyethylene (or polythene) (PE)

Polythene is one of the most widely used plastics and is available as low density (LDPE) and high density (HDPE).

LDPE has excellent resistance to water and oils, is tough, flexible and relatively transparent and is predominantly used in film applications. It is popular in applications where heat sealing is necessary. Applications include bags for dry cleaning, bread, frozen foods and fresh produce, and shrink-wrap coatings for paper cups, toys and squeezable bottles.

HDPE has excellent resistance to most solvents and has a high tensile strength compared to other forms of PE. It is a relatively stiff material with useful temperature capabilities. Applications include bottles for milk, juice, cosmetics, detergents and household cleaners, extruded pipes and conduit.

15.2.2.2 Polyvinyl chloride (PVC)

This is one of the oldest plastics and one of the most versatile, being available in many forms with a wide range of hardness and flexibility, and can be broadly divided into rigid and flexible materials. It is also one of the least expensive. PVC possesses good physical strength, durability and resistance to water and chemicals, is a good electrical insulator and is available in a wide range of colours.

Flexible packaging uses include bags, shrink wrap, meat wrapping, wire and cable installation and wall and floor coverings.

Rigid PVC is lightweight, stiff, hard and tough at room temperature and can be used outdoors when suitably stabilised. Applications include pipes, guttering and fittings, doors and windows, frames, bottles and containers.

15.2.2.3 Polystyrene (PS)

PS is a versatile plastic that can be rigid or foamed. High impact grades (HIPS) are produced by adding rubber or butadiene which increases the toughness and impact strength.

General-purpose PS is economical and is used for producing plastic model assembly kits, plastic cutlery, yogurt and cottage cheese containers, cups, CD 'jewel' cases and many other objects where fairly rigid, economical plastics are desired.

Expanded polystyrene foam (EPS) is usually white and made from expanded PS beads. It is lightweight, absorbs shocks and is a very good sound and heat insulator. It crumbles easily and burns readily unless flame proofed. Familiar uses include moulded packaging material to cushion fragile items inside boxes and as heat insulation in the construction industry. It is also widely used to make foam cups, plates, food containers, meat trays and egg cartons.

PS is melted by most solvents and by cyanacrylate (super) glues.

15.2.2.4 Polyethylene terephthalate (PET)

This is a thermoplastic polymer resin of the polyester family and is usually referred to as polyester. PET can be semi-rigid to rigid depending on its thickness and is very lightweight, clear, strong and impact resistant, has good gas- and moisture-resistant barrier properties and is widely used for soft drink bottles.

When produced as a thin film, it is often used in tape applications such as the carrier for magnetic tape and as a backing for pressure-sensitive tapes.

In the form of synthetic fibres, its major use is in carpet yarns and in textiles, as woven polyester fabrics used in the home furnishings such as bed and table sheets, curtains and draperies. Industrial polyesters are used in tyre reinforcement, ropes and fabrics for conveyor and safety belts.

15.2.2.5 Acrylonitrile-butadiene-styrene (ABS)

This material has exceptional impact strength and resistance to low temperatures and chemicals, and can be produced with an exceptionally high-gloss finish in a wide range of colours.

ABS is used in place of PS where greater toughness is required. Its uses include musical instruments, notably recorders, and clarinets, car roof boxes, rigid luggage and domestic appliance housings (e.g. food mixers) and children's toys (e.g. Lego building blocks).

15.2.2.6 Polymethyl methacrylate (PMMA)

Commonly called acrylic or by its familiar trade name Perspex, this is a transparent thermoplastic having optical clarity and is often used as an alternative to glass. PMMA possesses exceptional stability to outdoor weather resulting in wide use in signs, displays and light fittings. It is used as lenses for exterior car lights, for aircraft windows and as large flat panels and tunnels in aquariums. At room temperature it is rigid, can be easily machined and can be joined by cementing or with heat (melting). With the application of heat it becomes pliable and can be easily formed. Care must be taken when working PMMA as it splinters and scratches easily. Scratches can be removed by polish or by gently heating the surface of the material.

PMMA is available in various forms:

▶ sheet from 3 to 30 mm thick clear and in a range of colours and clear acrylic thick sheet from 35 to 150 mm thick;
▶ rod in various shapes and sizes, e.g. round, square, triangular and half-round;
▶ tube from 5 to 650 mm outside diameter and various wall thicknesses.

15.2.2.7 Polypropylene (PP)

PP is one of the lightest plastics available. It has good chemical resistance, is strong and has a high melting point, making it ideal for containers which have to be hot-filled. It has good resistance to fatigue and will withstand repeated bending, making it ideal as an integral ('living') hinge in flip-top bottles.

PP is widely used as containers for yogurt, margarine, take-away meals, microwavable meal trays, bottle tops and enclosures, and chairs. It is also widely used as a fibre in the production of carpets, rugs and mats for the home, for sacks and cloth and for ropes and twines.

This material is available as an extruded twinwall fluted PP sheet, under the trade name Correx, usually 2.5 and 4 mm thick in a range of colours. Uses include packaging and low-cost display, e.g. house for sale signs.

Expanded polypropylene (EPP) is a foam form of PP. It is lightweight with a very high strength to weight ratio. It is highly durable and can withstand impact without significant damage. Its applications include moulded seat cores and bumpers in the car industry and packaging, and as protective gear in the sport and leisure industry. Due to its impact characteristics, it is extensively used in radio-controlled model aircraft.

15.2.2.8 Nylon

Nylon is a member of the polyamide (PA) family of synthetic polymers first used commercially in a nylon-bristled toothbrush followed by ladies' stockings. Nylon is tough, has good resistance to oil, fuel and chemicals and has good thermal resistance. Nylon provides a good self-lubricating bearing surface and has a high melting point.

Nylon fibres are used in many applications including fabrics, carpets, musical instruments strings and ropes.

Nylon is non-toxic and is used in film form for food packaging, and coupled with toughness and its temperature resistance is used for boil-in-the-bag foods.

In its solid form it is used in many engineering applications including bushes, bearings, gears, wear pads and threaded parts.

Nylon is available as sheet, rod and tube in various sizes, usually cream or black coloured, and can be easily machined. To further enhance its properties, nylon is available in glass-filled, oil-filled and heat-stabilised versions.

15.2.2.9 Polytetrafluoroethylene (PTFE)

PTFE has an exceptional resistance to chemical attack, a very low coefficient of friction and excellent electrical insulating properties and can be used continuously at temperatures up to 260°C. It is well known under the brand name 'Teflon'.

Limitations of PTFE are its poor abrasion resistance, high coefficient of expansion and poor thermal conductivity. By adding filler materials such as glass fibre, carbon and finely powdered metal, these characteristics can be modified and improved.

PTFE is available as sheet, rod and tube in a variety of sizes, usually white in colour, but the filled varieties are available in blacks and browns.

The major application is in chemical engineering where its exceptional resistance to chemical attack makes it useful for handling highly corrosive chemicals and solvents and in pharmaceutical products as an alternative to glass such as beakers, bottles, evaporating dishes, funnels, jars and containers and syringes; in electrical engineering as an insulating material; and in mechanical engineering where its low frictional properties make it ideal for bearings, ball valve seatings, wear-strips and piston rings. It is also used in medical applications as it is stable and will not react to anything. It is widely used as a non-stick coating on pans and other cookware.

15.2.2.10 Polyetheretherketone (PEEK)

PEEK is regarded as one of the highest performing polymers. It is available as granules and powder for moulding and coating and as fibre and film. As sheet and rod, it is available in unfilled, glass and carbon fibre reinforced and wear-resistant grades.

PEEK offers excellent chemical resistance to common solvents including acids, salts and oil, and has very low moisture absorption and good wear, abrasion and electrical resistance. It has excellent strength and stiffness properties and low coefficients of friction and thermal expansion. It can be used continuously to 250°C and in hot water and steam without permanent loss in physical properties.

PEEK is lightweight compared to steel, titanium and aluminium alloys and is typically used as a replacement for machined metal parts in a wide variety of high-performance applications.

In the aerospace industry its uses include oil cooling and ventilation systems fans, fuel tank covers, door handles, radomes, fastening systems (nuts, bolts, inserts and brackets), pipes and convoluted tubing (a barrier against chemicals and moisture), wiring and cable clamps.

In the automobile industry PEEK has successfully replaced metal parts including oil pumps, camshaft bearings, turbocharger impeller, transmission seals, thrust washers, steering parts, seat adjustment gears, clutch friction ring, vacuum pump and brake parts.

In electronics uses include mobile phones, laser printers, audio speakers, circuit boards and in hard disk drives.

It is widely used in the oil and gas, food and beverage and semiconductor industries.

PEEK is used as a replacement for metal medical instruments that are exposed to extreme sterilisation where aggressive chemicals and radiation are used.

PEEK-based biomaterials are used in a variety of surgical implants.

15.3 Plastics recycling

This is the process of recovering scrap or waste plastics and reprocessing the material into useful products. It has been estimated that every tonne of plastics bottles recycled saves about 3.8 barrels of oil. Recycling is therefore seen as very eco-friendly.

In most cases, different types of plastics must be recycled separately so that the recycled material has the greatest value and potential use. In 1988 the American Society of the Plastics Industry Inc. (SPI) introduced a voluntary resin identification system which indicates the type of plastics that an item is made from. The primary purpose of the codes is to allow for efficient separation of different polymer types for recycling. The symbols used in the code consist of arrows that cycle clockwise to form a rounded triangle, enclosing a number from 1 to 7 usually with letters representing the polymer below the triangle as shown in Fig. 15.1. The number has no other meaning than identifying the type of polymer as follows:

1. PET
2. HDPE (polythene)
3. PVC
4. LDPE (polythene)
5. PP

6. PS
7. other plastics including acrylic, ABS, nylon, fibreglass and polycarbonate.

Figure 15.1 Resin identification codes

Local councils primarily target bottles and containers as they represent the largest usage and are easy to identify and separate, usually 1 and 2.

If you look at the base of a plastics bottle or container, you will see a triangle containing a number together with letters identifying the type of plastics material.

PET or number 1 is recycled as fibre for carpets, fleece jackets and containers for food and beverages and woven as fabrics for the clothing industry.

HDPE or number 2 is the most often recycled plastics and recycled items include bottles for non-food items such as shampoo and detergent, garden tables and benches, recycling bins and buckets.

15.4 Working in plastics

Various techniques are available within a conventional workshop for producing shapes in plastics materials without the need of expensive equipment.

The ease with which plastics can be moulded and formed to shape is one of the major advantages of this range of materials. It is not always practicable to produce the required shape, often for economical reasons; e.g. a small quantity would not justify the high cost of moulding equipment or large sizes. Small-scale casting or forming techniques may have to be used.

The moulding techniques used may not give the required accuracy. In this case a machining operation may have to be carried out.

Sheet plastics material may require to be fabricated, in which case sheets can be joined by welding.

15.5 Plastics welding

There are many different, cost-effective welding methods suitable for industrial mass production of plastics engineering components including ultrasonic, laser, spin, hot plate, linear and orbital vibration welding. Here we will deal with inexpensive manual methods for low quantities which can be carried out in small workshops or on site.

15.5.1 Hot-gas welding

This is a manual welding process which consists of heating and softening the two surfaces to be joined and a filler rod, usually of the same material, until complete fusion takes place. It is a similar process to welding metal, except that a naked flame is not used as this would burn the plastics material. Instead a stream of hot gas is used, directed from a special welding torch. The heat source may be electric or gas, at a temperature around 300 °C.

The surfaces to be welded have to be prepared to accept the filler rod. For butt joints the edges are chamfered to an included angle of about 60°. For fillet joints an angle of 45° is used, as shown in Fig. 15.2. The surfaces to be welded must be clean and free from grease. Filler rods are generally circular in section and for small work are usually 3 mm diameter.

The surfaces to be joined should be clamped together. The hot air is then directed at the surfaces and the filler rod, which is pressed

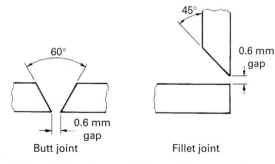

Figure 15.2 Butt- and fillet-weld preparation

into the joint as the area becomes tacky. The downward pressure of the filler rod makes the weld, which fuses and solidifies as welding proceeds, Fig. 15.3. Depending on the thickness of the material, more than one weld run may be necessary.

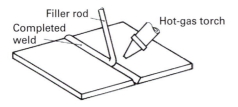

Figure 15.3 Hot gas welding

This method is used with good results to weld rigid PVC, PP and PE (polythene) sheet to fabricate tanks, vessels, pipes for all types of fluids, and ducting.

To speed up the welding process, speed tip welding can be used. A special nozzle is fitted which has a feed tube for the plastic filler rod. The speed tip heats the filler rod and the material being welded and presses a bead of softened plastic into the joint where it fuses the parts being joined.

15.5.2 Heat sealing

This method of welding uses a heated metal strip or bar at a temperature of between 180 °C and 230 °C. The heated strip is applied under pressure to the surfaces to be welded. To prevent the plastics being sealed sticking to the heated bars, a material such as PTFE is placed between them and the bar. This method is used on nylon and with PE sheet, e.g. in the manufacture of polyethene bags.

15.5.3 Solvent welding

Solvents can be used to soften thermoplastic materials, which, if placed together, will then completely fuse when the solvent evaporates. The main disadvantage of solvent welding is the risk of some of the solvent reaching surfaces other than those being joined and leaving a mark.

This technique is commonly used for connecting PVC and ABS pipes, for acrylic in the manufacture of display items and 'gluing' polystyrene and ABS model kits.

Care must be taken when using solvents, as many are flammable and give off toxic vapours.

15.6 Machining

Machining of plastics is carried out when the number of workpieces to be produced is small and to purchase expensive moulding equipment would be uneconomic. Alternatively, where accuracy greater than can be obtained from the moulding technique is required or where features such as tapped holes cannot be included, machining is essential.

Most plastics materials can be machined using metal-working tools and machines. Since plastics materials have a low thermal conductivity and a high coefficient of expansion, heat produced in cutting must be kept at an absolute minimum. To minimise this heat, it is advisable to use a cutting fluid and to grind all tools with larger clearance angles than are necessary when cutting metal. When machining plastics materials, take light cuts and use high cutting speeds with low feed rates.

The large range of plastics materials available makes it difficult to be specific, and the following is offered only as a general guide to the common machining techniques.

15.6.1 Sawing

In general with all soft materials, coarse-tooth hacksaw blades should be used to prevent clogging the teeth. With brittle materials such as acrylics (Perspex), it may be necessary to use a finer-tooth saw to avoid splintering the edges.

15.6.2 Drilling

Standard high-speed drills are satisfactory for use with plastics but must be cleared frequently to remove swarf. Slow-helix drills (20° helix) reduce the effect of swarf clogging in the flutes and, with a drill point of around 90°, give a better finish on breakthrough with the softer plastics. All drills should have an increased point clearance of 15° to 20°.

In drilling thin plastics sheet, a point angle as great as 150° is used on the larger diameter drills, so that the point is still in contact with the material when the drill starts cutting its full diameter.

Alternatively, the sheet can be clamped to a piece of waste material as with thin sheet metal.

Use cutting speeds of around 40 m/min with a feed rate of 0.1 mm/rev. Problems associated with drilling plastics materials were discussed in Section 8.7.

15.6.3 Reaming

Helical-flute reamers should always be used. The reamer must be sharp, otherwise the material tends to be pushed away rather than cut.

15.6.4 Turning

High-speed-steel cutting tools in a centre lathe can be used to turn plastics materials. Clearance angles should be increased to around 20°. A rake angle of 0° can be used on the more brittle plastics, while a 15° rake used on softer plastics aids the flow of material over the tool face. Cutting edges must be kept sharp. Cutting speeds of 150 m/min and higher are used, with feed rates between 0.1 and 0.25 mm/rev.

15.6.5 Milling

Milling plastics materials can be carried out using high-speed-steel cutters on a standard milling machine. The cutters should be kept sharp and in good condition.

To avoid distortion, the less rigid plastics materials should be supported over their complete area, due to the high cutting forces acting. Cutting speeds and feed rates similar to those for turning can be used.

15.6.6 Tapping and threading

Holes drilled or moulded in plastics materials can be tapped using high-speed-steel ground-thread taps. With softer plastics materials there is a tendency for the material to be pushed away rather than cut, and this may necessitate the use of special taps about 0.05 mm to 0.13 mm oversize. Alternatively, thread-forming and thread-cutting screws can be used during assembly. Thread-forming screws deform the plastics material, producing a permanent thread, and can be used in less brittle materials such as nylon. Thread-cutting screws physically remove material,

similar to a ground thread tap, and are used with the more brittle materials such as acrylic.

Threading can be carried out using single-point tools in a centre lathe with the same angles as for turning. High-speed-steel dies can also be used, but care must be taken to ensure that the thread is being cut and the material is not merely being pushed aside.

Threads cut directly in plastics materials will not withstand high loads and will wear out if screw fasteners are removed and replaced several times. Where high strength and reliability are required, threaded inserts are used.

Inserts, which have a tapped inner hole to accept a standard threaded screw, can be self-tapping, expansion, heat or ultrasonic and can be headed or unheaded.

Self-tapping inserts, Fig. 15.4(a), cut their own threads, are available in brass, case-hardened steel and stainless steel from M3 to M16 and are used in the more brittle thermoplastics and in thermosets.

Expansion inserts, Fig. 15.4(b), are pushed into a pre-drilled or pre-moulded hole and are locked in place during assembly as the screw expands the insert. They are used in the softer plastics which are ductile enough to allow the outer knurls to bite into the plastic. They are available in brass from M3 to M8.

Heat and ultrasonic inserts, Fig. 15.4(c), can only be used in thermoplastics materials and are available in brass from M3 to M8. Heat inserts are used for low-volume production. The insert is placed on the end of a thermal insert tool and heated to the correct temperature, depending on the type of plastics component and pressed into a pre-drilled or pre-moulded hole. The plastic adjacent to the insert is softened and flows into the grooves to lock the insert in place. The thermal insert tool is removed and the plastic re-solidifies.

(a)   (b)   (c)

Figure 15.4 Threaded inserts for plastics materials

For higher production rates, the above principles apply, but the heat source is generated by ultrasonic vibration.

15.7 Heat bending

The forming of thermoplastics can be conveniently carried out by applying heat, usually between 120°C and 170°C, and bending to shape. Care must be taken not to overheat, as permanent damage to the material can result. Provided no permanent damage has been done, a shaped thermoplastic sheet will return to a flat sheet on the application of further heat.

Simple bending is carried out by locally heating along the bend line, from both sides, until the material is pliable, using a strip heater. A strip heater can easily be constructed using a heating element inside a box structure, with the top made from a heat-resisting material. The top has a 5 mm wide slot along its centre, through which the heat passes, Fig. 15.5. When the material is pliable, it can be located in a former and bent to the required angle, e.g. in making a splash guard for a lathe, Fig. 15.6. The material can be removed from the former when the temperature drops to about 60°C. Formers can be simply made from any convenient material such as wood. For volume production large-scale industrial machines are available which can have multiple bending areas and advanced heat control.

Shapes other than simple bends can be carried out by heating the complete piece of material in an oven. To avoid marking the surface, the material can be placed on a piece of brown paper. The time in the oven depends on the type of material and its thickness, and time must be allowed for the material to reach an even temperature throughout.

Acrylic sheet material is easily worked at 170°C, 3 mm thickness requiring about 20 minutes and 6 mm thickness about 30 minutes in the oven. Again, a simple former can be used to obtain the required shape, e.g. in making a guard for a drilling machine, Fig. 15.7.

Drape forming (also called oven forming) is carried out commercially on small and large parts where the plastics material is pre-heated in an oven to the forming temperature, placed in a mould and held in place while it cools. Applications include

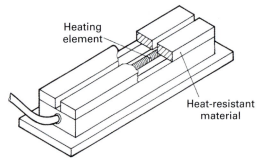

Figure 15.5   Strip heater

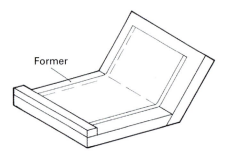

Figure 15.6   Bending a splash guard for a centre lathe

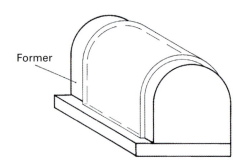

Figure 15.7   Forming a drill guard

windscreens, seat backs and bottoms, covers and doors.

15.8 Plastics moulding processes

As already stated, the shaping of plastics materials is achieved by the application of heat, pressure or both. There are many ways in which this can be done, depending on the nature of the polymer, the type and size of product and the quantity and dimensional accuracy required. The methods to be described here are moulding by compression, transfer, injection and low-pressure processes.

Metals have a definite melting point and in general tend to be free-flowing in a molten state. Polymers, on the other hand, have no definite melting point but are softened by the application of heat, which renders them 'plastic'. In this state they may be considered as very viscous fluids and, as a result, high pressures are required for moulding.

The viscosity of a polymer is reduced by the application of heat, but there is an upper temperature limit at which the polymer begins to break down in some way. This breakdown is known as degradation. All polymers are bad conductors of heat and are therefore susceptible to over-heating. If a polymer is subjected to excessive temperature or to prolonged periods in the mould, degradation will occur.

There is also a lower limit of temperature below which the polymer will not be soft enough to flow into the mould. The temperature for moulding must be between the upper and lower limits and will directly affect the viscosity of the polymer.

All the moulding processes require three stages:

1. application of heat to soften the moulding material;
2. forming to the required shape in a mould;
3. removal of heat.

The moulding techniques to be discussed essentially differ only in the way the moulding material is heated and delivered to the mould.

15.8.1 Types of moulding process

15.8.1.1 Compression moulding

Compression moulding is used chiefly with thermosetting plastics such as urea and melamine formaldehyde, epoxies and phenolics. The process is carried out in a hydraulic press with heated platens. The two halves of the moulding tool consist of a male and a female die, to give a cavity of the required finished shape of the product, and are attached to the platens of the press, Fig. 15.8. Depending on the size and shape of the product and on the quantity required, the moulding tool may contain a single cavity or a number of cavities, when it is known as a multi-cavity mould. The mould cavity is designed with an allowance for shrinkage of the moulding material and a draft angle of at least 1° to allow for the escape

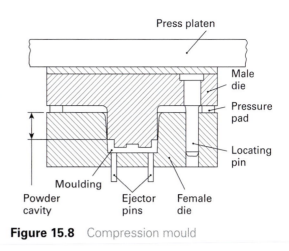

Figure 15.8 Compression mould

of gases and easy removal of the product after moulding. Moulding tools are manufactured from tool steel, hardened and tempered to give strength, toughness and a good hard-wearing surface. The male and female dies are highly polished and, once the tool has been proved, these surfaces are chromium-plated (around 0.005 mm thickness) to give a high surface gloss to the product and to facilitate its removal and protect the surfaces from the corrosive effects of the moulding material.

Moulding materials are normally in the form of loose powder or granules. These materials have a high bulk factor, i.e. the volume of the loose material is much greater than that of the finished product. The bulk factor is around 2.5:1 for granulated materials and can be as high as 4:1 for fine powder. To allow for this, a powder cavity is built into the female die attached to the bottom press platen. To prevent an excess of loose material being loaded into the tool, each charge is weighed, either on scales or by some automatic method.

Alternatively, loose powder material can be compressed to form a small pellet, or preform, of a size and shape to suit the mould cavity. This is done cold, so that no curing takes place, and is carried out in a special preforming or pelleting machine. The preform is easily handled and gives a consistent mass of charge. Depending on the material, these preforms may be preheated to around 85 °C in a high-frequency oven – this reduces the cycle time, since the preform is partially heated before it is loaded in the tool and requires less pressure in moulding.

The typical sequence of operations for compression moulding is as follows:

1. Load moulding material – as loose powder, granules or a heated preform – into the heated die cavity. Moulding temperatures vary, e.g. between 135 °C and 155 °C for urea powders and between 140 °C and 160 °C for melamine materials.
2. Close the split mould between the press platens. The pressure is around 30 N/mm² to 60 N/mm² (lower for preheated pellets). The combined effect of the moulding temperature and pressure causes the moulding material to soften and flow into the mould cavity. Further exposure to the moulding temperature causes the irreversible chemical reaction of cross-linking or curing. The curing time depends on the wall thickness, mass, moulding material, and moulding temperature; e.g. a 3 mm section of urea will cure in around 30 s at 145 °C and the same section in melamine in up to 2 min at 150 °C.
3. Open the split mould.
4. Eject the product from the mould. This may be done by hand or automatically, according to the complexity of the tool or product. Since the material is thermosetting and has cured, there is no need to wait for the moulding to cool and therefore it can be removed immediately while it is still hot.
5. Blow out the tool to remove any particles left behind by the previous moulding.
6. Lubricate the tool to assist the release of the next moulding.
7. Repeat the process.

Any material which escapes during the moulding process results in a feather edge on the moulding known as a flash. This can be removed by the operator during the curing cycle of the next moulding.

Compression moulding is used to produce a wide range of products, including electrical and domestic fittings, toilet seats and covers, bottle tops and various closures and tableware.

15.8.1.2 Transfer moulding

Transfer moulding is used for thermosetting plastics. This process is similar to compression moulding except that the plasticising and moulding functions are carried out separately. The moulding material is heated until plastic in a transfer pot from which it is pushed by a plunger through a series of runners into the heated split mould where it cures, Fig. 15.9. The two halves of the split mould are attached to the heated platens of a hydraulic press in the same way as for compression moulding.

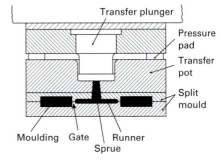

Figure 15.9 Transfer mould

The split mould is closed and the moulding material in the form of powder, granules, or a heated preform is placed in the transfer pot and pressure is applied. The pressure on the area of the transfer pot is greater than that in the compression-moulding process but, as the moulding material is plastic when it enters the mould cavity, the pressure within the cavity is much less. As a result of this, the process is suited to the production of parts incorporating small metal inserts. Intricate parts and those having variations of section thickness can be produced to advantage by this method. Cure times are less and greater accuracy is achieved than with compression moulding.

The main limitation of the process is the loss of material in the sprue, runners and gates – as thermosetting materials cure during moulding, this cannot be reused. Moulding tools are usually more complex and therefore more costly than compression-moulding tools.

Resin transfer moulding (RTM) is a low-pressure variation of transfer moulding where, instead of granules or preforms, a pre-mixed resin and catalyst (hardener) is injected into a closed mould containing dry glass, carbon or aramid fibre reinforcement. When the resin has cooled, the mould is opened and the composite part ejected.

The resins used include polyester, vinylester, epoxies, phenolics and methyl methacrylate combined with pigments and fillers. Applications include small complex aircraft and automotive components as well as automotive body parts, baths and containers.

15.8.1.3 Injection moulding

Injection moulding may be used for either thermoplastics or thermosetting plastics, but is most widely used for thermoplastics including polycarbonate, polythene, polystyrene, polypropylene, ABS, PEEK and nylon.

The major advantages of this moulding process are its high production rate, high degree of dimensional accuracy and its suitability for a wide range of products.

The moulding material is fed by gravity from a hopper to a cylindrical heating chamber where it is rendered plastic and then injected into a closed mould under pressure. The moulding solidifies in the mould. On solidification, the mould is opened and the moulding is ejected.

The type of injection-moulding machine most widely used is the horizontal type, Fig. 15.10. A hopper at the opposite end of the machine from the mould is charged with moulding material in the form of powder or granules which is fed by gravity to the heating chamber. Electric band heaters are attached to the outer casing of the heating chamber, inside which is an extruder-type screw similar to that of a domestic mincing machine. As

the screw rotates, it carries the material to the front of the heating chamber.

The band heaters and the frictional forces developed in the material by the rotating screw result in the material becoming plastic as it passes along.

When the plasticised material builds up in front of the screw, the screw moves axially backwards and its rotation is stopped. The amount of material to be injected into the mould – known as the shot size – is controlled by stopping the screw rotation at a predetermined position. At this stage, the mould, filled by the previous shot, is opened and the moulding is ejected. The mould is then closed and the stationary screw is moved axially towards the mould, pushing the plasticised material into the mould cavity under pressure. The screw then rotates, feeding more moulding material along the heating chamber to become plasticised, the material being continuously replaced from the hopper. The screw then moves axially backwards due to the build-up, and the sequence recommences. The heating-chamber temperature varies between 120 °C and 260 °C, depending on the type of moulding material and the shot size.

Moulding tools are of the same materials and finish as described for compression-moulding tools but without, of course, the need for a powder cavity. They are generally more expensive than compression-moulding tools but, because of the higher production rates of injection moulding, it is possible to use smaller, i.e. single-cavity, and hence cheaper tools and maintain the same

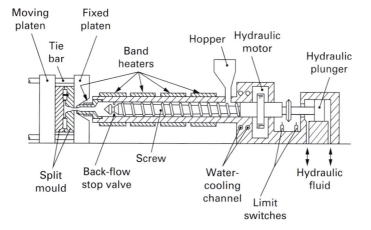

Figure 15.10 Injection-moulding machine

number of mouldings per hour as would be possible with a compression press.

The use of thermoplastics material requires that the mould be maintained at a constant temperature – usually around 75 °C to 95 °C – to cool and solidify the material within the mould before the moulding can be ejected. This is achieved by circulating water through the mould and makes the process much faster than compression moulding. Although material is used in the sprue and runners, material wastage is low since it can be reused.

Injection moulding of thermosetting materials is achieved in the same way as for thermoplastic materials except that the temperatures used are more critical. The temperature of the heating chamber is important – to avoid the moulding material curing before it enters and fills the mould cavity – and is between 95 °C and 105 °C for urea materials and between 100 °C and 110 °C for melamine materials. Moulding-tool temperature is also important, to ensure correct curing of the moulding material. Tool temperatures, between 135 °C and 145 °C for urea materials and between 145 °C and 155 °C for melamine materials are suitable for sections over 3 mm and are increased by 10 °C for sections below 3 mm. As the moulding material enters the mould at very near to the curing temperature, cycle times are low.

The range of injection moulded components is vast and includes toys, e.g. model kits, houseware, e.g. buckets and bowls, car components, mobile phones, computer enclosures, power tool housings, crates and containers.

15.8.2 Low-pressure moulding (LPM)

LPM is primarily a one-step process to encapsulate, seal and protect electronic assemblies. The process is also referred to as over-moulding or hot-melt moulding. The electronic assembly is placed in an aluminium mould which is then closed and a thermoplastic compound injected into the mould cavity encapsulating the assembly. The moulds are cooled by chilled water circulating through the mould platens to approximately 20 °C. When the part cools, it is removed from the mould. Cycle

times range from 15 to 45 seconds depending on the size and thickness of the component. Because of the low pressure (can be as low as 2 bar), the process offers very gentle encapsulation of fragile components. Figure 15.11 shows a USB stick before and after encapsulation. This process uses a high-performance polyamide hot-melt adhesive moulding compound to achieve the ultimate protection and seal against moisture, chemicals and vibration.

When used in cable and connector over-moulding, it offers a finger grip, strain relief, flex control and anti-vibration all-in-one, to prevent damage to cables during use (Fig. 15.12).

Figure 15.11 USB stick

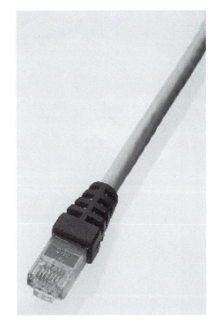

Figure 15.12 Modular connector

LPM is often used to replace cumbersome potting operations. Potting is a process of placing an electrical/electronic assembly in a pot and pouring in a two-part mix, usually epoxy resin, until the pot is full. The resin is allowed to cure and the process is complete. The transformer shown in Fig. 15.13 has been potted to provide a seal against moisture.

Figure 15.13 Transformer before and after potting

15.8.3 Insert moulding

The primary purpose of inserts is to strengthen relatively weak areas in the plastics material and to facilitate the joining of mouldings or the fastening of other parts to the moulding. Inserts are usually made from steel or brass, Fig. 15.14, and may be incorporated at the time of the moulding operation or be inserted after moulding into pre-moulded or drilled holes. The latter method was described in Section 15.6.6. When incorporated as part of the moulding process, the inserts can be loaded by automatic means or by the process operator. It is important to carefully consider the economics of including an insert at the time of moulding, due to the additional costs of the die to incorporate

Figure 15.14 Typical moulded-in insert

inserts and of the loading method. It is not good practice to allow the cycle times to vary from shot to shot (i.e. component to component) as this can have an adverse effect on the quality. This could happen with manual loading by the process operator.

Figure 15.15 shows moulded-in inserts in electrical components.

Figure 15.15 Inserts in electrical components

15.8.4 Safety in plastics moulding

Any machine with moving parts is a potential hazard and as such must be properly guarded. Moulding machines used with plastics are basically presses using high forces to lock the dies in position while the moulding operation is carried out. Adequate guarding must therefore be provided and used to prevent the operator coming in contact with moving parts and so eliminate the possibility of trapped fingers and hands. Most accidents are caused by:

▶ inadequate safeguarding fitted;
▶ removal or falling into disrepair of the safeguarding;
▶ overriding, usually for setting, of the safeguarding.

Heat is an essential part of any moulding process. Moulding materials have a high heat capacity and in their hot plastic state will stick on contact with the skin and are difficult to remove. Protection can be afforded by the use of suitable industrial gloves. Some of the materials used can cause dermatitis, which can be prevented by the use of gloves.

Harmful gases and vapours are given off by some plastics materials, so the moulding machine must be fitted with adequate extraction equipment.

Noise is often a problem, so the workplace and machinery must be designed to reduce noise exposure (see Chapter 1). Where appropriate, personal hearing protection must be provided and worn.

15

Review questions

1. Name two types of thermosetting plastics material.
2. Why are metal inserts often used in plastics components?
3. What is the definition of a thermosetting plastics material?
4. Name three types of plastics moulding process.
5. What is the definition of a thermoplastics material?
6. What precautions are necessary when machining plastics materials?
7. Name four types of thermoplastics materials.
8. State two types of welding used to join plastics materials.
9. Name two additives included in a plastics material and state their function.
10. What is a polymer composite and what are its advantages?
11. Give an advantage and a limitation of acrylic.
12. Give reasons why recycling plastics materials is eco-friendly and indicate the significance of resin identification codes.

CHAPTER 16

Primary forming processes

Most metal objects have at some stage in their manufacture been shaped by pouring molten metal into a mould and allowing it to solidify. On solidifying, the object is known as a casting if the shape is such that no further shaping is required – it may only require machining to produce the finished article. Castings are produced by various methods, e.g. sand casting, die-casting and investment casting. If, upon solidifying, the object is to be further shaped by rolling, extrusion, drawing or forging, it is known as an ingot, pig, slab, billet or bar – depending on the metal and the subsequent shaping process – and these are cast as simple shapes convenient for the particular forming process.

Most metals, with the exception of some precious ones, are found in the form of minerals or ores. The ores are smelted to convert them into metals; e.g. iron is obtained from iron ore (haematite), aluminium from bauxite and copper from copper pyrites.

Iron ore together with other elements is smelted in a blast furnace to give pig iron. According to the type of plant, the molten pig iron is cast as pigs or is transferred to a steel-making process. Cast pigs are refined in a cupola to give cast iron which is cast as notched ingots or bars of relatively small cross-section for ease of remelting in the foundry as required. Pig iron taken in a molten state to the steel-making process is made into steel which is cast as ingots or now more usually as slabs produced by the continuous casting process for subsequent rolling, drawing or forging.

Aluminium is extracted from bauxite by an electrolytic process. Commercially pure aluminium is soft and weak and it is alloyed to improve the mechanical properties. Aluminium alloys are available wrought or cast, in sections convenient for subsequent working by rolling, drawing, casting or extrusion.

Copper is extracted from copper pyrites and is refined by remelting, in a furnace or electrolytically. Copper is alloyed to produce a range of brasses and bronzes. These materials may then be rolled, drawn, cast or extruded.

16.1 Forms of supply of raw materials

Where the metal has to be remelted, as in the case of casting, it is usual to supply it to the foundry in the form of notched ingots or bars of relatively small cross-section. These can be broken into smaller pieces for ease of handling and loading into small furnaces.

Where the metal is to be subjected to further forming processes – e.g. rolling, extrusion, drawing or forging – it would be wasteful to remelt, both from an energy viewpoint and from the effect the remelting would have on the physical and mechanical properties. In this case the raw material would be supplied in the form most convenient for the process – i.e. slabs (width greater than three times the thickness) for rolling into sheet; blooms and billets (smaller section) for rolling into bar, sections, etc. and for forging and extrusion. Hot-rolled rod is supplied for cold drawing into rod, tube and wire.

16.2 Properties of raw materials

16.2.1 Fluidity

This property is a requirement for a metal which is to be cast. The metal must flow freely in a molten state in order to completely fill the mould cavity.

16.2.2 Ductility

A ductile material can be reduced in cross-section without breaking. Ductility is an essential property when drawing, since the material must be capable of flowing through the reduced diameter of the die and at the same time withstand the pulling force. The reduction in cross-sectional area aimed for in a single pass through the die is usually between 25% and 45%.

16.2.3 Malleability

A malleable material can be rolled or hammered permanently into a different shape without fracturing. This property is required when rolling and forging.

16.2.4 Plasticity

This is a similar property to malleability, involving permanent deformation without fracture. This property is required in forging and extrusion, where the metal is rendered plastic, i.e. made more pliable, by the application of heat.

16.2.5 Toughness

A material is tough if it is capable of absorbing a great deal of energy before it fractures. This property is required when forging.

16.3 Sand casting

Casting is the simplest and most direct way of producing a finished shape from metal. Casting shapes from liquid metals can be done by a variety of processes, the simplest of which is sand casting.

For the production of small castings, a method known as box moulding is employed. The box is made up of two frames with lugs at each end into which pins are fitted to ensure accurate alignment when the frames are placed together. The top frame is referred to as the cope and the bottom frame as the drag.

A removable pattern is used to create the required shape of cavity within the mould. The pattern is made in two parts, split at a convenient position for ease of removal from the mould – this position being known as the parting line. The two parts of the pattern are accurately aligned with each other using pins or dowels.

Consider the gear blank shown in Fig. 16.1(a), the pattern for which is shown in Fig. 16.1(b). One half of the pattern is placed on a board. The drag is then placed on the board such that the pattern half is roughly central. Moulding sand is then poured into the drag and is pressed firmly against the pattern, either manually or vibratory assisted. The drag is then filled completely and the sand is firmly packed using a rammer, Fig. 16.2(a). The amount of ramming should be sufficient for the sand to hold together but not enough to prevent the escape of gases produced during pouring of the molten metal. After ramming is complete, the

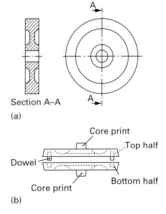

Section A–A
(a)

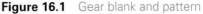

Core print
Top half
Dowel
Bottom half
Core print
(b)

Figure 16.1 Gear blank and pattern

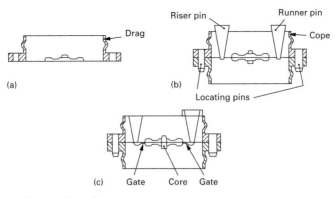

Figure 16.2 Basic steps in sand casting

sand is levelled off flush with the edges of the drag. Small vent holes can be made through the sand to within a few millimetres of the pattern, to assist the escape of gases.

The drag is then turned over and the second half pattern is located by means of the dowels. The upturned surface is then covered with a fine coating of dry 'parting sand', to prevent bonding between the sand in the cope and that in the drag. The cope is accurately positioned on the drag by means of the locating pins. To allow entry for the molten metal, a tapered plug known as a runner pin is placed to one side of the pattern. A second tapered plug known as a riser pin is placed at the opposite side of the pattern, Fig. 16.2(b) – this produces an opening which, when filled with molten metal during pouring, provides a supply of hot metal to compensate for shrinkage as the casting cools. The cope is then filled, rammed and vented as for the drag. The pins are then removed and the top of the runner hole is enlarged to give a wide opening for pouring the metal.

The cope is then carefully lifted off and turned over. Both halves of the pattern are carefully removed. Small channels known as gates are then cut from the bottom of the runner and riser to enable metal to fill the mould cavity. Any loose sand is blown away to leave a clean cavity in the mould. The core is placed in position in the bottom half of the mould box (the drag) and the top mould box (the cope) is carefully replaced in position with the aid of the locating pins and is clamped to prevent lifting when the metal is poured, Fig. 16.2(c). The core which is necessary to provide a hollow section, in this case the bore of the gear blank, is made separately.

The mould is now ready for pouring. When the metal has solidified, the mould is broken up to release the casting. The runner and riser are broken off, and the rough edges are removed by fettling (i.e. hand grinding).

For high-volume production of parts, automated sand casting systems are used which result in very low component costs.

16.3.1 Patterns

For the production of small quantities of castings, patterns are made from wood, smoothed, painted or varnished to give a smooth finish to the casting. Patterns are made larger than the finished part, to allow for shrinkage of the casting when it cools. A special rule, known as a contraction rule, is available to suit different metals – the pattern-maker makes the pattern using measurements from the contraction rule, which automatically gives the correct dimension of pattern with due allowance for shrinkage whatever the size of dimension.

Some surfaces of a casting may require subsequent machining, e.g. surfaces requiring a greater accuracy of size, flatness or surface finish than can be achieved by casting. Extra metal must be left on these surfaces, and the amount to be left for removal by machining must be allowed for on the appropriate surface of the pattern.

Where cores are to be incorporated in a casting (see Section 16.3.3), provision must be made on the pattern to provide a location seating in the mould. These sections added to the pattern are known as core prints.

To allow the pattern to be easily removed from the mould, a small angle or taper known as draft is incorporated on all surfaces perpendicular to the parting line.

Patterns can be reused indefinitely to produce new sand moulds.

16.3.2 Sand

Moulding sand must be permeable, i.e. porous, to allow the escape of gases and steam; strong enough to withstand the mass of molten metal; resist high temperatures and have a grain size suited to the desired surface of the casting.

Silicon sand is used in moulding, the grains of sand being held together in different ways.

In green-sand moulds the grains are held together by moist clay, and the moisture level has to be carefully controlled in order to produce satisfactory results.

Dry-sand moulds start off in the same way as green-sand moulds but the moisture is driven off by heating after the mould has been made to just above 100 °C. This makes the mould stronger and is suited to heavier castings.

With CO_2 (carbon dioxide) sand the silica grains are coated with sodium silicate instead of clay. When the mould is made, it is hardened by passing CO_2 gas through it for a short period of time. The sand 'sets' but is easily broken after casting.

Resin or no-bake moulding uses sand with organic and inorganic binders that strengthen the mould by chemically adhering to the sand, producing high-strength moulds having good dimensional control and providing an improved surface finish.

Specially prepared facing sand is used next to the pattern to give an improved surface to the casting. The mould can then be filled using a backing sand.

16.3.3 Cores

When a casting is to have a hollow section, a core must be incorporated into the mould. Cores can be made from metal or more commonly from sand, made separately in a core box, and inserted into the mould after the pattern is removed and before the mould is closed. They are located and supported in the mould in a seating formed by the core prints on the pattern. The core must be strong enough to support itself and withstand the flow of molten metal, and in some cases it may be necessary to reinforce it with wires to give added strength. The more complex cores are produced from CO_2 sand.

16.3.4 Chills

To control the solidification structure of the metal, it is possible to place metal plates (known as chills) in the mould. The rapid local cooling produces a finer grain structure forming a harder metal surface at these locations.

16.3.5 Advantages and limitations

The sand casting process can be used to economically produce small numbers of castings in both large and small sizes in a wide variety of metals.

The main advantages of sand casting are:

▶ adaptable to any quantity;
▶ complex shapes with intricate cores can be produced;
▶ low tooling costs;
▶ large ferrous and non-ferrous castings are achievable;
▶ reclaimable mould and cast material;
▶ easily adapted to high-production automated methods;
▶ suitable for a wide range of metals.

Metals which can be cast include cast irons, low-carbon and alloy steels, aluminium alloys and copper alloys (e.g. bronze and phosphor bronze).

Limitations include:

▶ low casting rate;
▶ high unit cost;
▶ cannot produce thin sections (minimum 4–5 mm);
▶ poor linear tolerances (e.g. 4 mm/m);
▶ rough surface finish;
▶ coarse grain size.

16.4 Vacuum casting

Vacuum casting is a variation of sand casting, is used for aluminium alloys and is known in the industry as V-casting. The process uses dry silica sand that contains no moisture or binders. The mould cavity holds the shape by means of a vacuum.

A specially vented pattern is placed on a hollow carrier plate. A heat-softened special plastic sheet (around 0.2 mm thick) is draped over the pattern while a vacuum is applied to draw the plastic tightly around the pattern. A special flask, with walls which are also a vacuum chamber, is placed over the pattern and filled with free-flowing sand (same as a cope in sand casting). The sand is vibrated to compact the sand and a sprue and pouring cup are formed. A second sheet of plastic is placed on top of the sand and a vacuum applied to draw out the air and strengthen the unbonded sand. The vacuum is then released from the pattern carrier plate and the top mould (cope) is stripped from the pattern. The bottom mould (drag) is made in exactly the same way. Any cores are set in place and the mould halves are closed. The molten metal is poured while the two halves are still under a vacuum, the plastic vaporises but the vacuum keeps the shape of the sand while the metal solidifies. When the metal solidifies, the vacuum is turned off and the sand runs out freely, releasing the casting.

The V-process does not require a draft angle since the plastic film allows easy removal from the pattern. This leads to a constant wall thickness, as small as 2.3 mm. The surface finish is very smooth, with excellent dimensional accuracy and good reproduction of detail which can reduce machining requirements. Pattern life is unlimited as the sand does not touch it. Use of dry sand gives lower sand costs, resulting in simplified sand control with no mixing, no reclamation, no waste sand, no shake out and no sand lumps, although simple screening is required to remove metal fragments.

16.5 Rolling

The cast ingots produced by the raw-material producers are of little use for manufacturing processes until they have been formed to a suitable shape, i.e. sheet, plate, strip, bar, sections, etc.

One of the ways in which shapes can be produced in order that manufacturing processes can

subsequently be carried out is by rolling. This can be performed as hot rolling or cold rolling – in each case, the metal is worked while in a solid state and is shaped by plastic deformation.

The reasons for working metals in their solid state are, first, to produce shapes which would be difficult or expensive to produce by other methods, e.g. long lengths of sheet, section, rods, etc., and, second, to improve mechanical properties.

The initial stage of converting the ingot to the required shape is by hot rolling. During hot working, the metal is in a plastic state and is readily formed by pressure as it passes through the rolls. Hot rolling has a number of other advantages.

▶ Most ingots when cast contain many small holes – a condition known as porosity. During hot rolling, these holes are pressed together and eliminated.
▶ Any impurities contained in the ingot are broken up and dispersed throughout the metal.
▶ The internal grain structure of the metal is refined, resulting in an improvement of the mechanical properties, e.g. ductility and strength.

Hot rolling does, however, have a number of disadvantages. Due to the high temperatures, the surface oxidises, producing a scale which results in a poor surface finish, making it difficult to maintain dimensional accuracy. Where close dimensional accuracy and good surface finish are not of great importance, e.g. structural shapes for construction work, a descaling operation is carried out and the product is used as-rolled. Alternatively, further work can be carried out by cold rolling.

When metal is cold rolled, greater forces are required, necessitating a large number of stages before reaching the required shape. The strength of the material is greatly improved, but this is accompanied by a decrease in ductility. Depending on the number of stages required in producing the shape, annealing (or softening) may have to be carried out between stages. Besides improving mechanical properties, cold rolling produces a good surface finish with high dimensional accuracy.

In the initial stage of converting it to a more suitable form, the ingot is first rolled into intermediate shapes such as blooms, billets or slabs. A bloom has a square cross-section with a minimum size of 150 mm square. A billet is smaller than a bloom and may have a square cross-section from 40 mm up to the size of a bloom. A slab is rectangular in cross-section from a minimum width of 250 mm and a minimum thickness of 40 mm. These are then cut into convenient lengths for further hot or cold working.

Most primary hot rolling is carried out in either a two-high reversing mill or a three-high continuous mill.

In the two-high reversing mill, Fig. 16.3, the metal is passed between the rolls in one direction. The rolls are then stopped, closed together by an amount depending on the rate of reduction required, and reversed, taking the material back in the opposite direction. This is repeated, with the rolls closed a little more each time, until the final size of section is reached. At intervals throughout this process, the metal is turned on its side to give a uniform structure throughout. Grooves are provided in the top and bottom rolls to give the various reductions and the final shape where appropriate, Fig. 16.4.

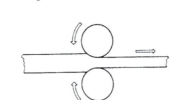

Figure 16.3   Two-high reversing mill

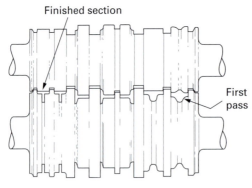

Figure 16.4   Rolls for producing a tee section

In the three-high continuous mill, Fig. 16.5, the rolls are constantly rotating, the metal being fed between the centre and upper rolls in one direction and between the centre and lower rolls in the

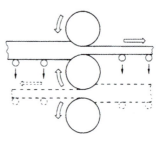

Figure 16.5   Three-high continuous mill

other. A platform is positioned such that it can be raised to feed the metal through in one direction or support it coming out from between the rolls in the opposite direction or be lowered to feed the metal back through or to support it coming out from between the rolls in the opposite directions.

In cold rolling, the roll pressures are much greater than in hot rolling – due to the greater resistance of cold metal to reduction. In this case it is usual to use a four-high mill, Fig. 16.6. In this arrangement, two outer rolls of large diameter are used as back-up rolls to support the smaller working rolls and prevent deflection.

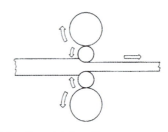

Figure 16.6   Four-high mill

When rolling strip, a series of rolls are arranged in line and the strip is produced continuously, being reduced by each set of rolls as it passes through before being wound on to a coil at the end when it reaches its final thickness.

16.6 Extrusion

Extrusion usually has to be a hot-working process, due to the very large reduction which takes place during the forming process. In operation, a circular billet of metal is heated to render it plastic and is placed inside a container. Force is then applied to the end of the billet by a ram which is usually hydraulically operated. This applied force pushes the metal through an opening in a die to emerge as a long bar of the required shape. The extrusion produced has a constant cross-section along its entire length. The die may contain a number of openings, simultaneously producing a number of extrusions.

There are a number of variations of the extrusion process, two common methods being direct extrusion and indirect extrusion.

16.6.1 Direct extrusion

This process, Fig. 16.7, is used for the majority of extruded products. A heated billet is positioned in the container with a dummy block placed behind it. The container is moved forward against a stationary die and the ram pushes the metal through the die. After extrusion, the container is moved back and a shear descends to cut off the butt end of the billet at the die face. The process is repeated with a new billet. As the outside of the billet moves along the container liner during extrusion, high frictional forces have to be overcome, requiring the use of high ram forces.

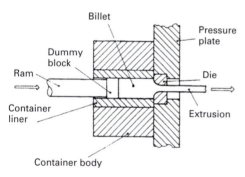

Figure 16.7   Direct extrusion

16.6.2 Indirect extrusion

In this process, Fig. 16.8, the heated billet is loaded in the container which is closed at one end by a sealing disc. The container is moved forward against a stationary die located at the end of a hollow stem. Because the container and billet move together and there is no relative movement between them, friction is eliminated. As a result,

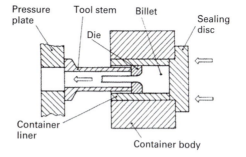

Figure 16.8   Indirect extrusion

longer and larger diameter billets can be used than with direct-extrusion presses of the same power. Alternatively, lower forces are required with the same size billets.

The indirect-extrusion process does have limitations, however. Metal flow tends to carry surface impurities into the extruded metal, and the billets have to be machined or chemically cleaned. The size of the extrusion is limited to the inside diameter of the hollow stem, and die changing is more cumbersome than with direct extrusion.

The important features of the extrusion process are:

▶ the complexity of shape possible is practically unlimited, and finished products can be produced directly, Fig. 16.9;
▶ a good surface finish can be maintained;
▶ good dimensional accuracy can be obtained;
▶ large reductions in cross-sectional area can be achieved;
▶ the metal is in compression during the process, so relatively brittle materials can be extruded;
▶ the mechanical properties of the material are improved.

The extrusion process is, however, limited to products which have a constant cross-section. Any holes, slots, etc. not parallel to the longitudinal axis have to be machined. Due to extrusion-press power capabilities, the size of shape which can be produced is limited. The process is normally limited to long runs, due to die costs, but short runs can be economical with simple die shapes.

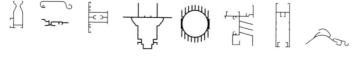

Figure 16.9   Examples of extruded shapes

Typical materials used are copper and aluminium and their alloys.

16.7 Drawing

The primary process of drawing is a cold-working process, i.e. carried out at room temperature. It is mainly used in the production of wire, rod and bar.

Wire is made by cold drawing a previously hot-rolled rod through one or more dies, Fig. 16.10, to decrease its size and improve the physical properties. The hot-rolled rod – usually around 10 mm in diameter – is first cleaned in an acid bath to remove scale, a process known as pickling. This ensures a good finish on the final drawn wire. The rod is then washed with water to remove and neutralise the acid. The end of the rod is then pointed so that it can be passed through the hole in the die and be gripped in a vice attached to the drawing machine. The rod is then pulled through the die to give the necessary reduction in section. Lubrication in the drawing process is essential for maintaining a good surface and long die life.

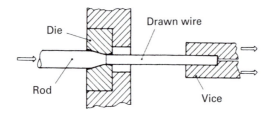
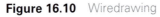

Figure 16.10   Wiredrawing

In continuous wire drawing, the wire is pulled through a series of progressively smaller dies until the final-size section is reached.

The strength of the material will limit the force which can be applied in pulling the wire through the die, while the ductility of the material limits the amount of reduction possible through each die. Typical materials used are steel, copper, aluminium and their alloys.

16.8 Forging

Forging is a hot-working process, heat being necessary to render the metal plastic in order that it may be more easily shaped. The oldest form of forging is hand forging as carried out by the blacksmith. Hand tools are used to manipulate the hot metal to give changes in section and changes in shape by bending, twisting, etc. Due to the hand operation, it is not possible to achieve high degrees of accuracy or extreme complexity of shape. This method is limited to one-off or small quantities and requires a high degree of skill.

When forgings are large, some form of power is employed. Mechanical and hydraulic hammers or a forging press is used, the process being known as open-sided forging. In this process, the hot metal is manipulated to the required shape by squeezing it between a vertically moving die and a stationary die attached to the anvil, Fig. 16.11. This method of forging is carried out under the direction of a forge-master who directs the various stages of turning and moving along the length until the finished shape and size are obtained. Again, great skill is required. This method is used to produce large forgings such as propeller shafts for ships.

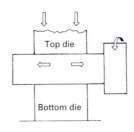

Figure 16.11   Open-die forging

When large quantities of accurately shaped products are required, these are produced by a process known as closed-die forging or drop forging. With this method, the hot metal is placed between two halves of a die each containing a cavity such that when the metal is squeezed into the cavity a completed forging of the required shape is produced, Fig. 16.12. The metal is subjected to repeated blows, usually from a mechanical press, to ensure proper flow of the metal to fill the die cavity. A number of stages

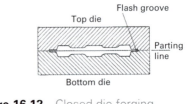

Figure 16.12   Closed-die forging

through a series of dies may be required, each stage changing the shape gradually until the final shape is obtained. The number of stages required depends on the size and shape of the part, how the metal flows and the degree of accuracy required.

One of the two halves of the die is attached to the moving part of the press, the other to the anvil. The cavity contained in the die is designed such that the parting line enables the finished forging to be removed and incorporates a draft in the direction of die movement in the same way as do the pattern and mould in sand casting. The size of cavity also allows for additional material on faces requiring subsequent machining.

Since it is impossible to judge the exact volume required to just fill the die cavity, extra metal is allowed for and this is squeezed out between the two die halves as they close. This results in a thin projection of excess metal on the forging at the parting line, known as a 'flash'. This flash is removed after forging by a trimming operation in which the forging is pushed through a correct-shape opening in a die mounted in a press.

Forging is used in the production of parts which have to withstand heavy or unpredictable loads, such as levers, cams, gears, connecting rods and axle shafts. Mechanical properties are improved by forging, as a result of the flow of metal being controlled so that the direction of grain flow

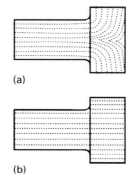
(a)

(b)

Figure 16.13 Direction of grain flow

increases strength. Figure 16.13 shows the difference in the grain flow between a shaft with a flange which has been forged up, Fig. 16.13(a), and one machined from a solid rolled bar, Fig. 16.13(b). Any form machined on the flange, such as gear teeth, would be much weaker when machined from solid bar. The structure of the material is refined due to the hot working, and the density is increased due to compression forces during the forging process.

The process of drop forging is normally restricted to larger batch quantities, due to die costs.

16.8.1 Upset forging

Upset forging increases the diameter of a workpiece by compressing its length. It is a

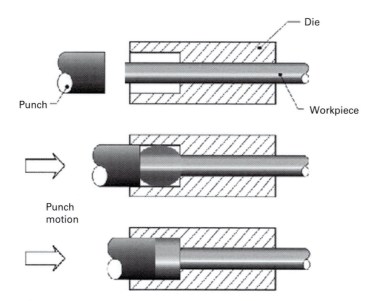

Figure 16.14 Upset forging

16

cold-working process carried out in special high-speed machines. The process is performed on bar stock which is fed into a split die which grips the bar. A punch is then forced against the workpiece deforming and enlarging the head of the workpiece into the shape of the die, Fig. 16.14. This produces a strong, tough part where the grain flow follows the head shape.

This process is used in the high-speed volume production of screws, bolts and other fasteners, engine valves and a multitude of similar headed parts.

16.9 Selection of a primary process

A number of factors have to be considered before a choice of process for a given component can

be made. For example, consideration would have to be given to the type of material to be used, the mechanical properties required, shape, accuracy, degree of surface finish and the quantity to be produced. Some of these factors are shown in Table 16.1 together with the appropriate primary processes, as an aid to the correct choice. For example, the requirement of variation of shape in three dimensions would eliminate rolling, extrusion and drawing; a high degree of accuracy would eliminate sand casting and forging and so on.

Table 16.1 Features of primary forming processes

	Sand casting	Rolling hot	Rolling cold	Extrusion	Drawing	Forging
Improved mechanical properties		✓	✓	✓	✓	✓
Three-dimensional shape variation	✓					✓
Constant cross-section		✓	✓	✓	✓	
Large quantities		✓	✓	✓	✓	✓
Good surface finish			✓	✓	✓	
High dimensional accuracy			✓	✓	✓	
High tool costs				✓		✓

Review questions

1. State two advantages of hot rolling.
2. What are the two types of extrusion process?
3. Name the five primary forming processes.
4. For what types of part is forging best suited and why?
5. Name five properties required of raw materials which make then suitable for primary forming processes.
6. What material is used to make patterns for sand casting and what is the main consideration when making patterns?
7. Name the type of machinery used in the rolling process.
8. Why are cores sometimes used in the sand casting process?
9. With the aid of a sketch describe the metal drawing process.
10. State four important characteristics of the extrusion process.

CHAPTER 17

Presswork

The term 'presswork' is used here to describe the process in which force is applied to sheet metal with the result that the metal is cut, i.e. in blanking and piercing, or is formed to a different shape, i.e. in bending.

The pressworking process is carried out by placing the sheet metal between a punch and die mounted in a press. The punch is attached to the moving part – the slide or ram – which applies the necessary force at each stroke. The die, correctly aligned with the punch, is attached to the fixed part or bedplate of the press.

The press used may be manually operated by hand or by foot and used for light work or it may be power operated, usually by mechanical or hydraulic means, and capable of high rates of production.

The time to produce one component is the time necessary for one stroke of the press slide plus load/unload time or time for feeding the material. Using a power press, this total time may be less than one second.

It is possible to carry out a wide range of operations in a press, and these include blanking, piercing and bending.

Blanking is the production of an external shape, e.g. the outside diameter of a washer.

Piercing is the production of an internal shape, e.g. the hole in a washer.

Bending – in this case simple bending – is confined to a straight bend across the metal sheet in one plane only.

17.1 Presses

17.1.1 The fly press

Blanking, piercing and bending of light work where the required force is small and the production rate is low may be carried out on a fly press.

A hand-operated bench-type fly press is shown in Fig. 17.1. The body is a C-shaped casting of rigid proportions designed to resist the forces acting during the pressworking operation. The C shape gives an adequate throat depth to accommodate a range of work sizes. (The throat depth is the distance from the centre of the slide to the inside

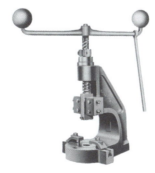

Figure 17.1   Fly press

Workshop Processes, Practices and Materials, Fifth Edition. 9781138784727.
© 2015 Bruce J. Black. Published by Taylor & Francis. All rights reserved.

face of the body casting.) The bottom part of the casting forms the bed to which tooling is attached. The top part of the casting is threaded to accept the multi-start square-threaded screw which carries the handle at its top end and the slide at its bottom end. An adjustable threaded collar is fitted at the top end of the screw and can be locked at a required position to avoid overtravel of the screw during operation. The slide contains a hole and a clamping screw to locate and secure the punch. The horizontal portion of the handle is fitted with ball masses which produce a flywheel effect when larger forces are required. These masses are fitted on spikes, and one or both may be removed when smaller forces are required.

In operation, the vertical handle is grasped and the handle is partially rotated. This provides, through the multi-start thread, a vertical movement of the slide. A punch and die fitted in the slide and on the bed, and correctly aligned with each other, are used to carry out the required pressworking operation.

The fly press can be set up easily and quickly for a range of blanking, piercing and bending operations, and the manual operation gives a greater degree of sensitivity than is often possible with power presses. This type of press may also be used for operations such as pressing dowels and drill bushes into various items of tooling. Due to the manual operation, its production rates are low.

17.1.2 Power presses

Power presses are used where high rates of production are required. A power press may be identified by the design of the frame and its capacity – i.e. the maximum force capable of being delivered at the work, e.g. 500 kN ('50 tons'). The source of power may be mechanical or hydraulic. Different types of press are available in a wide range of capacities, the choice depending on the type of operation, the force required for the operation and the size and type of tooling used.

One of the main types of power press is the open-fronted or gap-frame type. This may be rigid or inclinable. The inclinable feature permits finished work to drop out the back by gravity. A mechanical open-fronted rigid press is shown in Fig. 17.2. The model shown has a capacity of

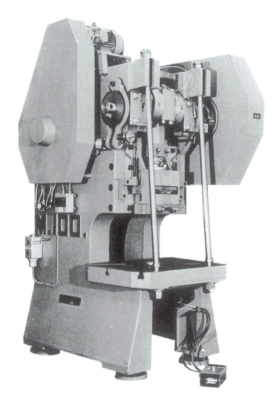

Figure 17.2 Mechanical open-fronted rigid power press

1000 kN ('100 tons') and operates at 60 strokes per minute (it is shown without guards for clarity). The open nature of this design gives good accessibility of the tools and allows the press to be operated from either side or from the front.

The limitation of the open-fronted press is the force which can be applied. High forces have a tendency to flex the frame and so open the gap between the slide and the bedplate. This flexing of the frame can be overcome on the large-capacity presses by fitting tie-rods between the bedplate and the top of the frame as shown in Fig. 17.2.

As the capacity of a press is increased, it becomes necessary to increase the strength and rigidity of its frame – for the reason already outlined. This is achieved in the other main type of power press – the straight-sided or column type – since the large forces are taken up in a vertical direction by the side frames. A mechanical straight-sided press of 3000 kN ('300 tons') capacity is shown in Fig. 17.3.

Straight-sided presses sacrifice adaptability and accessibility to gain frame rigidity and are best

17

With mechanical presses, the maximum force is available at the bottom of the slide stroke. In blanking and piercing operations, the work is done very near to the bottom of the stroke. However, where a part is blanked and then taken further down the stroke, e.g. to form a bend, less force will be available for the blanking operation which will have been carried out some distance above the bottom of the stroke. For example, a 500 kN press with a stroke of 120 mm will exert a force of only 120 kN half-way down its stroke.

17.1.2.2 *Hydraulic presses*

Hydraulic power presses derive their power from high-pressure hydraulic pumps which operate the ram. The load applied is completely independent of the length and stroke, i.e. full load can be applied at any point in the stroke. The applied load is controlled by a relief valve which gives automatic protection against overload. Again by means of valves, the ram can be made to approach the work rapidly and then be shifted to a lower speed before contacting the work, thus prolonging the life of the tool but still giving fast operating speeds. Rapid ram reversal can also be controlled. Switches are incorporated to determine the positions at which these controls become effective, thereby increasing productivity by making tool setting faster and by keeping the actual working stroke to a minimum.

Figure 17.3 Mechanical straight-sided power press

The number of moving parts are few and these are fully lubricated in a flow of pressurised oil, leading to lower maintenance costs. Fewer moving parts and the absence of a flywheel reduce the overall noise level of hydraulic presses compared to mechanical presses.

suited to work on heavy-gauge metals and on large surfaces. Since the sides are closed by the side frames and are open at the front and the back, these presses are limited to operation from the front only.

Longer strokes are available than with mechanical presses, giving greater flexibility of tooling heights.

17.1.2.1 *Mechanical presses*

A typical hydraulic press of 100 kN ('10 ton') capacity is shown in Fig. 17.4. This model has a ram advance speed of 475 mm/s, a pressing speed of 34 mm/s and a return speed of 280 mm/s. The model shown is fitted with a light-screen guard operated by a continuous curtain of infrared light.

Mechanical power presses derive their energy for operation from a constantly rotating flywheel driven by an electric motor. The flywheel is connected to a crankshaft through a clutch which can be set for continuous or single stroking. In the single-stroke mode, the clutch is automatically disengaged at the end of each stroke and the press will not restart until activated by the operator. A connecting rod is attached at one end to the crankshaft and at its other end to the slide. Adjustment is provided to alter the position of the slide, and in some presses the length of stroke can also be adjusted by means of an eccentric on the crankshaft. A brake is fitted to bring the crankshaft to rest at the correct position.

17

Figure 17.4 Hydraulic straight-sided power press

17.1.3 Safety

Power presses are among the most dangerous machines used in industry and have resulted in many serious injuries to operators and tool setters, mainly fingers and hands amputated or crushed between closing tools during loading/unloading or when setting up.

The primary objective of the Use of Work Equipment Regulations 1998 PUWER is to ensure that no work equipment, including power presses, gives rise to risks to health and safety regardless of the work equipments, age, condition and origin.

PUWER Part IV, however, contains specific requirements for power presses, although employers must still comply with the whole of the Regulations (see Chapter 1).

PUWER Part IV applies only to mechanically driven presses which are power driven, have a flywheel and clutch and which are wholly or partly used to work metal. The Regulations do not apply to presses which do not have a clutch mechanism (e.g. pneumatic and hydraulic presses).

17.1.3.1 Employers' duties under PUWER

Employers must ensure that:

▶ power presses and all their guards, the control systems and ancillary equipment (e.g. automatic feeder systems) are maintained so that they do not put people at risk;
▶ maintenance work on power presses is carried out safely, i.e. machinery is shut down and isolated, and done by people who have the right skills and knowledge;
▶ training is provided for 'appointed persons' to help them fulfil their role;
▶ adequate health and safety information and, where appropriate, written instructions are made available to everyone using the process or supervising and managing their use;
▶ an inspection and test is done by the appointed person within 4 hours of the start of each shift or day that the press is used, or after tool setting or adjustment, and the certificate signed to confirm that the press is safe to use;
▶ any existing maintenance log is kept up to date;
▶ the presses and safety devices are thoroughly examined by a competent person at the required intervals.

Preventative maintenance is needed to identify potential failures before employees are put at risk of injury. Worn or defective parts need to be repaired or replaced and adjustments need to be made at set intervals to ensure the press will continue to work safely, in particular those parts that could cause danger if they failed or deteriorated such as brakes, clutches, guards and safety-related parts of the control system.

Every employer must appoint a person to inspect and test the guards and safety devices on each press:

▶ every day they are in use (within the first 4 hours of each working period);
▶ after setting, resetting or adjustment of the tools.

The appointed person has to be adequately trained and competent to do the work.

The appointed person must sign the daily log certificate kept beside each machine which is an indication that the press is fit for use. Without the signed certificate, the press cannot be operated.

Before a power press is put into service for the first time, it must be thoroughly examined and tested by a competent person to ensure the press and its safeguards are installed safely and are safe to operate. The competent person needs to have enough practical or theoretical knowledge and experience to detect defects or weaknesses and assess how these will affect safe operation.

For the purpose of ensuring that health and safety conditions are maintained, every power press is thoroughly examined at least every 12 months where it has fixed guards, and at least every 6 months in other cases (e.g. interlocked guarding or automatic guarding).

A signed report of the thorough examination and test has to be provided to and kept by the employer.

17.1.4 Power-press mechanisms

Press mechanisms such as the clutch, brake, connecting rod and flywheel journals, as well as guards and guard mechanisms, must be subjected to the requirements of the Regulations. Many presswork operations involve feeding and removing workpieces by hand, and every effort must be made to ensure that this can be done safely without the risk that the press will operate inadvertently while the operator's hands are within the tool space. The most obvious way of avoiding accidents to the operator is to design the tools in such a way as to eliminate the need for the operator to place his fingers or hands within the tool space for feeding or removal of work. This can often be done by providing feeding arrangements such that the operator's hands are outside the working area of the tools.

All dangerous parts of the press must be guarded, and the principal ways of guarding the working area are:

▶ enclosed tools, where the tools are designed in such a way that there is insufficient space for entry of fingers;
▶ fixed guards, which prevent fingers or hands reaching into the tool space at any time;
▶ interlocked guards, which allow access to the tools but prevent the clutch being engaged until the guards are fully closed – the guards cannot be opened until the cycle is complete,

the clutch is disengaged, and the crankshaft has stopped;
▶ automatic guards, which push or pull the hand clear before trapping can occur;
▶ light-screen guards – operated by a continuous curtain of infrared light which, if broken, stops the machine.

17.2 Press-tool design

The tools used in presses are punches and dies. The punch is attached to the press slide and is moved into the die, which is fixed to the press bedplate. In blanking and piercing, the punch and die are the shape of the required blank and hole and the metal is sheared by passing the punch through the die. In bending, the punch and die are shaped to the required form of the bend and no cutting takes place. In each case, the punch and die must be in perfect alignment. The complete assembly of punch and die is known as a press tool.

17.2.1 Blanking and piercing

In blanking and piercing operations the work material, placed in the press tool, is cut by a shearing action between the adjacent sharp edges of the punch and the die. As the punch descends on to the material, there is an initial deformation of the surface of the material followed by the start of fracture on both sides, Fig. 17.5. As the tensile strength of the material is reached, fracture progresses and complete failure occurs.

The shape of the side wall produced by this operation is not straight as in machining operations and is shown greatly exaggerated in Fig. 17.6. The exact shape depends on the amount of clearance between the punch and the die. Too

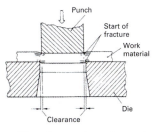

Figure 17.5 Shearing action of punch and die

265

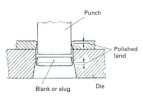

Figure 17.6 Characteristics of sheared edge

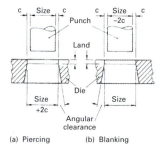

(a) Piercing   (b) Blanking

Figure 17.7 Clearance on punch and die for piercing and blanking operations

large a clearance leads to a large angle of fracture and a large burr, while too small a clearance will result in premature wear of the tools and a risk of tool breakage.

The clearance of the space between the punch and the die is quoted as a percentage of the work-material thickness per side. Establishment of the correct clearance to be used for a given blanking or piercing operation is influenced by the required characteristics of the cut edge and by the thickness and properties of the work material. The values given in Table 17.1 are offered only as a general guide.

Table 17.1 Typical values of clearance for press-tool design

Work material	Clearance per side (% of work-material thickness)
Low-carbon steel	5–7
Aluminium alloys	2–6
Brass:	
annealed	2–3
half hard	3–5
Phosphor bronze	3.5–5
Copper:	
annealed	2–4
half hard	3–5

In blanking and piercing operations, the punch establishes the size of the hole and the die establishes the size of the blank. Therefore in piercing, where an accurate size of hole is required, the punch is made to the required hole size and the clearance is made on the die, Fig. 17.7(a). Conversely, in blanking, where an accurate size of blank is required, the die is made to the required blank size and the clearance is made on the punch, Fig. 17.7(b). On this basis, the material punched through during piercing is scrap and the material left behind in the die during blanking is scrap.

17.2.2 Example 17.1

It is required to produce 50 mm diameter blanks from 2 mm thick low-carbon steel. If a clearance of 6% is chosen, then

diameter of die = diameter of blank = 50 mm

clearance per side = 6% of 2 mm = 0.12 mm

clearance on diameter is therefore
2 × 0.12 mm = 0.24 mm

Thus the diameter of the punch is smaller than the die by this amount; therefore

diameter of punch = 50 mm − 0.24 mm
= 49.76 mm

17.2.3 Example 17.2

It is required to punch 20 mm diameter holes in 1.5 mm thick copper. If a clearance of 4% is chosen, then

diameter of punch = diameter of hole = 20 mm

clearance per side = 4% of 1.5 mm = 0.06 mm

clearance on diameter is therefore
2 × 0.06 mm = 0.12 mm

Thus the diameter of the die is larger than the punch by this amount; therefore

diameter of die = 20 mm + 0.12 mm = 20.12 mm

To prevent the blanks or slugs removed by the punch from jamming in the die, it is usual to provide an angular clearance below the cutting edge of the die as shown in Fig. 17.7(a). Thus the pieces can fall through the die, through a hole in the bedplate, into a bin. A land equal in width to approximately twice the metal thickness may be provided, which enables a large number of regrinds to be carried out on the top die face to maintain a sharp cutting edge.

17

17.2.4 Stripping

As the punch enters the work material during a blanking or piercing operation, it becomes a tight fit in the material. To withdraw the punch without lifting the material along with the punch, it is necessary to provide a method of holding the material while the punch is withdrawn on the upward stroke. This is known as stripping and can be done by providing a fixed stripper or a spring-type stripper.

A fixed stripper, Fig. 17.8, is used when the work material is in the form of strip, fed across the top of the die. The stripper in this case is a flat plate screwed to the top of the die and contains a hole through which the punch passes. When the punch is on the upward return stroke, the plate prevents the material from lifting and strips it off the punch.

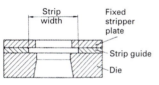

Figure 17.8 Fixed stripper

When the work material is not in the form of strip and has to be loaded in the tool by hand, the spring-pad type of stripper is used, Fig. 17.9. The stripper pad is set in advance of the punch, holding the material on the die face by means of springs and keeping it flat while the operation is being carried out. Stripper bolts keep the pad in position. When the punch is on the upward return stroke, the pad holds the material against the die and strips it off the punch.

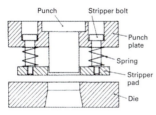

Figure 17.9 Spring-pad stripper

17.2.5 Bending

Bending as described here refers to simple bending, confined to straight bends across the work material in one plane only. In bending operations, no metal cutting takes place: the previously cut material is placed between a punch and a die and force is applied to form the required bend.

Strip or sheet metal should, wherever possible, be bent in a direction across the grain of the material rather than along it. The direction of the grain is produced in the rolling process, and by bending across it there is less tendency for the material to crack. Keeping the bend radius as large as possible will also reduce the tendency for the material to crack.

Bends should not be positioned close to holes, as these can be pulled into an oval shape. It is generally accepted that the distance from the centre of the bend to the edge of a hole should be at least two-and-a-half times the thickness of the work material.

In bending operations, the length of the blank before bending has to be calculated. Any metal which is bent will stretch on the outside of the bends and be compressed on the inside. At some point between the inside and outside faces, the layers remain unaltered in length and this point is known as the neutral axis. For bends of radius more than twice the material thickness, the neutral axis may be assumed to lie at the centre of the material thickness, Fig. 17.10(a). For sharper bends of radius less than twice the material thickness, the neutral axis shifts towards the inside face. In this case, the distance from the neutral axis to the inside face may be assumed to be 0.33 times the material thickness, Fig. 17.10(b).

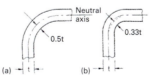

Figure 17.10 Position of neutral axis

The length of the blank is determined by calculating the lengths of the flat portions either side of the radius plus the stretched out length of the bend radius (known as the bend allowance).

17.2.6 Example 17.3

Determine the blank length of the right-angled bracket shown in Fig. 17.11(a).

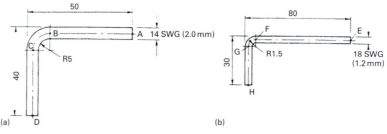

Figure 17.11 Bending examples

Length AB = 50 mm − inside radius
                − material thickness

        = 50 mm − 5 mm − 2 mm

        = 43 mm

Length CD = 40 mm − 5 mm − 2 mm

        = 33 mm

Since the radius is greater than twice the material thickness t, we can assume that the distance from the inside face to the neutral axis is $0.5t$.

∴ radius to neutral axis = 5 mm + (0.5 × 2 mm)

                       = 6 mm

Since it is a 90° bend, length BC equals a quarter of the circumference of a circle of radius 6 mm

∴ length BC = $\dfrac{2\pi R}{4} = \dfrac{\pi R}{2} = \dfrac{6\pi}{2}$ = 9.4 mm

∴ blank length = 43 mm + 33 mm + 9.4 mm

                  = 85.4 mm

17.2.7 Example 17.4

Determine the blank length of the right-angled bracket shown in Fig. 17.11(b).

Length EF = 80 mm − 1.5 mm − 1.2 mm

           = 77.3 mm

Length GH = 30 mm − 1.5 mm − 1.2 mm

           = 27.3 mm

Since the radius in this case is less than twice the material thickness t, we can assume that the distance from the inside face to the neutral axis is $0.33t$.

∴ radius to neutral axis = 1.5 mm + (0.33 × 1.2 mm)

                       = 1.9 mm

∴ length FG = $\dfrac{\pi R}{2} = \dfrac{1.9\pi}{2}$ = 2.98 mm, say 3 mm

∴ blank length = 77.3 mm + 27.3 mm + 3 mm

                = 107.6 mm

Another factor which must be considered in bending operations is the amount of spring-back. Metal that has been bent retains some of its original elasticity and there is some elastic recovery after the punch has been removed. This is known as springback. In most cases this is overcome by overbending, i.e. bending the metal to a greater extent so that it will spring back to the required angle – see also section 17.3.3.

17.2.8 Die-sets

The punch is held in a punch plate and in its simplest form has a step as shown in Fig. 17.12. The step prevents the punch from pulling out of the punch plate during operation.

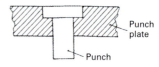

Punch plate

Punch

Figure 17.12 Location of punch in punch plate

In order to ensure perfect alignment in the press, the punch plate and the die are secured in a die-set. Standard die-sets are available in steel or cast iron, and a typical example is shown in Fig. 17.13. The top plate carries a spigot which is located and held in the press slide. The bolster contains slots for clamping to the press bedplate. The bolster has two guide pins on which the top plate slides up and down through ball bushes which reduce friction and ensure accurate location between the two parts.

The punch plate is fixed to the underside of the top plate and the die is fixed to the bolster, the punch and die being accurately aligned with each other.

The die-set is mounted in the press as a complete self-contained assembly and can be removed and

Figure 17.13   Standard die-set

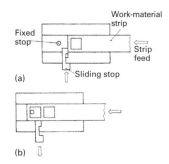

Figure 17.15   Use of stops in simple blanking

replaced as often as required in the knowledge that accurate alignment of punch and die is always maintained. Change-over times are low, since the complete die-set is removed and replaced by another die-set complete with its punches and dies for a different component.

17.3 Blanking, piercing and bending operations

17.3.1 Simple blanking

As previously stated, blanking is the production of an external shape from sheet metal. In its simplest form, this operation requires one punch and die.

For simple blanking from strip fed by hand, the press tool consists of a die on top of which are attached the strip guides and the stripper plate. The punch is held in the punch plate. The arrangement is shown in section in Fig. 17.14.

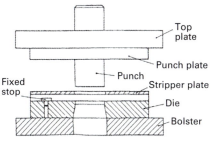

Figure 17.14   Simple blanking tool

To assist in setting up and for subsequent operation, stops are required. For setting up, a sliding stop is pushed in and the strip of work material is pushed against it by hand, Fig. 17.15(a). The punch descends and blanks the first part, the blank falling through the die opening and out through the bedplate. On the upward stroke, the work material is stripped from the punch by the stripper plate. The sliding stop is then retracted and the work material is pushed up to the fixed stop locating in the hole produced in the blanking operation, Fig. 17.15(b). This maintains a constant pitch between blankings. The punch then descends and produces another blank; the punch is raised; the strip is moved forward against the fixed stop; and the operation is repeated. Thus a blank is produced each time the punch descends.

17.3.2 Blanking and piercing

Where the required workpiece is to have both an external and an internal shape – e.g. a washer – the two operations, i.e. blanking and piercing, can be done by the same press tool. This type of press tool is known as a follow-on tool.

The principle is the same as for simple blanking but the punch plate has two punches fitted – one for blanking and one for piercing – and there are two holes of the required shape in the die, Fig. 17.16. The work material is again in the form of strip, fed by hand.

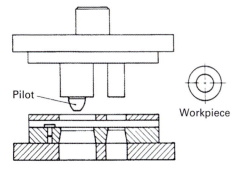

Figure 17.16   Follow-on tool

17

For setting up, two sliding stops are required. The first sliding stop is pushed in and the strip of work material is pushed against it, Fig. 17.17(a). The punches descend and the hole for the first workpiece is pierced. The punches are then raised. The first sliding stop is then retracted, the second sliding stop is pushed in, and the work material is pushed against it, Fig. 17.17(b) – again to maintain a constant pitch. At this stage the pierced hole is now positioned under the blanking punch. The punches once more descend, the blanking punch producing a completed workpiece and at the same time the other punch piercing a hole. The second sliding stop is withdrawn and the work material is now moved forward against the fixed stop, Fig. 17.17(c), again positioning the pierced hole under the blanking punch. The operation is then repeated, a completed workpiece being produced at each stroke of the press.

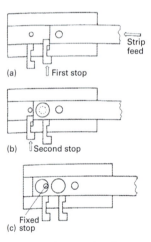

Figure 17.17 Use of stops in piercing and blanking

Greater accuracy of the inner and outer profiles of a workpiece can be obtained by fitting a pilot in the blanking punch (Fig. 17.16). In this case the fixed stop is used for approximate positioning and is arranged so that the work material is drawn slightly away from it as the pilot engages in the pierced hole.

17.3.3 Bending

Two bending methods are commonly used, one known as vee bending and the other as side bending.

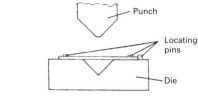

Figure 17.18 Vee-bending tool

Vee-bending tools consist of a die in the shape of a vee block and a wedge-shaped punch, Fig. 17.18. The metal to be bent is placed on top of the die – suitably located to ensure that the bend is in the correct position – and the punch is forced into the die. To allow for springback, the punch is made at an angle less than that required of the finished article. This is determined from experience – e.g. for low-carbon steel an angle of 88° is usually sufficient to allow the metal to spring back to 90°.

Side-bending tools are more complicated than those employed in vee bending but give a more accurate bend. The metal to be bent is placed on top of the die and is pushed against the guide block, which determines the length of the bent leg. Where the leg length is short, location pins can be used. As the punch plate descends, the pressure pad contacts the surface of the metal in advance of the punch and holds it against the die while the punch forms the bend. The guide plate prevents the punch from moving away from the work material during bending and helps the punch to iron the material against the side of the die, so preventing springback. The arrangement of this type of tool is shown in Fig. 17.19.

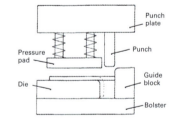

Figure 17.19 Side-bending tool

17.4 Blanking layouts

When parts are to be blanked from strip material, it is essential that the blank is arranged within the strip to gain the greatest economical use of the material by minimising the amount of scrap produced. The final layout will determine the width of strip, which in turn determines the general design and dimensions of the press tool.

The layout may be influenced by subsequent operations such as bending. In this case it is necessary to consider the direction of grain flow, as previously outlined.

It is also necessary to consider the minimum distance between blanks and between blanks and the edge of the strip – this distance must be large enough to support the strip during blanking. Insufficient distance results in a weakened strip which is subject to distortion or breakage, leading to misfeeding. The actual distance depends on a number of variables, but for our purpose a distance equal to the work-material thickness is acceptable.

The material utilisation can be calculated from the area of the part divided by the area of strip used in producing it, given as a percentage. The area of strip used equals the strip width multiplied by the feed distance. An economical layout should give at least a 75% material utilisation.

By virtue of their shape, some parts are simple to lay out whereas others are not so obvious.

Consider a 30 mm × 20 mm blank to be produced from 14 SWG (2.0 mm) material. This would simply be laid out in a straight line as shown in Fig. 17.20.

$$20\,\text{mm} + 2\,\text{mm} = 22\,\text{mm}$$

$$\text{Material utilisation} = \frac{\text{area of part}}{\text{area of strip used}} \times 100\%$$

$$= \frac{\text{area of part}}{\text{strip width} \times \text{feed distance}} \times 100\%$$

$$= \frac{30\,\text{mm} \times 20\,\text{mm}}{34\,\text{mm} \times 22\,\text{mm}} \times 100\% = 80\%$$

Now consider the blank shown in Fig. 17.21, to be produced from 19 SWG (1.0 mm) material. The simple layout would be as shown in Fig. 17.22(a). This gives a strip width of 32 mm and a feed of 31 mm, allowing 1 mm between blanks and between blanks and the edges of the strip. Thus the area of strip used per blank would be 32 mm × 31 mm = 992 mm².

Area of blank = 675 mm²

$$\therefore \text{ material utilisation} = \frac{675\,\text{mm}^2}{992\,\text{mm}^2} \times 100\% = 68\%$$

An alternative layout with the blanks turned through 45° is shown in Fig. 17.22(b). In this

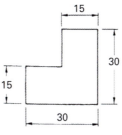

Figure 17.21 Blank

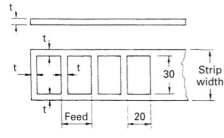

Figure 17.20 Blank layout

Since the work-material thickness is 2 mm, the distance between blanks and between blanks and the edges will be assumed to be 2 mm; therefore

strip width = 30 mm + 2 mm + 2 mm = 34 mm

and the distance the strip must feed at each stroke of the press in order to produce one blank is

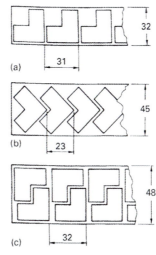

Figure 17.22 Blank layout

case the strip width is 45 mm and the feed 23 mm. Thus the area of strip used per blank would be 45 mm × 23 mm = 1035 mm².

Area of blank = 675 mm²

∴ material utilisation = $\frac{675 \text{ mm}^2}{1035 \text{ mm}^2} \times 100\% = 65\%$

A further alternative layout is shown in Fig. 17.22(c), with the blanks in an alternating pattern. This gives a strip width of 48 mm and a feed of 32 mm, which in this case produces two blanks. Therefore the feed per blank is half this amount, i.e. 16 mm. Thus the area of strip used per blank would be 48 mm × 16 mm = 768 mm².

Area of blank = 675 mm²

∴ material utilisation = $\frac{675 \text{ mm}^2}{768 \text{ mm}^2} \times 100\% = 88\%$

This gives the most economical use of the material and would therefore be the obvious choice. However, with this layout the strip would have to be worked in two passes. On the first pass the bottom row would be blanked; the strip would then be turned round and passed through again for the other row to be blanked.

Review questions

1. What is the limitation of an open-fronted press and how can it be overcome?
2. Name four types of guarding used to prevent operator access to dangerous parts of a power press.
3. State the major advantage in using a hand-operated fly press.
4. What is meant by the term 'presswork'?
5. To what types of production process is a power press best suited?
6. Name the two parts of a tool required to carry out a blanking operation.
7. State four characteristics of a hydraulic power press.
8. State which tool requires clearance for a blanking and for a piercing operation and give the reasons for this clearance.
9. Explain the use of a standard 'die-set'.
10. Explain the importance of the layout of blanks in strip material used in the blanking and piercing process.

CHAPTER 18

Investment casting, lost foam casting and shell moulding

Investment casting, lost foam casting and shell moulding are employed where greater accuracy is required than can usually be achieved by sand casting, especially with small intricate castings and those made from materials otherwise difficult to work with. The moulds in each process are expendable, as in sand casting, and, in comparison with sand casting, these processes are often referred to as 'precision casting'.

Equipment and materials are more expensive than those required in sand casting and therefore require the production of larger quantities to be economic. The higher costs for small quantities can, however, often be offset by savings as a result of the ability to produce a high degree of dimensional accuracy and surface finish, which can reduce or eliminate subsequent machining operations.

Any metal capable of being cast can be cast by these methods, although advantage is most readily gained when using those difficult to work by other methods. Metals for investment casting are the subject of British Standard BS 3146.

18.1 Investment casting

The investment-casting process – also known as the lost-wax process – is one of the oldest casting processes, having been practised for thousands of years. Today highly sophisticated plant, equipment and materials are employed to produce a large variety of components in materials ranging from high-temperature nickel- or cobalt-based alloys – known as 'super-alloys' – to the non-ferrous aluminium and copper alloys.

The process starts with the production of an expendable wax pattern of the shape required which is then coated with a refractory material to produce the mould, which is allowed to dry. The mould is heated, melting the wax, which is allowed to run out and so produce a cavity. Further heat is applied to fire the mould before pouring the casting metal to fill the cavity left by the melted wax. When the molten metal has solidified, the refractory shell is broken away to release the casting. The casting is cut away from any runners and is dressed or fettled and finished as required.

The expendable wax pattern, exactly the shape and size of the required casting, with allowances for pattern contraction and contraction of the casting during solidification, is produced in a split pattern die, Fig. 18.1. The wax is injected into the pattern die in a plastic state, under pressure, and solidifies quickly. It is then ejected from the pattern die, Fig. 18.2. Preformed ceramic or water-soluble cores can be introduced into the wax pattern for the production of castings which

Workshop Processes, Practices and Materials, Fifth Edition. 9781138784727.
© 2015 Bruce J. Black. Published by Taylor & Francis. All rights reserved.

18

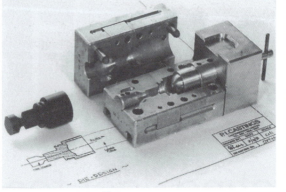

Figure 18.1 Split aluminium die and wax pattern

Figure 18.3 Wax patterns attached to a runner system

The completed wax-pattern assembly is first given a primary coat or investment by dipping it in a ceramic slurry of very fine particle size to make a complete one-piece mould, Fig. 18.4. This first coat is extremely important, as it determines the surface finish and final quality of the casting. When the primary coat is dry, the secondary investment or back-up coats are applied to build up the shell to the required thickness. This is done by successive dipping in slurry followed by coating with dry granular refractory or stucco of large particle size – a process known as 'stuccoing'. A completed

Figure 18.2 Wax-pattern removal

require hollow or complex interior forms. When the wax patterns are small, a number of them are joined by heat welding to a wax runner-and-riser system to produce a pattern assembly for convenience of casting, Fig. 18.3.

Pattern dies are made from a variety of materials, depending on the number of patterns required, their complexity, and the required dimensional accuracy. The die materials used include cast low-melting-point alloys, epoxy resin and aluminium alloys.

Technology is now available which enables the production of a three-dimensional model in wax directly from 3D CAD data and therefore provides the optimum route to develop new products using the investment casting process. It takes a fraction of the time it would take to produce models or tooling using the more traditional methods.

Figure 18.4 Dipping completed wax-pattern assembly

shell is built up of six to nine such back-up coats, dried between each coat, to give a final thickness of about 6 mm to 12 mm.

After the shell has been formed round the wax pattern and dried, it is dewaxed by heating to about 150 °C and allowing the wax to run out, leaving the required shape of cavity inside the mould, Fig. 18.5. The mould is then fired at about 1000 °C to attain its full mechanical strength, and the casting metal is poured as soon as the mould is removed from the firing furnace.

Figure 18.6   Refractory removed from cast metal

Figure 18.5   Dewaxing mould shells

Casting may be carried out in air, in an inert atmosphere, or in a vacuum. Vacuum casting is preferred for casting nickel- and cobalt-base 'super-alloys', not only to avoid contamination of the alloy but also to ensure perfect filling of the mould and to reduce oxidation of the component after casting.

When the cast metal has solidified, the mould is broken up to release the castings, Fig. 18.6, which are cut away from their runners. The castings are then finished by fettling, finishing using a moving abrasive belt, or vapour blasting where loose abrasive is forced on to the surface under pressure. Any cores present in the casting are removed by using a caustic solution.

A typical range of investment castings is shown in Fig. 18.7.

The dimensional accuracy of investment casting will vary depending upon pattern contraction, mould expansion and contraction,

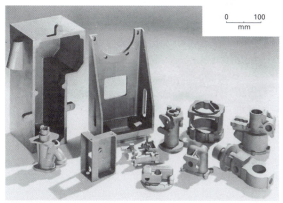

Figure 18.7   Typical investment castings

and contraction of the cast metal. In general, a tolerance of ±0.13 mm per 25 mm can be achieved on relatively small castings. A surface finish comparable with that of machining can be achieved by the transfer of the excellent finish of the pattern to the surface of the mould cavity by the investment technique.

The majority of investment castings are from a few grams to 10 kg, although castings up to 150 kg have been produced. A wall-section thickness of 1.5 mm can be achieved on production runs, although thinner wall sections are possible.

Although the process can be used for prototype castings, quantities usually produced are within a range from 50 to 50 000.

A patented process has been developed known as counter gravity low pressure air melt (CLA) casting process, which is the same as has been described, up to the metal pouring stage.

Once the fully dried mould has been dewaxed and fired to 1000 °C it is placed inside a casting chamber and sealed. The casting chamber is then located in position and vertically lowered towards the furnace and the molten metal so that the open neck is below the liquid surface, Fig. 18.8. A vacuum is then applied to the casting chamber forcing molten metal up into the mould and so filling it at a consistent and even rate. After a pre-determined cycle time the vacuum is released leaving the solidified components intact but allowing the runner, which is still in a molten state, to return to the furnace for use on the next mould.

Due to this vacuum technology there are a number of additional advantages over gravity pouring.

▶ Sections as thin as 0.4 mm can be achieved.
▶ Since the runner returns to the furnace less metal is used.
▶ Oxidation is eliminated as the molten metal does not come in contact with atmosphere while casting.
▶ As the metal is drawn up from below the slag line the chance of non-metallic inclusions is reduced.
▶ Casting temperatures can be greatly reduced resulting in a finer grain structure so giving superior mechanical properties.

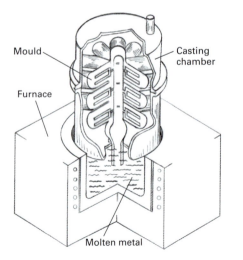

Figure 18.8 Casting chamber in furnace

▶ Molten metal always fills at a constant rate thus eliminating inconsistencies in the casting process.

18.1.1 Advantages of investment casting

▶ High degree of dimensional accuracy and surface finish.
▶ Accurate reproduction of fine detail.
▶ Virtually unlimited freedom of design enabling intricate shapes, both external and internal, to be produced. Sub-assemblies can often be replaced by a single casting.
▶ Metals can be cast which are difficult or even impossible to work by other methods.
▶ Reduction or elimination of finishing or machining operations, especially on metals which are difficult to machine, resulting in savings in material and machining costs.
▶ Comparatively small quantities can be produced, as well as large quantities.
▶ High production rates.

18.2 Metals for investment casting

Investment castings can be made from an extremely wide range of ferrous and non-ferrous metals.

The ferrous range covers low-alloy, carbon and tool steels as well as corrosion- and heat-resistant steels. The non-ferrous range covers aluminium and zinc alloys and the copper-based alloys, brasses, bronzes and gunmetal.

18.3 Lost foam casting

The lost foam casting process originated in 1958 but due to strong patents that covered the process early development was restricted. Only when the patent expired around 1980 was free development possible and during the 1990s much work was done and the lost foam casting process is now a technical and commercial success.

The lost foam process consists of first making a foam pattern having the shape of the finished metal part. The foam patterns are created from polystyrene beads, similar in size and shape to sand granules, expanded to the desired shape using aluminium tooling. More complex shapes can be created by gluing a number of patterns together. The assemblies are then attached to a central foam piece or tree. Depending on the size, multiple patterns can be produced on a single tree.

After a short stabilising period, the completed pattern is strengthened by dipping in a refractory material which coats the foam pattern leaving a thin heat-resistant layer which is then air dried. This ceramic coating also provides a good surface finish for the finished casting.

When drying is complete the coated foam is suspended in a steel container which is vibrated while traditional green sand, with no binders or other additives, is poured around the coated pattern providing mechanical support to the thin ceramic layer. During the sand pouring operation, the vibration causes a layer of sand near the pattern surface to fluidise ensuring all voids are filled, and shapes supported, while the whole mass of sand is compacted.

After compaction, molten metal is poured into the mould causing the foam to burn up and vaporise as the molten metal replaces the foam pattern, exactly duplicating all the features of the original pattern. Like other investment casting methods, for each new casting produced, a new pattern must be made; hence the name lost foam casting.

After solidification, the container is tipped over and the unbonded sand flows out together with the castings. Because there are no binders or other additives the sand is reclaimable. The castings are then subjected to further operations, e.g. removal of sprue, gate, risers, any machining required, heat treatment, etc. similar to other casting processes.

Generally all ferrous and non-ferrous metals can be cast using this method but it is unsuitable for low- and medium-carbon steels due to the possibility of carbon pick-up. Because the foam pattern has to decompose to produce the casting, metal pouring temperatures above 540 °C are usually required so that casting of low-temperature metals is limited.

Characteristics of lost foam casting
▶ Design freedom to produce complex shapes.
▶ No parting line.
▶ No cores are needed.

▶ Low draft angles can be used.
▶ Close dimensional tolerances can be achieved.
▶ Uses dry unbonded sand which is reusable.
▶ High production rates are possible.
▶ Machining can be eliminated.
▶ Minimum finishing operations are needed, e.g. shot blasting, grinding.
▶ Expensive tooling restricts the process to long production runs.

The process is used in the automotive industry in the production of cylinder heads, engine blocks, crankshafts and exhaust manifolds, in the electrical industry for electric motor frames, and for mains water valves to name but a few.

18.4 Shell moulding

The shell-moulding process uses a thin expendable mould or 'shell' made from a fine silica or zircon sand bonded with a thermosetting resin. The fineness of the sand influences the surface of the finished casting.

The shell is produced by heating a metal pattern of the shape to be cast to a temperature of about 230 °C to 260 °C and covering it with the sand-and-resin mixture. This is done by locating the heated pattern on the open end of a dump box, mounted on trunnions, which contains the sand-and-resin mixture, Fig. 18.9(a). When the dump box is inverted, the mixture falls on to the heated pattern which melts the resin through a layer of sand approximately 10 mm thick, Fig. 18.9(b). The dump box is then returned to its upright position and surplus mixture falls back into the dump box, leaving the pattern covered with a layer of sand bonded with resin. This layer or shell is then hardened or 'cured' by the further application of heat, before being removed from the pattern. The shell is removed with the aid of ejector pins which are pushed to lift the shell clear of the pattern, Fig. 18.9(c).

Patterns suitable for shell moulding can be made from any material capable of withstanding oven temperatures of about 400 °C and of transferring heat to the moulding material. The materials most widely used are cast iron and steel. These metal patterns are more expensive than the wooden patterns used with green-sand casting and so, to

18

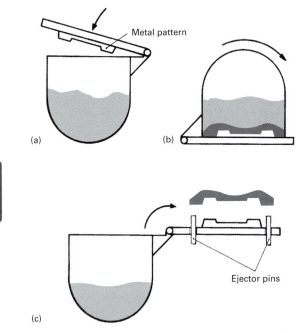

(a)

(b)

(c)

Figure 18.9 Stages in producing a shell mould

Figure 18.10 Shell-moulding pattern, mould, cores and casting

be economic, the process requires larger numbers of castings – seldom less than 200.

When the shells required to form a complete mould have been cured, they are placed together and secured ready for metal pouring. They can then be used immediately or stored for later use.

Before pouring, large shells may have to be supported to avoid distortion under the heavy mass of molten metal.

Molten metal is poured into the prepared shell mould to form the casting, the surface of which will be free of blemishes as air and gases can readily escape due to the permeability of the shell. When the cast metal has solidified, the shell is broken away to reveal the casting. Figure 18.10 shows the metal pattern, shell mould, cores and resulting casting.

Shell moulding can generally be expected to produce small castings more accurately and with smoother surfaces than other forms of sand casting. It is comparable with gravity die-casting in aluminium but not as good as pressure die-casting or investment casting.

An accuracy within 0.25 mm can be maintained on dimensions up to 100 mm in the same mould half. Accuracies of about 0.4 mm could be expected for similar dimensions across the mould parting line.

These accuracies will often eliminate machining operations, with consequent cost savings.

The smooth surface finish provides a good base for paint finishes or an improved appearance if left in the as-cast condition.

Shell moulding allows intricate shapes to be produced with minimum taper or draft angle. An angle of $\frac{1°}{2}$ to 1° is usually sufficient, compared with the 2° to 3° needed with green-sand casting.

Although the sand used is more expensive than green sand, the volume used is less, due to the thin shell, and this can show savings in the amount purchased, stored and handled.

It has been estimated that, on average, a relatively simple shell-moulded casting will cost around 10% to 15% more than the same casting produced in green sand. However, by taking advantage of the greater accuracy and reducing or eliminating machining operations, the use of a shell-moulded casting can often result in a cheaper finished component.

The shell-moulding process can be used to produce castings in any metal capable of being cast. Materials commonly used include bronze, aluminium, magnesium, copper and ferrous alloys. Although the patterns and sand are relatively expensive, the use of this process can usually be justified, in comparison with green-sand casting, for complex castings where high accuracy and definition are required on reasonable-size repeat batches.

18.4.1 Advantages of shell moulding

▶ Can be fully automated.
▶ Lower capital costs compared with the mechanised green-sand casting.
▶ Lightweight moulds easy to handle and store.
▶ Shells are easily broken at knockout stage.
▶ Lower finishing costs (e.g. machining and fettling).

▶ Improved surface finish and dimensional accuracy than green-sand castings.

18.4.2 Limitations

▶ The gating system must be part of the pattern, making it expensive.
▶ Raw materials are relatively expensive (however – not much is used).
▶ Size and weight range of castings is limited.
▶ Process generates fumes which must be extracted.

Review questions

1. From what material is the expendable mould made in the shell moulding process?
2. State four characteristics of the lost foam process.
3. Briefly describe the lost wax casting process.
4. How is the foam pattern strengthened when used with the lost foam process?
5. What are the requirements of the materials used to make patterns for shell moulding?

6. Briefly describe the lost foam casting process.
7. State four advantages of investment casting.
8. State four additional advantages of using the CLA casting process over the traditional investment casting process.
9. Briefly describe the shell moulding process.
10. What tolerance can be expected on relatively small investment cast products?

CHAPTER 19

Die-casting and metal injection moulding

Die-casting is the name given to the production of castings by processes which make use of permanent metal moulds or 'dies'. The term 'permanent' is used since the moulds can be used to produce thousands of castings before they require replacement, unlike those used in sand casting, where each mould is destroyed to remove the casting.

The basic principle of all die-casting is for the die, which contains a cavity of the required shape, to be filled with molten metal. The die is usually in two parts, but may often have movable pieces or 'cores', depending on the complexity of shape being produced.

With the two die halves securely locked in the closed position, molten metal is introduced into the die cavity. When the molten metal has solidified, the die halves are opened and the casting is removed or ejected. The die halves are then closed and the operation is repeated to produce the next die-casting. With fully automatic machines this complete cycle can be done extremely quickly.

Each fill of metal is known as a 'shot'. It is the way in which filling is done which distinguishes the different die-casting methods.

19.1 Gravity die-casting

Gravity die-casting is the simplest and most versatile of the die-casting methods. The molten metal is poured into the die, using the force of gravity to ensure that the die cavity is completely filled in the same manner as for sand casting. The die contains the necessary cavity and cores to produce the required shape of casting, together with the runners, gates, risers and vents needed to feed the metal and allow air and gases to escape, Fig. 19.1. Apart from the dies, very little additional equipment is required, the dies being arranged near to the furnaces where the metal is melted and from where it can be transferred by ladle.

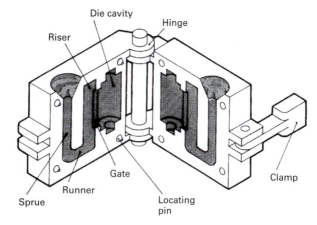

Figure 19.1 Gravity die-casting

Workshop Processes, Practices and Materials, Fifth Edition. 9781138784727.
© 2015 Bruce J. Black. Published by Taylor & Francis. All rights reserved.

This process provides castings of greater accuracy than sand casting, due to the use of metal moulds which can be used to produce thousands of castings before they require replacement. The most commonly used material for the manufacture of dies for gravity die-casting is close-grained cast iron. Any cores required are usually made of heat-resisting steel. The dies are more expensive than sand-casting moulds but cheaper than those required with high-pressure die-casting methods.

To allow for runners and risers, the volume of material required can amount to over twice the volume of the actual component. Although the surplus can be recovered, it is still wasteful in its removal, dressing the component (or fettling using a hand-held grinder), and the energy required in remelting.

The output from gravity die-casting varies with casting size and may be as high as 25 shots/h, but the process requires greater manpower in operation and filling than other die-casting methods. It is generally economical for batch sizes 500–2500 in aluminium alloys, zinc alloys and copper-based alloys. The minimum wall thickness which can be achieved is between 3 mm and 5 mm depending on workpiece size, which is usually up to around 23 kg but can be up to 70 kg. Tolerances of 0.4 mm up to 150 mm sizes and 0.05 mm per 25 mm above this size can be achieved.

19.2 Low-pressure die-casting

The low-pressure die-casting process involves holding the molten metal to be cast in a crucible around which are arranged the heating elements necessary to keep the metal molten. The crucible and elements are contained inside a sealed pressure vessel in the top cover of which is a riser tube whose lower end is immersed in the molten metal. Low-pressure air (0.75 bar to 1.0 bar) is introduced into the crucible above the surface of the molten metal and forces the metal up the riser tube and out through the top end. This top end is attached to the bottom or fixed half of the die, which is designed to allow air to escape as the molten metal enters. When the metal has solidified in the die, which is air-cooled, the air pressure is released and the remaining molten metal in the riser tube drops.

The casting is removed by raising (usually by means of a hydraulic cylinder) the moving platen to which the top or moving half of the die is attached. The dies are designed so that the moving half contains the male shape, and so, due to shrinkage of the metal, the casting clings to this half. When the moving platen nears the end of its stroke, striker pins on the ejector mechanism contact the underside of the fixed top platen, pushing the ejector pins which release the casting from the die. In some cases a stripper ring is used to release the casting, but the principle is the same.

The moving platen is then reversed to close the two halves of the die ready for the next casting to be produced.

A diagrammatic view of the machine set up with the dies in the closed position is shown in Fig. 19.2.

The dies are cheaper than for high-pressure methods since they can usually be made of cast iron, but they are sometimes more expensive than for gravity die-casting. The capital outlay for machines is higher than for gravity but lower than for the high-pressure method.

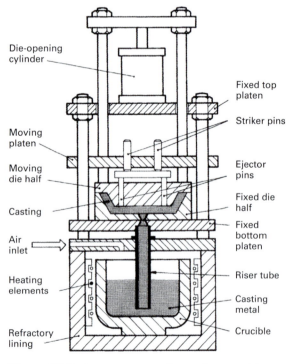

Figure 19.2 Low-pressure die-casting machine

Due to the need for only a single feed, less metal is used and therefore less dressing (or fettling) is required. This results in minimal distortion and makes the process suitable for the production of large flat castings. Components produced by this method can be both lighter and thinner than with gravity die-casting, as the molten metal is introduced under pressure rather than relying solely on gravity to fill the die.

While low-pressure die-casting is not necessarily quicker than gravity die-casting, it does give a more dense-structured casting and lends itself to automation, where one man can operate more than one machine with a consequent higher output per man. This process is ideal where the component is symmetrical about an axis of rotation leading to its use in the manufacture of car wheels. Low-pressure die-casting can be economical for batch sizes greater than 1000 most commonly in aluminium alloys for castings ranging from 5 kg to 25 kg although larger sizes can be achieved. Minimum wall thickness is between 2 mm and 5 mm. Surface finish and wall thickness are better than can be achieved with gravity die-casting but poorer than high-pressure die-casting.

Castings in excess of 60 kg can be produced by this method.

19.3 High-pressure die-casting

Two principal types of machine are used in the production of die-castings by high-pressure methods: hot-chamber and cold-chamber machines.

19.3.1 Hot-chamber machine

A section through a machine of this type is shown in Fig. 19.3. As shown, the furnace and the crucible containing the molten metal are contained within the machine.

The casting cycle starts with the moving machine platen, to which one die half is attached, being pushed forward, usually by a pneumatic cylinder, until it locks against the other die half attached to the fixed machine platen. Only when the dies are completely locked does the safety system allow the plunger to be depressed.

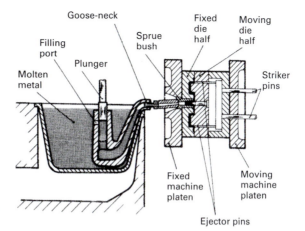

Figure 19.3 High-pressure hot-chamber die-casting machine

As the plunger is depressed, it covers up the filling port and so prevents molten metal escaping back into the crucible. The molten metal is forced through the goose-neck into the sprue bush and fills the die cavity. To enable air to escape from the die cavity, small channels known as 'vents' are machined from the cavity across the die face. These vents must be deep enough to allow the air to escape but shallow enough to prevent metal escaping from the die. Usually a depth of 0.13 mm to 0.2 mm is sufficient.

When the die is filled, the plunger is returned to the 'up' position, the molten metal in the nozzle flows back into the goose-neck, and the injection cylinder refills through the filling port.

Meanwhile the molten metal in the die has solidified, due to the lower temperature of the die. The die is maintained at the correct temperature by circulating cold water through it.

When the plunger reaches the 'up' position, the mechanism to unlock the dies is operated and the moving machine platen and the die half attached to it are withdrawn. The dies are constructed so that the male form and cores are on the moving die half, and the casting shrinks on to this half and comes away with it.

As the moving machine platen continues to open, striker pins bearing on the ejector plate strike an ejector block on the machine which pushes the ejector pins forward, releasing the casting. The casting falls through an opening in the machine base, down a chute, and into a container or quenching tank. The dies are then closed and the cycle is repeated.

This cycle can take place as rapidly as 500 times per hour, or more on high-production machines.

The casting as ejected from the machine is known as the 'spray' and contains the component, sprue, runner and overflow as shown in Fig. 19.4.

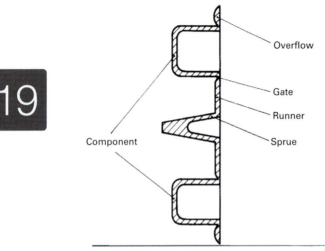

Figure 19.4 Spray from high-pressure hot-chamber die-casting machine

The 'runner' is a channel in the die through which the molten metal fills the die cavity. The term is also applied to the metal that solidifies in this channel. The 'gate' is that part of the die through which the metal enters the die cavity from the runner.

The 'sprue' is the metal attached to the runner that solidifies in the sprue bush. The 'overflow' is a recess with entry from the die cavity, to assist in the production of sound castings and ensure complete filling of the cavity.

Metal in the region of the gate and to the overflow is made thin so that the runner and overflow can be easily and quickly broken off the casting. These are placed in such a position as not to impair the finished appearance of the casting.

Hot-chamber machines are most widely used with zinc alloy at pressures in the region of 100 bar. The metal temperature in the crucible is between 400 °C and 425 °C, and the die temperature, which should be the lowest that will give castings of good quality, is usually from 180 °C to 260 °C. Casting size can vary from 0.01 kg to 25 kg.

The cost of dies, which are made from heat-resisting steel, is high and makes this process uneconomical for small numbers of components. Annual requirements of 20 000 or more are generally necessary to make production economical, but this would depend on the complexity of the component and whether more than a single cavity could be incorporated in the die. Dies with more than one cavity are known as multi-impression dies.

This process is highly automated to produce rapid cycle times. A very high surface finish can be obtained on dimensionally accurate castings requiring little or no machining. Thin sections can be produced down to 1 mm or even 0.5 mm, which result in lighter castings and a reduction in the amount of metal required.

19.3.2 Cold-chamber machine

A section through a machine of this type is shown in Fig. 19.5. Unlike in the hot-chamber type, the metal is melted and held in a furnace away from the machine.

Again, one die half is attached to the fixed machine platen and the other to the moving machine platen.

With the dies locked in the closed position, a measured quantity of molten metal is fed from

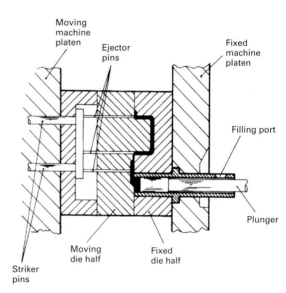

Figure 19.5 High-pressure cold-chamber die-casting machine

the holding furnace into the injection cylinder. This operation can be automated but is usually done by hand, using a ladle of a size appropriate to the shot size required for the casting. The plunger is pushed forward, usually by means of a hydraulic cylinder, forcing the molten metal into the die cavity, where it quickly solidifies. Pressure on the metal varies between 350 bar and 3500 bar. The dies are opened and the casting is released by the ejector mechanism in the same way as in the hot-chamber machine.

The plunger is then withdrawn and the dies are closed ready for the next cycle. The cycle time is not as fast as for the hot-chamber machine.

In this machine there is no sprue, since the die is mounted at the end of the cylinder. Instead, the metal not used in filling the cavity remains in the cylinder and is attached to the casting when the spray is removed from the die. This excess metal attached to the runner is known as the 'slug', Fig. 19.6.

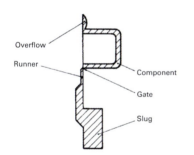

Figure 19.6 Spray from high-pressure cold-chamber die-casting machine

Cold-chamber machines are more commonly used with the higher melting-point aluminium alloys poured at about 650 °C to 670 °C and with magnesium alloys poured at about 680 °C.

Rates of production in the region of 80 to 100 shots/h can be achieved with handfilled machines, but these can be increased by about 40% by using mechanised metal-feed systems.

Zinc alloys containing aluminium and copper, typically Al 27% Cu 2% and Al 11% Cu 1%, are now used extensively in cold-chamber die-casting for thin-walled, intricate castings where strength, corrosion resistance and non-sparking properties (e.g. in petroleum plants and underground) are required.

19.4 Die-casting metals

Although techniques for pressure die-casting in steel have been developed, the metals most widely used are certain non-ferrous alloys which have relatively low melting points.

19.4.1 Zinc alloys

The designation of zinc alloys is set out in BS EN 1774. The alloy (a number) comprises the letters ZL denoting zinc alloy, followed by four numbers. The first two numbers denote the aluminium content, the third number the copper content and the fourth number any other main element (if less than 1%, the number is 0). The alloys are also given a short designation.

These alloys are shown in Table 19.1.

Table 19.1 Zinc alloys

Alloy symbol	Alloy number	Short designation
ZnAl4	ZL0400	ZL3
ZnAl4Cu1	ZL0410	ZL5
ZnAl4Cu3	ZL0430	ZL2
ZnAl6Cu1	ZL0610	ZL6
ZnAl8Cu1	ZL0810	ZL8
ZnAl11Cu1	ZL1110	ZL12
ZnAl27Cu2	ZL2720	ZL27

Alloys ZL3 and ZL5 are most suitable, the compositions being similar except for the copper content in ZL5.

ZL3 is a general-purpose alloy used in hot-chamber high-pressure die-casting for the engineering and automotive industry, household equipment and utensils, office equipment, locks and toys. It is easily machined, polished, lacquered and electroplated for decorative and functional purposes.

ZL5 is used for castings which require moderately higher strength and hardness than the ZL3 alloy. The copper content makes this alloy more expensive and its use is therefore restricted to applications requiring the higher properties.

Zinc alloys can be easily cast since they are extremely fluid even at relatively low temperatures, which allows the molten metal to flow into the most intricate shapes and permits thin sections to be cast.

ZL3 solidifies at 382 °C and is generally cast at temperatures from 400 °C to 425 °C. This relatively low temperature enables smaller tolerances to be maintained and leads to longer die life, since the dies are not subjected to repeated high thermal shock.

The low casting temperature of these alloys makes them particularly suitable for use with hot-chamber high-pressure machines which are fully automatic and allow high rates of production.

Zinc alloys have good resistance to corrosion, and die-castings can be left as-cast. However, in situations which are particularly corrosive, chromating or anodising is used for added protection. Where functional and decorative finishes are required, the most widely used is chromium plating, and the best results are obtained when an initial layer of copper is followed by layers of nickel and chromium. Other finishes such as painting and plastics coating can also be applied.

Zinc alloys possess good machining qualities and, although zinc die-casting can be produced to fine tolerances, machining can be easily carried out where necessary.

19.4.2 Aluminium alloys

The major advantages of aluminium alloys are lightness and high electrical and thermal conductivity. Aluminium alloys are about 40% as dense as zinc alloys and, although not generally as strong, many aluminium alloys can be heat-treated to give comparable mechanical properties. Section thicknesses, however, have generally to be thicker to obtain equivalent strengths.

The melting temperature is about 600 °C, with casting temperatures about 650 °C. Because of the higher melting temperature required, aluminium-alloy die-castings are less close in tolerance than zinc die-castings but can be used at higher working temperatures.

The compositions of aluminium alloys are specified in BS EN 1706. The two alloys most widely used in pressure die-casting are LM6 and LM24. The excellent casting characteristics of LM24 make it suitable for most engineering applications. LM6 has a higher silicon content and is preferred to LM24 where high resistance to corrosion is the primary requirement and for thin and intricate castings. It is also suitable for gravity die-casting. Other aluminium alloys such as LM5, LM9, LM25 and LM27 are used in gravity die-casting as well as in sand casting.

Due to the wider tolerances required with aluminium-alloy die-castings, it is usual to machine in order to obtain acceptable limits of size. Machining can readily be carried out using conventional machine tools and cutting tools, but with the higher cutting speeds associated with aluminium and its alloys.

Finishing of aluminium-alloy die-castings is usually for decorative purposes and includes anodising, which can be dyed a wide variety of colours; painting and electroplating. The more usual electroplating material is chromium, which involves fairly complicated pre-treatment to prevent the formation of an oxide film which would make it difficult for the plated metal to bond to the aluminium.

19.4.3 Magnesium alloys

Magnesium alloys are the lightest of the casting metals, being about 60% as dense as aluminium alloy, and are used where weight saving is important, e.g. for portable saws, cameras, projectors and automobile components including specialised car wheels. The specification of these alloys is contained in BS EN 1753.

Because of choking of the nozzle due to the disturbance of the flux cover on the molten metal, this material is not normally used in hot metal machines. Alloys are available with a very low manganese content which minimises the formation of the aluminium–manganese precipitation which causes the 'sludge' to form. This material is, however, used satisfactorily with the cold-chamber machines, where it is poured at about 680 °C.

Magnesium alloys machine extremely well, but care must be taken not to allow the cutting edge of the tool to become dull, as the heat generated may be sufficient to ignite the chips.

Magnesium alloys do not retain a high lustre like the aluminium alloys and cannot be electroplated with the same ease as the zinc alloys. Finishing is usually achieved by painting or lacquering. Chromate treatments are used to protect against corrosion, but have little decorative value.

19.4.4 Copper alloys

Of the many copper alloys, only 60/40 brass and its variants with a melting temperature of about 900 °C combine acceptable mechanical properties with a casting temperature low enough to give reasonable die life. The higher zinc content lowers the casting temperature and gives essential hot ductility. Small additions of silicon or tin improve fluidity; tin also improves corrosion resistance. Aluminium is added to form a protective oxide film to keep the molten metal clean and reduce the attack on die materials.

Apart from its high injection temperature of about 950 °C, brass is an extremely good die-casting alloy, reproducing fine detail and flowing readily into exceptionally thin sections. By keeping all the sections thin, heat transfer to the die can be very greatly reduced and die life be extended.

19.4.5 Lead and tin alloys

These low-melting-point alloys were the first alloys to be die-cast, originally for printer's type. Lead alloyed with antimony, sometimes with small additions of tin, has a melting point of about 315 °C and can be cast to very close tolerances and in intricate shapes. The castings have low mechanical properties and are used mainly for their density, e.g. car-wheel balance masses, and corrosion-resistance, e.g. battery-lead terminals.

Several tin-based alloys, usually containing lead, antimony and copper, with a melting point of about 230 °C, are also die-cast where the very highest accuracy is required and great strength is not of importance. Their excellent corrosion-resistance to moisture makes them suitable for such components as number wheels in gas and water meters, and they have also found use in small complex components of electrical instruments.

19.5 Special features of die-castings

Of all the methods used in manufacturing, die-casting represents the shortest route from molten metal to finished part. To take greatest advantage of the special features of die-casting – such as accuracy, rate of production, thinness of section, lack of machining and fine surface finish – great attention must be given to the design of parts intended for die-casting. It is advisable for the designer to consult the die-caster at an early stage of design in order that those features which permit ease of production can be incorporated.

19.5.1 Section thickness

Sections should be as thin as possible consistent with adequate strength. Thin sections reduce metal cost and allow the casting to solidify faster in the die, thus shortening the production cycle. They also result in lighter components. Small zinc die-castings, for instance, can be produced with wall sections down to 0.5 mm, though with larger zinc die-castings a thickness of 1 mm is more general. Additional strength of a thin section can be achieved by providing ribs in the required position.

The sections of a die-casting should always be as uniform as possible – sudden changes of section affect the metal flow and lead to unsound areas of the casting. Differences in the rate of cooling between thick and thin sections produce uneven shrinkage, causing distortion and stress concentrations.

Bosses are sometimes required to accommodate screws, studs, pins, etc. and, if designed with a section thicker than an adjacent thin wall, will also cause unequal shrinkage. This can be minimised by making the variation in thickness as small and as gradual as conditions permit.

19.5.2 Die parting

The die parting is the plane through which the two halves of the die separate to open and close. It is usually across the maximum dimension. The designer should visualise the casting in the die and design a shape which will be easy to remove.

When the two die halves are closed, there is always a small gap at the two faces, into which metal will find its way due to pressure on the metal. This results in a small ragged edge of metal known as a 'flash', which has subsequently to be removed, usually by means of a trimming tool. The position of the flash has to be arranged so that it can be removed efficiently without leaving an unsightly blemish on the finished casting.

19.5.3 Ejector pins

Ejector pins are used in the moving die to release the casting. The ejector pins will leave small marks on the surface of the casting and should be positioned so that these will not appear on a visible face of the finished casting. If a casting has a face which is to be machined, then, where possible, the ejector-pin marks should be arranged on that face, to be removed by the machining.

19.5.4 Draft angle

To allow the casting to release easily from the die, a wall taper or draft angle, normally between 1° and 2° per side, is provided. With shallow ribs more taper is required, about 5° to 10°.

19.5.5 Undercuts

A part which contains a recess or undercut requires slides or movable cores in the die, otherwise the casting cannot be ejected. These slides and movable cores greatly increase die costs and slow down the rate of production. As a general rule, the design of a part should avoid undercut sections.

19.5.6 Corners

Sharp internal corners on castings are always a source of weakness, and should be avoided by the use of a blend radius or fillet. For example, it is common practice with high-pressure casting of zinc alloys to have a minimum radius of 1.6 mm on inside edges. A slight radius on the outside corners of castings reduces die cost.

The provision of radii within the die cavity is beneficial to the flow of molten metal and the production of sound castings.

19.5.7 Lettering

When lettering, numerals, trademarks, diagrams or instructions are required to be cast on a die-casting, they should be designed as raised from the casting surface. This reduces die cost, as it is easier to cut the design into the die surface than to make a raised design on the die surface.

19.5.8 Threads

Threads can be cast on high-pressure machines but should be specified only where their use reduces cost over that for machine-cut threads. Cast internal threads under 20 mm in diameter are rarely economical.

19.5.9 Inserts

It is sometimes desirable to include inserts to obtain features in a casting which cannot be obtained from the cast metal. This may be done to provide

▶ additional strength;
▶ locally increased hardness;
▶ bearing surfaces;
▶ improved electrical properties;
▶ passages otherwise difficult to cast;
▶ passages intended to carry corrosive fluids;
▶ facilities for soldered connections;
▶ means of easier assembly.

Usually inserts are cast in place, but there are instances in which they are applied after casting, in holes cast for that purpose. The object of casting the insert in place is either to anchor it securely or to locate it in a position where it could not be placed after casting.

The insert material must be able to withstand the temperature of the molten metal and is usually steel, brass or bronze.

Inserts which are cast in position have to be manufactured with a small tolerance, otherwise they will not locate accurately in the die or in the casting. In some instances they may be quite expensive to manufacture, and their use tends to slow down the casting process. They should therefore be used only where distinct advantages can be obtained.

Very small inserts are difficult to place in the die and should be avoided. Large inserts can lead to distortion resulting from different coefficients of expansion and should be used with caution.

When inserts are cast in place they are located in the mould cavity, molten metal flows around them and solidifies, and the insert becomes part of the casting. However, there is little or no bond between the casting and the insert other than the mechanical effect of casting shrinkage, so the insert must contain some feature to provide positive anchorage. The simplest methods make use of knurling, holes, grooves or flats machined on the insert. Some examples are shown in Fig. 19.7.

Figure 19.7(a) shows a location pin anchored by means of a diamond knurl on the outside diameter.

Figure 19.7(b) shows a flat electrical connector anchored by its shape and by molten metal flowing through the hole.

Figure 19.7(c) shows a threaded bush anchored due to its hexagonal shape and the groove round the outside.

Figure 19.7(d) shows a plain circular bush anchored due to the two flats machined in the outside diameter.

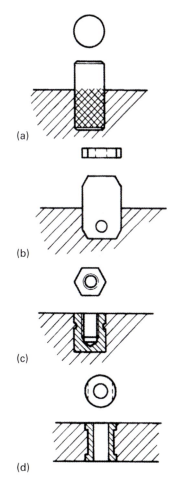

(a)

(b)

(c)

(d)

Figure 19.7 Inserts used in die-casting

19.6 Advantages and limitations of die-castings

19.6.1 Advantages

▶ Die-castings can be produced to good dimensional accuracy. Actual values depend on the die-casting method, size of dimension and material being cast; e.g. a 25 mm dimension pressure-die-cast in zinc alloy can be produced to a tolerance of 0.1 mm if the dimension is in one die half or 0.26 mm if the dimension is across the die parting.

▶ Due to good dimensional accuracy, machining operations can be reduced or eliminated. Holes, recesses, chamfers, etc. can be cast in position.

▶ Where machining has to be carried out, the consistent size and shape of die-castings enable accurate location in jigs and fixtures.

▶ Complex-shaped die-castings can be produced, giving designers a certain freedom to achieve attractive styling, and can be shaped to give localised strength where required.

▶ The ease with which complex parts can be produced enables die-castings to replace parts which are extremely difficult to produce by machining and to replace a number of assembled parts with a single casting and so save on assembly costs.

▶ Thin walls can be produced, resulting in
  i) a lighter casting;
  ii) material-cost savings, since less material is used;
  iii) reduced cost per casting, since thin walls allow higher rates of production.

▶ Smooth surfaces of die-castings from metal moulds reduce or eliminate pre-finishing operations such as buffing and polishing prior to finishing operations such as electroplating and painting.

▶ The cost of die-castings can be reduced by using multi-impression dies.

▶ A wide range of finishes can be given to die-castings. Choice depends on the die-cast materials and includes electroplating, vacuum metallising, painting, lacquering, plastics coating, anodising and chromating.

▶ Die-castings can be produced with cast-in inserts of other metals. Inserts such as tubes,

289

heater elements, fasteners and bushes are widely used.

▶ The cost of die-castings can be low, due to very rapid fully automated production over long periods. High-pressure machines, e.g. can operate at rates in excess of 500 shots/h.

▶ A wide range of sizes of die-castings can be produced, from a few grams to over 60 kg depending on the die-casting method.

19.6.2 Limitations

▶ Large quantities of castings are required to offset the high cost of dies and casting machines.

▶ Die-castings can be produced only from a limited range of the lower melting-point non-ferrous alloys.

▶ There are limitations on the maximum size of die-casting, due to metal temperatures and pressures.

19.7 Choice of a die-casting process

There are a number of factors which influence the choice of a die-casting method. These are set out in Table 19.2 opposite each of the die-casting methods.

For example, if the component under consideration is to be made from aluminium alloy and only 2000 per year are required, the choice would be limited to gravity die-casting. If, however, the quantity required was 50 000 per year then the high-pressure cold-chamber method would be appropriate.

Use of Table 19.2 is intended only as a guide, as it is difficult to give precise values. For instance, casting masses depend on the material being cast, accuracy depends on size, and so on.

19.8 Metal injection moulding

Metal injection moulding, or MIM, combines the technologies of powder metallurgy and injection moulding enabling the production of high integrity metal parts in complex shapes, yielding close tolerances, smooth surface finish and finely reproduced detail. This process virtually eliminates machining and is ideal for small components at medium to high quantities.

Fine metal powder is mixed with a binder system to create a feedstock. The binder is a mix of wax and plastics and makes up roughly 50% by volume of the feedstock. The feedstock, in the form of granules, is injected under high pressure into a mould cavity using moulding machines

Table 19.2 Factors influencing choice of die-casting process

Die-casting method		Economic run per year	Accuracy	Casting mass	General minimum thickness of section	Type of metal cast	Rate of prodn	Surface finish	Die costs	Capital cost of equipment
Gravity		1000	Good	Up to about 23 kg or more	3–5 mm	Al alloys Cu alloys Mg alloys	Medium	Good	High	Nil to very little
Low-pressure		5000	Good	Up to about 25 kg or more	2–3 mm	Al alloys	Low	Very good	Usually greater than for gravity	High
High-pressure	Hot	20 000	Excellent	Less than 1 g to about 25 kg	1 mm	Zn alloys	Very high	Excellent	Usually 2 to 3 times greater than for low pressure	Very high
	Cold				2 mm	Al alloys Cu alloys Mg alloys	High			

very similar to plastics injection moulding machines. The mould consisting of two halves, is securely fitted to the fixed and moving platens of the machine and when closed together the material can be injected. The machine injection unit consists of a screw which transports the compound and compresses it so that it is free of bubbles, a heating system which controls the temperature of the compound, and the nozzle out of which the compressed and heated material is injected under pressure into the mould. The mould temperature is also controlled as it must be low enough to ensure that the compact, which the moulded item is now called, is rigid when it is removed. The compact, because it is still in a fragile state is called a 'green' compact.

Once the feedstock, after a predetermined cycle, has hardened, the mould is opened, the compact ejected and picked up by a robotic handling unit to avoid any damage to the 'green' compact.

Removal of the binder, which can be 50% by volume of the 'green' compact, is a key stage of the process and one that requires most careful control. This process, called debinding, consists of heating up the 'green' compact to cause the binder to melt, decompose and/or evaporate. During this process, the strength of the compact decreases markedly and great care is necessary in handling these 'brown' compacts as they are now called.

The 'brown' compacts are now sintered, the name given to the heating process in which the separate parts fuse together and provide the necessary strength in the finished product. Sintering is carried out in a controlled atmosphere furnace, sometimes in a vacuum, at a temperature below the melting point of the metal. Large shrinkages occur and the temperature must be very closely controlled in order to retain the final shape and size. The finished part will have a typical density of 98% and with mechanical properties, which are not significantly, if at all, below those of wrought metal of the same composition.

Depending on the part, this can be the final step, unless there are secondary operations to be performed, e.g. tapping threads.

Almost any metal that can be produced in a suitable powder form can be processed by metal injection moulding. Aluminium is an exception because the adherent oxide film that is always present on the surface inhibits sintering. Metals used include low-alloy steels, stainless steels, magnetic alloys, low-expansion alloys (Invar and Kovar) and bronze. Metal injection moulding becomes advantageous when producing small complex metal parts, e.g. parts with a volume of less than a golf ball or deck of playing cards, in relatively high volume around 10 000–15 000 pieces/year.

Typical tolerances that can be achieved are ±0.08 mm/25 mm, i.e. on a 25 mm diameter the achievable tolerance would be ±0.08 mm.

Review questions

1. State four advantages of the die-casting process.
2. Give four reasons why it may be desirable to use inserts in a die-casting.
3. Describe briefly a die-casting process.
4. What type of metal is unsuitable for metal injection moulding and why?
5. Name three types of die-casting process.
6. Describe briefly the metal injection moulding process.
7. State two limitations of the die-casting process.
8. What is the name given to the mix of powder and binder in the metal injection moulding process?
9. Give two reasons for keeping wall thickness as thin as possible in a die-casting.
10. State the criteria under which it becomes advantageous to use the metal injection moulding process.

CHAPTER 20

Moving loads

More than a third of over-three-day injuries reported each year are associated with manual handling. Manual handling is defined as the transporting or supporting of loads by hand or bodily force, i.e. human effort as opposed to mechanical handling such as a crane or fork-lift truck. Introducing mechanical assistance, e.g. a sack truck, may reduce but not eliminate manual handling since human effort is still required to move, steady or position the load.

The load may be moved or supported by the hands or any other part of the body, e.g. the shoulders.

A load is defined as a distinct movable object.

Common injuries are sprains or strains often to the back which arise from incorrect application and/or prolonged use of body force though hands, arms and feet are also vulnerable. Some of these injuries can result in what are known as musculoskeletal disorders or MSDs which can affect the body's muscles, joints, tendons, ligaments and nerves. Typically MSDs affect the back, neck, shoulders and upper limbs. Health problems range from discomfort, minor aches and pains to more serious medical problems.

It is now widely accepted that an ergonomic approach will remove or reduce the risk of manual handling injury. Ergonomics can be described as 'fitting the job to the person rather than the person to the job'. This ergonomic approach looks at manual handling taking account of a whole range of relevant factors including the nature of the task, the load, the working environment and the capability of the individual.

20.1 The Manual Handling Operations Regulations 1992 (as amended)

The Regulations apply to a wide range of manual handling activities including lifting, lowering, pushing, pulling or carrying and seek to prevent injury not only to the back, but also to any other part of the body. Account is taken of physical properties of loads which may affect grip or cause injury by slipping, roughness, sharp edges, extremes of temperature.

The Regulations require that where there is the possibility of risk to employees from the manual handling of loads, the employer should take the following measures, in this order:

1. avoid the need for hazardous manual handling so far as is reasonably practicable;
2. assess risk of injury from any hazardous manual handling that cannot be avoided; and
3. reduce the risk of injury from hazardous manual handling so far as is reasonably practicable.

In taking steps to avoid hazardous manual handling operations, first check whether the load needs to be moved or whether the following operation can be carried out next to the load. Consideration should be given to automation, particularly if new processes are involved and can be incorporated at an early stage, or mechanisation such as the use of a fork-lift truck. However, care must be taken to avoid creating new hazards which may arise through the use of automation or mechanisation.

In assessing and reducing the risk of injury, the employer needs to carry out a risk assessment. The problems to be examined are:

1. the tasks and what they involve;
2. the nature of the loads;
3. the working environment;
4. the individual's capacity to carry out the task;
5. handling aids and equipment;
6. work organisation.

The risk assessment should result in a number of ways of reducing the risk of injury. The task – can you

▶ use a lifting aid;
▶ improve workplace layout to improve efficiency;
▶ reduce the amount of twisting and stooping;
▶ avoid lifting from floor level or above shoulder height;
▶ reduce carrying distances;
▶ avoid repetitive handling;
▶ vary the work pattern, allowing one set of muscles to rest while another is used;
▶ push rather than pull?

Nature of the load – can you make the load

▶ lighter or less bulky;
▶ easier or less damaging to grasp;
▶ more stable?

Working environment – can you

▶ remove obstructions to free movement;
▶ provide better flooring;
▶ avoid steps and steep ramps;
▶ prevent extremes of hot and cold;
▶ improve lighting;
▶ consider less restrictive clothing or personal protective equipment?

Individual's capacity – can you

▶ pay particular attention to those who have a physical weakness;
▶ give your employees more information, e.g. about the range of tasks likely to be faced;
▶ provide more training?

Handling aids and equipment – can you

▶ provide equipment that is suitable for the task;
▶ carry out planned preventative maintenance to prevent problems;
▶ change wheels/tyres and/or flooring so that equipment moves easily;
▶ provide better handles and handle grips;
▶ make equipment brakes easier to use, reliable and effective?

Work organisation – can you

▶ change tasks to reduce monotony;
▶ make more use of workers' skills;
▶ make workloads and deadlines more achievable;
▶ encourage good communications and teamwork;
▶ involve workers in decisions;
▶ provide better training and information?

Training should cover:

▶ manual handling risk factors and how injuries can occur;
▶ how to carry out safe manual handling including good handling technique;
▶ appropriate systems of work for the individual tasks and environment;
▶ use of mechanical aids;
▶ practical work to allow the trainer to identify and put right anything the trainee is not doing safely.

Employees have duties too. They should:

▶ follow appropriate systems of work laid down for their safety;
▶ make proper use of equipment provided for their safety;
▶ co-operate with employer on health and safety matters;
▶ inform the employer if they identify hazardous manual handling activities;
▶ take care that their activities do not put themselves or others at risk.

Duties of employees are already in place under HSWA and the Management of Health and Safety at Work Regulations, and the Manual Handling Operations Regulations supplement these general duties as they apply to manual handling.

Although the Regulations do not set out specific requirements, such as weight limits, they do give numerical guidelines to reduce the risk and assist with assessment. Guidelines for lifting and lowering are shown in Fig. 20.1. This shows guidelines for men taking into consideration vertical and horizontal position of the hands as they move during the handling operation. Compare the activity with the diagram and if the lifter's hand enters more than one box during the lifting operation, use the smallest value. For example, 10 kg if lifted to shoulder height at arm's length or 5 kg if lifted to full height at arm's length. This assumes that the load can be easily grasped with both hands, with a good body position and in reasonable working conditions. The guidelines for a woman are proportionally less.

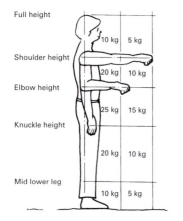

Figure 20.1 Lifting and lowering loads

20.1.1 Good handling techniques for lifting

The development of good handling techniques is no substitute for the risk reduction steps already outlined but is an important addition which requires training and practice. The following should form the basic lifting operation.

20.1.1.1 Stop and think

Plan the lift. Organise the work to minimise the amount of lifting necessary. Know where you are going to place the load. Use mechanical assistance if possible. Get help if the load is too heavy; there is a difference between what people can lift and what they can safely lift. Make sure your path is clear. Don't let the load obstruct your view. For a long lift, i.e. from floor to shoulder height, consider a rest midway on a bench in order to adjust your grip. Alternatively lift from floor to knee then from knee to carrying position – reverse this method when setting the load down.

20.1.1.2 Place your feet

Keep your feet apart to give a balanced and stable base for lifting (see Fig. 20.2). Your leading leg should be as far forward as is comfortable. Avoid tight clothing and unsuitable footwear.

Figure 20.2 Placing the feet

20.1.1.3 Adopt a good posture

At the start of the lift, slight bending of the back, hips and knees is preferable to fully flexing the back (stooping) or fully flexing the hips and knees (squatting), Fig. 20.3. Don't flex the back any further during lifting. Keep shoulders level and facing in the same direction as the hips, i.e. don't twist your body.

Figure 20.3 Adopting a good posture

20.1.1.4 Get a firm grip

Lean forward a little over the load to get a good grip. Try to keep your arms within the boundary formed by the legs. Balance the load using both hands. A hook grip is less tiring than keeping the fingers straight. Wear gloves if the surface is rough or has sharp edges. Take great care if load is wrapped or slippery in any way. Look ahead, not down at the load, once it has been held securely.

20.1.1.5 Don't jerk

Carry out the lifting operation smoothly, keeping control of the load. Move the feet: don't twist your body if you turn to the side.

20.1.1.6 Keep close to the load

If the load is not close when lifting, try sliding it towards you. Keep the load close to your body for as long as possible. Keep the heaviest side of the load next to your body (see Fig. 20.4).

Figure 20.4 Keep load close to body

20.1.1.7 Put down

Putting down the load is the exact reverse of your lifting procedure. If precise positioning of the load is required, put it down, then slide it to the desired position. Whenever possible make use of mechanical assistance involving the use of handling aids.

Although an element of manual handling is still present, body forces are applied more efficiently reducing the risk of injury. Levers can be used which lessen the body force required and also remove fingers from a potentially dangerous area. Hand or power operated hoists can be used to support a load and leave the operator

free to control its positioning. A trolley, sack truck, roller conveyer or ball table can reduce the effort required to move a load horizontally while chutes using gravity can be used from one height to the next. Hand-held hooks and suction pads can be used where loads are difficult to grasp.

As a general rule, loads over 20 kg need the assistance of lifting gear.

20.1.2 Good handling techniques for pushing and pulling

The risk of injury may also be reduced if lifting can be replaced by controlled pushing or pulling. The following points should be observed when loads are pushed or pulled:

- ▶ trolleys and trailers should have handle heights that are between waist and shoulder;
- ▶ handling aids should be well maintained with large-diameter wheels that run smoothly;
- ▶ push rather than pull, providing you can see and are able to control steering and stopping;
- ▶ get help to negotiate a slope or ramp;
- ▶ keep feet well away from the load;
- ▶ travel no faster than walking pace.

As a rough guide the force that needs to be applied to move a load over flat level ground is at least 2% of the load, e.g. a load of 500 kg will require 10 kg to move the load. Greater forces will be required if the handling aid is poorly maintained or the surface is soft or uneven.

The task is within guidelines if the following values are not exceeded:

	Men	Women
Force to stop and start a load (kg)	20	15
Sustained force to keep load in motion (kg)	10	7

20.1.3 Transporting loads

Light loads can be transported with the aid of a variety of hand trucks and trailers largely depending on the shape of the articles.

20.1.3.1 Hand trucks (Fig. 20.5)

With two wheels, these have a capacity of around 250 kg and can be used to transport sacks and

Figure 20.5 Hand truck

boxes, or special types are available for transporting oil drums, gas cylinders and similar items.

20.1.3.2 *Flat trolley or trailers (Fig. 20.6)*

With four wheels, these are more stable and capable of moving heavier loads with a typical capacity of between 500 and 800 kg depending on the design. These are used where uniform shapes such as crates and boxes have to be transported. Where this can be done safely, the load can be stacked, but care must be taken not to stack too high or the load may become unstable and dangerous.

Figure 20.6 Flat trailer

Palletised loads can be easily moved with the aid of hand pallet trucks as shown in Fig. 20.7. The forks are simply pushed under the pallet, a hydraulic lifting system raises the pallet and the easy-going wheels enable the load to be placed where required. These trucks have a loading capacity of up to 3000 kg.

There are a number of do's and do not's to observe regarding safe stacking either in storage areas or during transportation:

Figure 20.7 Hand pallet truck

Do:

▶ always wedge objects that can roll, e.g. drums, tubes, bars;
▶ keep heavy objects near floor level;
▶ regularly inspect pallets, containers, racks for damage;
▶ stack palletised goods vertically on a level floor so they won't overbalance.
▶ 'key' stacked packages of uniform size like a brick wall so that no tier is independent of another;
▶ use a properly constructed rack secured to a wall or floor.

Do not:

▶ allow items to stick out from racks or bins into the gangway;
▶ climb racks to reach upper shelves – use a ladder or steps;
▶ lean heavy objects against structural walls;
▶ de-stack by climbing up and throwing down from the top or pulling out from the bottom;
▶ exceed the safe working loading of racks, shelves or floors.

20.1.4 **Power lifting**

Where loads are too heavy to be manually lifted, some form of lifting equipment is required.

Lifting equipment in use should be:

▶ of good construction, sound material, adequate strength, free from patent defect and be properly maintained;
▶ designed so as to be safe when used;
▶ proof tested by a competent person before being placed into service for the first time and after repair;
▶ thoroughly examined by a competent person after proof testing;

297

- ▶ certified on the correct test certificate including proof load and the safe working load (SWL);
- ▶ marked with its SWL and with a means of identification to relate to its certificate;
- ▶ thoroughly examined by a competent person within the periods specified in the Regulations and a record kept

20.2 Lifting Operations and Lifting Equipment Regulations 1998 (LOLER)

These Regulations aim to reduce risks to people's health and safety from lifting equipment provided for use at work. In addition to the requirements of LOLER, lifting equipment is also subject to the requirements of the Provision and Use of Work Equipment Regulations 1998 (PUWER).

Lifting equipment includes any equipment used at work for lifting and lowering loads, including attachments used for anchoring, fixing or supporting the equipment. The Regulations cover a wide range of equipment including cranes, fork-lift trucks, lifts, hoists, mobile elevating work platforms and vehicle inspection platform hoists. The definition also includes lifting accessories such as chains, slings, eyebolts, etc.

Generally, the Regulations require that lifting equipment provided for use at work is:

- ▶ sufficiently strong, stable and suitable for the particular use and marked to indicate SWL;
- ▶ positioned and installed to minimise any risks, e.g. from the equipment or load falling or striking people;
- ▶ used safely, i.e. the work is planned, organised and performed by competent people;

Where appropriate, before lifting equipment and accessories is used for the first time, it is thoroughly examined. Lifting equipment may need to be thoroughly examined in use at periods specified in the Regulations (i.e. at least 6 monthly for accessories and, at a minimum, annually for all other equipments) or at intervals laid down in an examination scheme drawn up by a competent person.

Following a thorough examination of any lifting equipment, a report is submitted to the employer to take any appropriate action.

20.2.1 Types of lifting equipment

Lifting equipment can be classified according to its power source: manual, electrical, pneumatic, hydraulic or petrol or diesel.

20.2.1.1 Manual

Muscle power is restricted to the manual effort required to operate portable lifting equipment. An example of this is the chain block shown in Fig. 20.8 capable of lifting loads up to around 5000 kg. The mechanical advantage is obtained through the geared block thereby reducing the manual effort required in operating the chain.

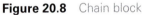

Figure 20.8 Chain block

Small hydraulically operated portable cranes (Fig. 20.9) are available which have an adjustable jib with a lifting capacity typically from 350 to 550 kg on the smaller models and up to 1700–2500 kg on the larger models. The smaller figure is the load capable of being lifted with the jib in its most extended position. In this case the

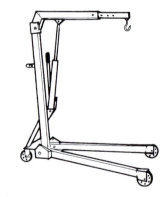

Figure 20.9 Portable crane

mechanical advantage is obtained through the hydraulic system, again reducing the manual effort.

20.2.1.2 Electrical

Electric motors are used to power chain blocks, hoists such as vehicle lifting platforms and large overhead gantry cranes capable of lifting substantial loads.

20.2.1.3 Pneumatic

Air chain hoists are available operated by air motors driven by compressed air at either 4 or 6.3 bars. Safe working loads range from 0.25 to 50 tonnes. These may also be linked to air-driven trolleys where hoists are used in conjunction with an overhead gantry.

Exhausting air from a system is commonly used in lifting. Powered vacuum lifters are used to lift sheet material, plastics sacks and cartons in a range of industries.

20.2.1.4 Hydraulic

Hydraulic systems use pressurised liquid, usually oil. This can be in the form of a hydraulic cylinder moving a heavy load over a short distance or through a pump to a hydraulic motor linked to a mechanical means of movement, e.g. cables or screws.

20.2.1.5 Petrol or diesel

This power source is not used in a factory environment but is common for on-site work in hoists used for raising and lowering materials and equipment. They are totally portable and independent.

20.2.2 Lifting gear accessories

20.2.2.1 Hooks

Are forged from hardened and tempered high-tensile steel and are rated according to their SWL. A range of types are available including those with a safety latch as shown in Fig. 20.10. Form and dimensions are set out in BS EN 1677-5 which states that all hooks shall be free from patent defect and shall be cleanly forged in one piece. After manufacture and heat treatment

each hook shall be subjected to a proof load. After proof testing each hook is stamped to allow identification with the manufacturer's certificate of test and examination.

Figure 20.10 Hook

20.2.2.2 Slings

Slings are manufactured from a number of materials, generally man-made fibre, wire rope and chain.

Belt slings made from a high-tenacity polyester webbing are commonly used. Flat webbing or belt slings made from high-tenacity polyester webbing are commonly used but are also available made from polypropylene and polyamide (nylon) in a range of sizes and capacities. Fibre round slings with a woven tubular sleeve are also available. These are colour coded for easy identification of their SWL.

SWL (tonnes)	1	2	3	4	5	6	8	10	Over 10
Colour	Violet	Green	Yellow	Grey	Red	Brown	Blue	Orange	Orange

Belt slings are available in standard lengths from 1 to 10 m and in widths from 25 to 450 mm. They are available up to 12 tonnes SWL and may be endless, fitted with two soft-eye lifting loops for general-purpose lifting or lightweight metal fittings for improved protection from pressure at the lifting point (see Fig. 20.11).

Figure 20.11 Belt sling

If a lift is vertical (i.e. the lifting hook on one end and the load vertically below on the other end), the total load that may be lifted is that marked on the sling. However, if the method of lifting differs, the working load limit (WLL) will alter as shown in Table 20.1. The figures relate to a sling with a 1 tonne (1000 kg) SWL.

Table 20.1

			α	α
Vertical lift	Choke hitch	Basket hitch	Basket hitch 0–45°	Basket hitch 45–60°
SWL 1000 kg	WLL 800 kg	WLL 2000 kg	WLL 1400 kg	WLL 1000 kg

As shown in Table 20.1, a sling with a SWL of 1 tonne used in a basket hitch lift, has a WLL of 2 tonnes.

Wire rope and chain slings may be endless, single leg or have two, three or four legs with various eyes and fittings at the ends, such as shackles, links, rings and hooks. Figure 20.12(a) shows a three-leg wire rope sling and Fig. 20.12(b) a two-leg chain sling.

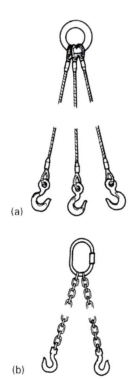

(a)

(b)

Figures 20.12 (a) Three-leg wire rope sling
(b) Two-leg chain sling

Slings are proof tested, identified and marked with the SWL and provided with a dated test certificate.

It is against the law to use lifting equipment or accessories for a load greater than the SWL

except in the case of a multi-legged sling, where a table showing the load limits at different angles of legs must be posted in prominent positions in the factory.

Multi-legged slings are rated at a uniform working load for angles between 0° and 45° to the vertical and between 45° and 60° to the vertical.

The WLL of wire rope and chain slings vary with the number of legs and the angle of the legs in operation.

Table 20.2 shows the working load limits for a variety of chain slings whose links are made from 10 mm diameter material and operating at various angles.

Table 20.2

Size	Single leg	Two leg		Three and four leg	
		0–45°	45°–60°	0–45°	45°–60°
		α	α	α	α
mm 10	kg 3150	kg 4250	kg 3150	kg 6700	kg 4750

Table 20.3 shows the working load limits for a variety of wire rope slings made from 10 mm diameter wire and operating at various angles (see BS EN 13414).

Table 20.3

Rope dia.	Single leg	Two leg		Three and four leg	
		0°–45°	45°–60°	0–45°	45°–60°
		α	α	α	α
mm 10	kg 1050	kg 1500	kg 1050	kg 2250	kg 1600

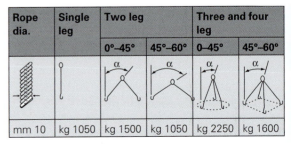

20.2.2.3 Collar eyebolt (Fig. 20.13)

Some larger components are fitted with a drilled and tapped hole to accept an eyebolt to simplify lifting. Collar eyebolts, manufactured to BS 4278: 1984 are available with threads from 6 to 52 mm with larger sizes available to order. Imperial threads are available for replacement on older equipment. Typical SWL of an eyebolt having an M45 thread is 8 tonnes.

Figure 20.13 Collar eyebolt

Care must be taken to avoid mismatching threads, i.e. screwing a metric eyebolt into an imperial threaded hole and vice versa. When fitted, the eyebolt should be screwed against the face of the collar which should sit evenly on the contacting surface. If a single eyebolt is used for lifting a load which is liable to revolve or twist, a swivel-type hook should be used to prevent the eyebolt unscrewing.

Collar eyebolts may be used up to the SWL for axial lifting only. Eyebolts with a link (Fig. 20.14) offer considerable advantages over collar eyebolts when loading needs to be applied at angles to the axis (Fig. 20.15). Their SWL is relatively greater than those of the plain collar eyebolt and the load can be applied at any angle.

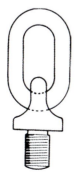

Figure 20.14 Eyebolt with link

The maximum recommended working load at various angles (as indicated in Fig. 20.15) is set out in Table 20.4 for collar eyebolts and in Table 20.5 for eyebolts with a link.

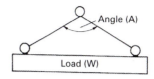

Figure 20.15 Loading at angle using eyebolt with link

Table 20.4

Single eyebolt vertical SWL	Maximum load (W) lifted by pair of collar eyebolts when angle (A) between sling is		
	up to 30°	30°–60°	60°–90°
2000 kg	2500 kg	1600 kg	1000 kg

Table 20.5

Single eyebolt vertical SWL	Maximum load (W) lifted by pair of eyebolts with links when angle (A) between sling is		
	up to 30°	30°–60°	60°–90°
2000 kg	4000 kg	3200 kg	2500 kg

20.2.2.4 Shackles (Fig. 20.16)

Made from alloy steel and manufactured to BS 3551: 1962, shackles are used in conjunction with lifting equipment and accessories. There are four types of pin which can be fitted but the most common is the screwed pin with eye and collar as shown. The range of pin diameters vary from 16 mm up to 108 mm. A range of pin diameters and their SWL is shown in Table 20.6.

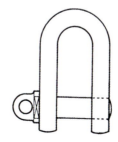

Figure 20.16 Shackle

Table 20.6

Diameter of pin (mm) SWL	16	25	32	38	48	60	70	83	108
(Tonnes)	1.1	4.5	7.5	10.5	16.8	27	35	50	80

20.2.2.5 Chains

Chains with welded links in alloy steel are manufactured in accordance with British Standards (BS) covering the designation of size, material used and its heat treatment and dimensions, e.g. material diameter, welds and dimensions of links. The welds should show no fissures, notches or similar faults and the finished condition should be clean and free from any coating other than rust preventative.

All chain is tested, proof loaded and marked to comply with the BS requirements and issued with a certificate of test and examination together with particulars of the heat treatment which the chain was subjected to during manufacture.

20.2.2.6 Special-purpose equipment

Where production items are regularly lifted, lifting and spreader beams can be used, specially designed for the purpose. Standard lifting beams (Fig. 20.17a) or spreader beams (Fig. 20.17b) are used to give a vertical lift where the lifting points are too far apart to use slings.

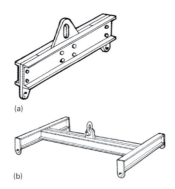

Figure 20.17 (a) Lifting beam (b) Spreader beam

For lifting items such as steel drums and pipes, chain slings can be fitted with special hooks. Figure 20.18 shows pipe hooks, which are used in pairs.

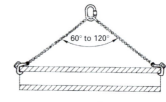

Figure 20.18 Pipe hooks in use

Where large plates are to be lifted, plate clamps (Fig. 20.19) are available which are used to vertically lift plate up to 130 mm thick and having a SWL of 30 tonnes.

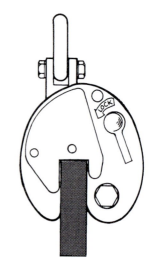

Figure 20.19 Plate clamp for vertical lift

20.2.3 Safe lifting

There are a range of safety rules to observe when carrying out any lifting operation.

Always:

▶ Only use equipment marked with its SWL for which there is a current test certificate.
▶ Plan the lift, establish the weight of the load and prepare the landing area ensuring that it will take the weight.
▶ Check slings and equipment are free of damage, use slings/slinging methods suitable for the load and protect slings from sharp edges and corners.
▶ Attach the sling securely to the load or appliance and position hooks to face outwards.
▶ Ensure the load is balanced and will not tilt or fall.
▶ Make a trial lift and trial lower.

- ▶ Keep fingers, toes, etc., clear when tensioning slings and when landing loads.
- ▶ Ensure that the load is free to be lifted.
- ▶ Use the established code of signals.
- ▶ Ensure that people and loads can't fall from a high level.

Never:

- ▶ Exceed the SWL or rated angle.
- ▶ Lift a load if you doubt its weight or adequacy of the equipment.
- ▶ Use damaged slings or accessories.
- ▶ Twist, knot or tie slings.
- ▶ Hammer slings into position.
- ▶ Snatch a load.
- ▶ Trap slings when landing the load.
- ▶ Drag slings over floors, or attempt to pull trapped slings from under loads.
- ▶ Allow people to pass under or ride on loads.
- ▶ Leave a load hanging unnecessarily for any period of time.

20.3 The Health and Safety (Safety Signs and Signals) Regulations 1996

These Regulations cover various means of communicating health and safety information. As well as the use of signs, they set out minimum requirements for acoustic signals, verbal communication and hand signals.

Hand signals, when moving loads, must be precise, simple, expansive, easy to make and to understand and clearly distinct from other such signals.

Where both arms are used at the same time they must be moved symmetrically and used for giving one sign only.

The person giving the sign, or signalman, who must be competent and trained in their correct use, uses arm/hand movements to give manoeuvring instructions to an operator and must be able to monitor all manoeuvres visually without being endangered. The signalman's duties must consist exclusively of directing manoeuvres and ensuring the safety of workers in the vicinity.

The operator must interrupt the ongoing manoeuvre in order to request new instructions when he is unable to carry out the orders he has received, with the necessary safety guarantees.

The operator must be able to recognise the signalman without difficulty. The signalman must wear one or more brightly coloured distinctive items such as jacket, helmet, sleeve or armbands or carry bats.

The set of coded signals are shown in Fig. 20.20.

| DANGER - emergency stop

Both arms point upwards with the palms facing forwards | START - Attention. Start of command

Both arms are extended horizontally with the palms facing forwards | STOP - Interruption. End of movement

The right arm points upwards with the palm facing forwards |
| END - of the operation

Both hands are clasped at chest height | RAISE

The right arm points upwards with the palm facing forward and slowly makes a circle | LOWER

The right arm points downwards with the palm facing inwards and slowly makes a circle |

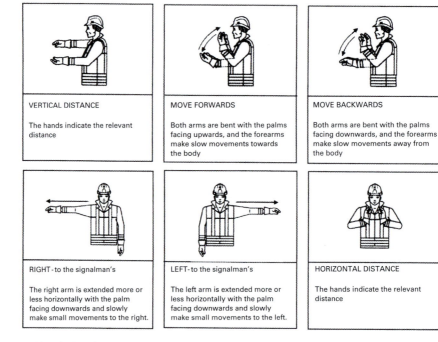

Figure 20.20 Hand signals

Review questions

1. State four steps to take in the working environment to reduce the risk of injury.
2. List four safety requirements of lifting equipment in use.
3. State the three measures to be taken to reduce the risk to employees under the Manual Handling Operations Regulations.
4. List four observations which should be made for safe stacking of objects.
5. Define 'ergonomics'.
6. State four duties of an employee in avoiding risk to himself or others.
7. List four safety rules to be observed when carrying out any lifting operation.
8. State five requirements for lifting equipment under the Lifting Operations and Lifting Equipment Regulations.
9. List the sequence to be observed when manually lifting a load.
10. By means of a sketch show the hand signal for emergency stop.

CHAPTER 21

Drawings, specifications and data

The technical drawing is the means by which designers communicate their requirements. A technical drawing is a pictorial representation of a component with dimensions and other data added. Together with various specifications, the drawing sets out a detailed description of the part which is to be made. The arrangement of dimensions shown on the drawing determine the shape and, together with various characteristics, will have an influence on how the part is made, e.g. whether the part is to be produced from a casting or forging, whether it is a fabricated or welded structure or machined from solid material.

The material specification will include the type of material required together with any standard parts such as nuts, bolts, washers, etc.

Types of tools to be used in the manufacture of the part may also be determined, e.g. if the drawing specifies an 8 mm reamed hole. There may be a requirement for special equipment necessary for installation or assembly.

Specifications such as tolerances, surface finish, heat treatment or protective finishes such as painting or plating all influence the sequence of operations for manufacture.

You can see from this that a technical drawing and its associated specification, communicate a large amount of information.

Drawings are a fundamental part of engineering and are used from the moment design work starts on a new product, through all its stages, to the final finished product.

21.1 Standardisation

In order to create the necessary uniformity in communicating technical information, a system of standardisation has to be adopted. The standardisation can be the information contained on a drawing, the layout of the drawing, the standardisation of sizes or of parts.

This standardisation can be adopted within an individual firm or group of firms, or within an industry, e.g. construction, shipbuilding, etc. nationally by adopting British Standards (BS), throughout Europe (EN) or internationally through ISO. British Standard BS 8888 provides the unified standards through which all this information can be presented clearly. Guides to the use of BS 8888 are available for schools and colleges and for further and higher education as PP 8888 parts 1 and 2.

Adopting a system of standard specification, practices and design results in a number of advantages:

▶ reduction in design costs;
▶ reduction in cost of product;

Workshop Processes, Practices and Materials, Fifth Edition. 9781138784727.

- ▶ redundant items and sizes are eliminated;
- ▶ designs are more efficient;
- ▶ level of interchangeability is increased;
- ▶ mass-production techniques can be adopted;
- ▶ purchasing is simplified;
- ▶ control of quality is enhanced;
- ▶ spares can be easily obtained;
- ▶ costing can be simplified;
- ▶ overheads are reduced.

Except for written notes, technical drawings have no language barriers. They provide the universal language for design, for the craftsman and the technician in manufacture, assembly and maintenance, for the sales team as an aid to selling and for the customer before buying or indeed servicing after purchase.

21.2 Communicating technical information

Many different methods are used to communicate technical information. The method chosen will depend on how much information has to be dispensed and its complexity. Whichever method is chosen the all-important factor is that the information is simply presented, easy to understand and unambiguous.

21.2.1 Technical drawings

Technical drawings can vary from thumbnail sketches to illustrate a particular piece of information through pictorial drawings in isometric or oblique projection to major detail and assembly drawings. These are covered later in this chapter.

21.2.2 Diagrams

Diagrams are used to explain rather than represent actual appearances. For example, an electrical circuit diagram shows the relationship of all parts and connections in a circuit represented by lines and labelled blocks without indicating the appearance of each part. Figure 21.1 is a diagram to explain the route of oil circulation in a car engine but does not go into detail of the engine itself.

21.2.3 Exploded views

Exploded views are used where it is necessary to show the arrangement of an assembly in three

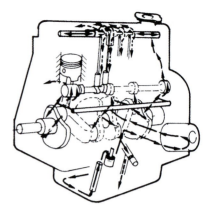

Figure 21.1 Engine oil circulation diagram

dimensions. They are used for assembly purposes and in service or repair manuals where reference numbers of the parts and the way in which they fit together are shown. Figure 21.2 shows an exploded view of a hand riveting tool, listing the spare parts, their identification numbers and their relative position within the finished product.

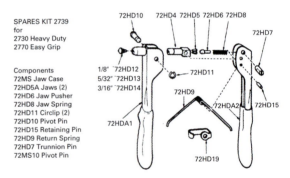

Figure 21.2 Exploded view of hand riveting tool

21.2.4 Operation sheet

The purpose of an operation sheet is to set out the most economic sequence of operations required to produce a finished object or process from the raw material. Although the main purpose of operation sheets is to set out the sequence of operations, they also serve a number of other very important functions:

- ▶ They determine the size and amount of material required. From this information, the material can be ordered in advance and appropriate material stock levels can be maintained.
- ▶ Any tooling such as jigs, fixtures and gauges can be ordered or manufactured in advance so that it will be available when required.

- Knowing the machines or plant which are to be used enables machine-loading charts to be updated so that delivery dates to customers will be realistic.
- The sequence of operations listed will enable work to be progressed through the factory in an efficient manner.
- The inclusion of estimated times for manufacture on the operation sheet enables a cost of manufacture and hence selling price to be established.

Figure 21.3 shows a fitted bolt with a hexagon head which is to be produced in one operation on a capstan lathe. The sequence of operations is shown in Operation sheet 1.

Operation sheet 1 Material: 17 A/F × 42 mm. M.S. hexagon bar. Name: fitted bolt

Op. no.	Machine	Operation	Tooling	Position
1	Capstan lathe	Feed to stop	Adjustable stop	Turret 1
		Turn 10 mm thread diameter	Roller box tool	Turret 2
		Turn 12 mm diameter	Roller box tool	Turret 3
		Face end and chamfer	Face and chamfer tool	FTP
		Form undercut	Undercut tool	FTP
		Die thread	Self-opening diehead	Turret 4
		Chamfer head	Chamfer tool	FTP
		Part off	Parting-off tool	RTP

21.2.5 Data sheets

Data sheets are available where a range of information is available and where you can obtain

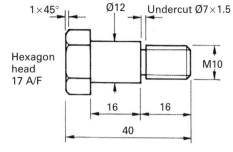

Figure 21.3 Hexagon-headed fitted bolt

the information required for your application. Table 21.1 shows data relating to a range of tin/lead base solders and their use.

There are many examples of data sheets, some of which are made up as wall charts for easy reference.

21.2.6 Tables

Tables are used to show a range of options or sizes available and help with a choice for a particular set of circumstances. Table 21.2 shows the selection of a hacksaw blade with the correct number of teeth for the task, given the type and thickness of material to be cut.

Tables are also available setting out tapping and clearance drill sizes for a range of thread sizes. Table 21.3 shows such a table for ISO metric coarse threads. These are also available as wall charts for easy reference.

21.2.7 Graphs

Graphs are a useful means of communication where there is a relationship between sets of technical data. Figure 21.4 shows the relationship between tempering temperature and hardness of a particular type of steel. From the graph it can be

Table 21.1

Alloy N°	Composition (%)		Melting range (°C)	Uses
	Sn	Pb		
1	64	36	183–185	Mass soldering of printed circuits
2	60	40	183–188	General soldering
3	50	50	183–220	Coarse tinman's solder
5	40	60	183–234	Coating and pre-tinning
7	30	70	183–255	Electrical cable conductors
8	10	90	268–302	Electric lamp bases

Sn – chemical symbol for tin; Pb – chemical symbol for lead.

Table 21.2

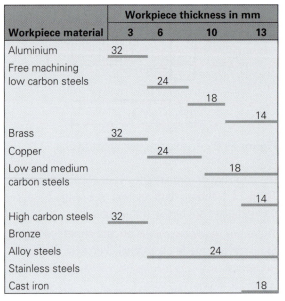

Workpiece material	Workpiece thickness in mm			
	3	6	10	13
Aluminium	32			
Free machining low carbon steels		24		
			18	
				14
Brass	32			
Copper		24		
Low and medium carbon steels			18	
				14
High carbon steels	32			
Bronze				
Alloy steels			24	
Stainless steels				
Cast iron				18

Table 21.3

ISO metric coarse thread			
Diameter (mm)	Pitch (mm)	Tapping drill (mm)	Clearance drill (mm)
2.0	0.40	1.60	2.05
2.5	0.45	2.05	2.60
3.0	0.50	2.50	3.10
4.0	0.70	3.30	4.10
5.0	0.80	4.20	5.10
6.0	1.00	5.00	6.10
8.0	1.25	6.80	8.20
10.0	1.50	8.50	10.20
12.0	1.75	10.20	12.20

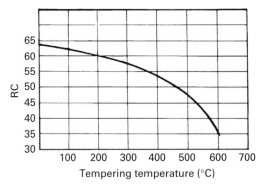

Figure 21.4 Temperature/hardness graph for tempering

seen that to achieve a final hardness of 60 RC, a part made from this steel would be tempered at 200 °C.

20.2.8 Storing technical information

Storing large amounts of original drawings in paper form is very bulky and takes up a lot of space. Original drawings were stored on microfilm by photographically reducing them on to film which was then mounted and suitably filed. When required they were re-enlarged and printed for use on the shop floor. Care must be taken when preparing drawings which are to be microfilmed otherwise some detail may be lost in the reducing process.

Libraries and spares departments carry complete documents or lists of spares on microfiche systems. The information is stored on a large film which is divided into a number of frames or grids identified by letters along one edge and numbers along the other. The information is viewed on a screen by moving to the required grid reference, e.g. B12.

Although microfiche systems are still available most technical drawing and design is now done on computer-aided draughting or design (CAD) systems where the computer is capable of generating and storing much technical information. This information can be stored in the computer itself on the hard disk or on separate microdisks for retrieval at any time. Any modification can be easily carried out. Information can be easily retrieved and viewed on a visual display unit and when required can be downloaded to a printer or plotter. Furthermore, the information can also be accessed by others in the same factory through computer networks or sent using a modem, through the telephone system to other factories in other parts of the country. In the case of computer numerical control (CNC) machines, the information can be sent directly to the machine control system.

CDs and DVDs are increasingly being used in technical, sales and education sectors. Again the information contained can be easily accessed and sent anywhere in the country without the need and expense of actual machines or equipment being present.

Vast amounts of information can now be accessed very quickly and easily on the worldwide web.

21.3 Interpreting drawings

Presenting information in the form of a drawing is the most appropriate method of communication. Drawings provide a clear and concise method of conveying information to other individuals and other departments. They give a permanent record of the product and enable identical parts to be made at any time and if modifications are required these can be easily incorporated and recorded. Finally, as has already been stated, they do not represent a language barrier.

When an initial design is decided, it is broken down into a series of units for assembly or sub-assembly, which are further broken down into individual parts referred to as detail drawings.

21.3.1 Assembly drawing

An assembly drawing shows the exact relative position of all the parts and how they fit together. Dimensions of individual parts are not shown although indication of overall dimensions relative to, and important for, the final assembly may be included. All individual parts are identified by an item number. A parts list is shown giving information such as: item number, part description, part number and quantity required. The parts list would include details of bought-out items such as nuts, bolts, screws, bearings, electric motors, valves etc.

As the name implies these drawings are used by engineers to fit together, or assemble, all the parts as the final product and by maintenance engineers during servicing, repair or replacement of worn parts. An example of an assembly drawing is shown in Fig. 21.5.

1	Flange	7643	1
2	Bearing B/o	A302	1
3	Pin	5781	1
4	Shaft	4352	1
5	Pulley	2113	1
6	Bearing B/o	A201	1
7	Bush	5874	1

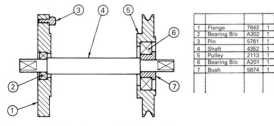

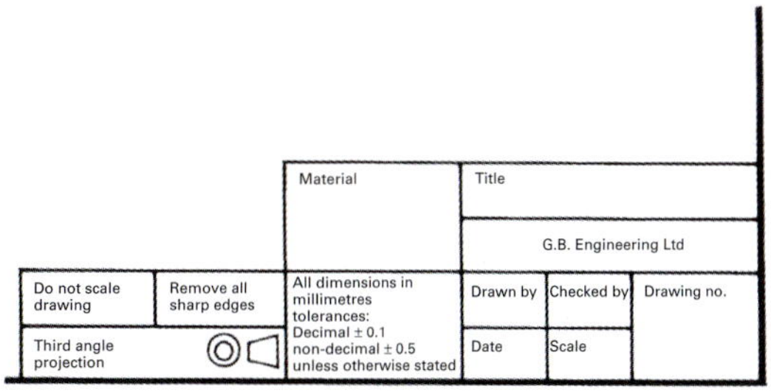

Figure 21.5 Assembly drawing

21.3.2 Detail drawing

A detail drawing is done to give production departments all the information required to manufacture the part. Every part required in an assembly, except standard or bought-out items will have its own detail drawing.

The information given on a detail drawing must be sufficient to describe the part completely and would include:

- dimensions;
- tolerances;
- material specification;
- surface finish;
- heat treatment;
- protective treatment;
- tool references;
- any notes necessary for clarity.

All detail drawings should have a title block usually located at the bottom right-hand corner. For ease of reference and filing of its drawings, a company should standardise the position and content of a title block and this is pre-printed on the drawing sheet.

The information contained will vary from one company to another but would be likely to include most of the following:

- name of the company;
- drawing number;
- title;
- scale;
- date;
- name of draughtsman/woman;
- unit of measurement;
- type of projection;
- general tolerance notes;
- material specification.

An example of a title block is shown in Fig. 21.6.

Figure 21.6 Example of a title block

21.3.3 Block plan

Block plans are used in the construction industry to identify a site and locate the outlines of a construction in relation to a town plan or other context.

21.3.4 Location drawing

Location drawings are used in the construction industry to locate sites, structures, buildings, elements, assemblies or components.

21.3.5 Projection

The two common methods used to represent a three-dimensional object on a flat piece of paper are orthographic projection and pictorial projection.

21.3.5.1 Orthographic projection

This is the most frequently used method of presenting information on a detail drawing. It is usually the simplest and quickest way of giving an accurate representation, and dimensioning is straightforward. It is only necessary to draw those views that are essential and it is simple to include sections and hidden detail.

Orthographic projection may be first angle or third angle and the system used on a drawing should be shown by the appropriate symbol.

An example of first angle projection is shown in Fig. 21.7 each view showing what would be seen by looking on the far side of an adjacent view.

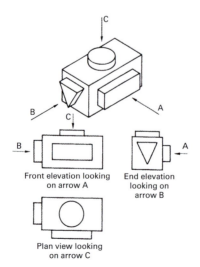

Figure 21.7   First angle projection

An example of third angle projection is shown in Fig. 21.8 each view showing what would be seen by looking on the near side of an adjacent view.

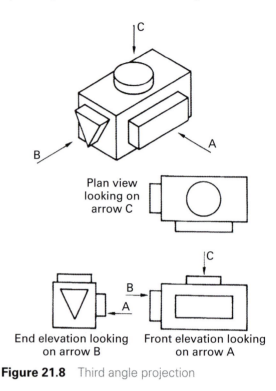

Figure 21.8   Third angle projection

21.3.5.2 Pictorial projection

This is a method of showing, on a single view, a three-dimensional picture of an object. It is therefore easier to visualise and is useful for making rough sketches to show someone what you require.

Two kinds of pictorial projection are used: isometric projection and oblique projection.

21.3.5.2.1 Isometric projection

In isometric projection, vertical lines are shown vertical but horizontal lines are drawn at 30° to the horizontal on each side of the vertical. This is shown by the rectangular box in Fig. 21.9 and in Figs. 21.7 and 21.8.

With this method, a circle on any one of the three faces is drawn as an ellipse. In its simplest form, all measurements in orthographic views may be scaled directly onto the isometric view.

Figures 21.7 and 21.8 show an isometric view and the corresponding orthographic views.

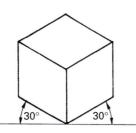

Figure 21.9   Isometric projection

21.3.5.2.2 Oblique projection

The main difference between isometric and oblique projection is that with oblique projection one edge is horizontal. Vertical edges are still vertical which means two axes are at right angles to each other. This means that one face can be drawn as its true shape. The third edge can be drawn at any angle but is usually 30° or 45°. This is shown by the rectangular box in Fig. 21.10.

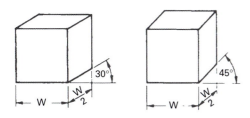

Figure 21.10   Oblique projection

To avoid a distorted view, the dimensions along the receding edges are drawn half full size.

The rule to follow when drawing oblique projection is to chose the face with most detail as the front face so it can be drawn as its true shape, e.g. a circle remains a circle and can be drawn using a pair of compasses. This will make use of the advantage that oblique has over isometric and will show the object with minimum distortion.

Figure 21.11 shows an oblique view and two corresponding orthographic views.

21.4 Sectional views

A part with little internal detail can be satisfactorily represented by orthographic projection. However, where the internal detail is complicated and are represented by hidden detail lines, the result may be confusing and difficult to interpret correctly. In such cases, the interior detail can be exposed by

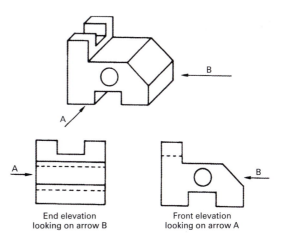

End elevation
looking on arrow B

Front elevation
looking on arrow A

Figure 21.11   Oblique and corresponding orthographic views

'cutting away' the outside and showing the inside as full lines instead of hidden detail. The view thus drawn is known as a sectional view.

There are a range of sectional views which can be done.

Figure 21.12 shows a single plane sectional view, i.e. an imaginary slice taken straight through the object called the 'cutting plane'. The cutting plane is shown by long chain lines, thickened at the ends and labelled by capital letters. The direction of viewing is shown by arrows resting on the thickened lines. The area which is sectioned is hatched by thin lines at 45° to the part profile.

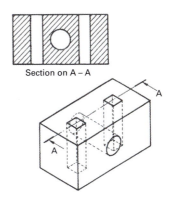

Section on A – A

Figure 21.12   Sectional view

21.5 Standard conventions

Because an engineering drawing is the means of communication it is important that all drawings are uniform in their layout and content. British Standards sets out to standardise

21

conventions used in engineering drawings which include: layout, lines, systems of projection, sections, conventional representation and dimensions.

21.5.1 Layout of drawings

Layouts should use the 'A' series of sheet sizes A4, A3, A2, A1 and A0 the dimensions of which are shown in Table 21.4. The drawing area and title block should be within a frame border.

Table 21.4

Name	Size in mm
A4	210 × 297
A3	297 × 420
A2	420 × 594
A1	594 × 841
A0	841 × 1189

21.5.2 Lines and line work

All lines should be uniformly black, dense and bold. Types of line are shown in Table 21.5.

Table 21.5

Type of line	Description of line	Application
————	Thick, continuous	Visible outlines and edges
————	Thin, continuous	Dimension and leader lines; projection lines; hatching
– – – – –	Thin, short dashes	Hidden outlines and edges
–·–·–·–	Thin, chain	Centre lines
⌐·–·–⌐	Chain, (thick at ends and at changes of direction, thin elsewhere)	Cutting planes of sections

21.5.3 Systems of projection

The system of projection used on a drawing should be shown by the appropriate symbol as shown in Fig. 21.13.

21.5.4 Conventional representation

Standard parts which are likely to be drawn many times result in unnecessary waste of time and

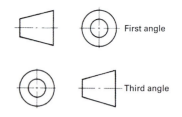

Figure 21.13   Symbols for system of projection

therefore cost. Conventional representation is used to show common engineering parts as simple diagrammatic representations, a selection of which is shown in Fig. 21.14.

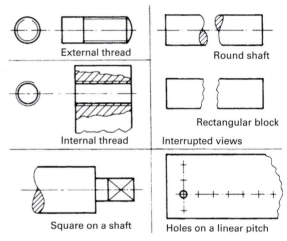

Figure 21.14   Conventional representation

21.5.5 Dimensioning

All dimensions necessary for the manufacture of a part should be shown on the drawing and should appear once only. It should not be necessary for a dimension to be calculated or for the drawing to be scaled.

Dimensions should be placed outside the outline of the view wherever possible. Projection lines are drawn from points or lines on the view and the dimension line placed between them. Dimension lines and projection lines are thin continuous lines. There should be a small gap between the outline and the start of the projection line, and the projection line should extend for a short distance beyond the dimension line. The dimension line has an arrowhead at each end which just touches the projection line.

Dimension lines should not cross other lines unless this is unavoidable.

Leaders are used to show where dimensions or notes apply and are thin continuous lines ending in arrowheads touching a line or as dots within the outline of the object. These principles are shown in Fig. 21.15.

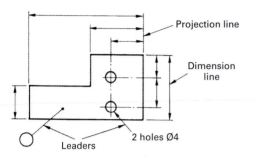

Figure 21.15 Principles of dimensioning

Difficulties can arise if the dimensioning of a part is not done correctly, which may lead to errors of the finished part.

In Fig. 21.16(a) the part is dimensioned from the datum in sequence, i.e. cumulatively, known as chain dimensioning. Each dimension is subject to the tolerance of ±0.15 so that a cumulative error of ±0.45 in the overall length is possible.

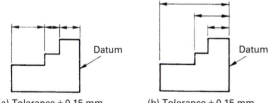

(a) Tolerance ± 0.15 mm   (b) Tolerance ± 0.15 mm

Figure 21.16 Dimensioning to avoid cumulative error

By dimensioning the part as shown in Fig. 21.16(b) the cumulative error is avoided.

Difficulties also arise if a part is dimensioned from more than one datum. The shouldered pin shown in Fig. 21.17(a) has been dimensioned from each end, i.e. two separate datums. This could result in a possible maximum error of ±0.6 mm between the shoulders A and B.

If the dimension between shoulders A and B is important, then it would have to be individually dimensioned relative to one datum and subject to a tolerance of ±0.2 mm as shown in Fig. 21.17(b).

Difficulties may also arise where several parts are assembled and an important dimension has to be maintained. The datum chosen must reflect this important dimension. Figure 21.18(a) shows an assembly where A is the important dimension. The chosen datum face is indicated. Thus the dimensioning of each individual part of the assembly reflects this datum as shown in Fig. 21.18(b)–(d).

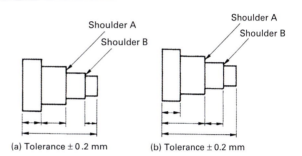

(a) Tolerance ± 0.2 mm   (b) Tolerance ± 0.2 mm

Figure 21.17 Dimensioning to maintain important size

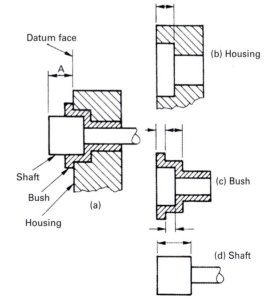

Figure 21.18 Dimensioning to maintain correct assembly

Review questions

1. What is the purpose of an assembly drawing?
2. By means of a sketch show three views of an object in first angle projection.
3. State eight advantages for adopting a system of standard specification.
4. List eight items of information which would be included in a detail drawing.
5. State the objectives for creating an operation sheet.
6. By means of a sketch show where dimensions would create a cumulative error and how this would be overcome.
7. Give an example of a data sheet in your workplace with which you are familiar listing the important information it contains.
8. Under what circumstances would a graph be useful?
9. Why is it desirable to use standard conventions when creating a drawing?
10. Why would it be necessary to show a sectional view on a drawing?

21

APPENDICES

Screw-thread forms

Appendix 1: basic form for ISO metric threads, Fig. A1

Major diameter in millimetres (first-choice sizes)	Pitch in millimetres	
	Coarse-pitch series	Fine-pitch series
1.0	0.25	0.2
1.2	0.25	0.2
1.6	0.35	0.2
2.0	0.4	0.25
2.5	0.45	0.35
3.0	0.5	0.35
4.0	0.7	0.5
5.0	0.8	0.5
6.0	1.0	0.75
8.0	1.25	0.75
8.0	–	1.0
10.0	1.5	0.75
10.0	–	1.0
10.0	–	1.25
12.0	1.75	1.0
12.0	–	1.25
12.0	–	1.5
16.0	2.0	1.0
16.0	–	1.5
20.0	2.5	1.0
20.0	–	1.5
20.0	–	2.0
24.0	3.0	1.0
24.0	–	1.5
24.0	–	2.0

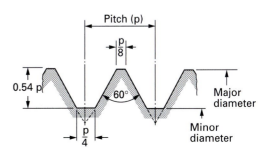

Figure A1 Basic form for ISO metric threads

Appendix 2: basic form for Unified threads (UNC and UNF), Fig. A2

Major diameter in inches	Threads per inch		Pitch in inches	
	UNC	UNF	UNC	UNF
1/4	20	28	0.05	0.036
5/16	18	24	0.055	0.042
3/8	16	24	0.062	0.042
7/16	14	20	0.071	0.05
1/2	13	20	0.077	0.05
9/16	12	18	0.083	0.055
5/8	11	18	0.091	0.055
3/4	10	16	0.100	0.062
7/8	9	14	0.111	0.071
1″	8	12	0.125	0.083

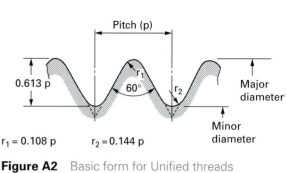

$r_1 = 0.108\,p \qquad r_2 = 0.144\,p$

Figure A2 Basic form for Unified threads (UNC and UNF)

Appendix 3: basic form for Whitworth threads (BSW, BSF, and BSPF), Fig. A3

Major diameter in inches	Number of threads per inch (t.p.i.)			Pitch in inches			Outside diameter in inches
	BSW	BSF	BSPF	BSW	BSF	BSPF	BSPF
1/8	40	–	28	0.025	–	0.036	0.383
3/16	24	32	–	0.042	0.031	–	–
7/32	–	28	–	–	0.036	–	–
1/4	20	26	19	0.050	0.038	0.053	0.518
9/32	–	26	–	–	0.038	–	–
5/16	18	22	–	0.055	0.045	–	–
3/8	16	20	19	0.062	0.050	0.053	0.656
7/16	14	18	–	0.071	0.055	–	–
1/2	12	16	14	0.083	0.062	0.071	0.825
9/16	12	16	–	0.083	0.062	–	–
5/8	11	14	14	0.091	0.071	0.071	0.902
11/16	11	14	–	0.091	0.071	–	–
3/4	10	12	14	0.100	0.083	0.071	1.041
7/8	9	11	14	0.111	0.091	0.071	1.189
1″	8	10	11	0.125	0.100	0.091	1.309

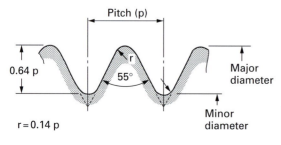

$r = 0.14\,p$

Figure A3 Basic form for Whitworth threads (BSW, BSF and BSPF)

Appendix 4: basic form for British Association threads (BA), Fig. A4

BA designation number	Major diameter in millimetres	Pitch in millimetres
0	6.0	1.00
1	5.3	0.90
2	4.7	0.81
3	4.1	0.73
4	3.6	0.66
5	3.2	0.59
6	2.8	0.53
7	2.5	0.48
8	2.2	0.43
9	1.9	0.39
10	1.7	0.35

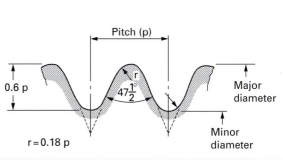

Figure A4 Basic form for British Association threads (BA)

Index

Note: Page numbers in **bold** are for figures, those in *italics* are for tables.